T0205653

Practical Molecular Virology

Methods in Molecular Biology

John M. Walker, SERIES EDITOR

1. **Proteins,** edited by ***John M. Walker,*** *1984*
2. **Nucleic Acids,** edited by ***John M. Walker,*** *1984*
3. **New Protein Techniques,** edited by ***John M. Walker,*** *1988*
4. **New Nucleic Acid Techniques,** edited by ***John M. Walker,*** *1988*
5. **Animal Cell Culture,** edited by ***Jeffrey W. Pollard and John M. Walker,*** *1990*
6. **Plant Cell and Tissue Culture,** edited by ***Jeffrey W. Pollard and John M. Walker,*** *1990*
7. **Gene Transfer and Expression Protocols,** edited by ***E. J. Murray,*** *1991*
8. **Practical Molecular Virology:** *Viral Vectors for Gene Expression,* edited by ***Mary K. L. Collins,*** *1991*

Methods in Molecular Biology • 8

Practical Molecular Virology

Viral Vectors for Gene Expression

Edited by

Mary K. L. Collins

Chester Beatty Laboratories, Institute of Cancer Research, London, UK

Humana Press • Totowa, New Jersey

999 Riverview Dr., Suite 208
Totowa, NJ 07512

Printed in the United States of America. 9 8 7 6 5 4 3 2

Library of Congress Cataloging in Publication Data

Practical molecular biology: viral vectors for gene expression/
edited by Mary Collins.
p. cm. – (Methods in molecular biology : 8)
Includes index.
ISBN 0-89603-191-8 (hardcover) ISBN 0-89603-299-X (paperback)
1. Genetic vectors. 2. Viruses. I. Collins, Mary II. Series:
Methods in molecular biology : v. 8.
[DNLM: 1. Gene Expression Regulation, Viral. 2. Genetic Vectors.
3. Viruses. W1 ME9616F v. 8 / QW 160 P895]
QH442.2.P73 1991
574.87'322–dc20
DNLM/DLC
for Library of Congress 91-7049
CIP

Preface

Recombinant viruses provide an efficient mechanism for the transfer and expression of DNA in eukaryotic cells. First, the transfer of DNA by viral infection—utilizing specific cell surface receptors and cellular internalization mechanisms—occurs much more readily than DNA transfer via uptake induced by such physical methods as calcium phosphate coprecipitation or electroporation. Second, the novel strategies employed by the virus to express its own genes can then be "hijacked" in the recombinant virus to express the researcher's gene of interest.

The purpose of *Practical Molecular Virology* is thus to compile a collection of readily repeatable gene transfer and expression methods from workers expert in the use of a variety of recombinant viral vectors. These include those designed for the production of recombinant antigens, such as poliovirus and yeast Ty-VLPs; those giving very high levels of recombinant protein expression, for example, baculovirus, vaccinia virus, and SV40; and finally viral vectors used for efficient, stable gene transfer to eukaryotic cells, such as retroviruses and herpesviruses.

The first chapter describes the viral life cycle for each virus, and explains how this can be adapted to allow construction of recombinant vectors. Subsequent chapters deal with methods for producing and characterizing recombinant viruses. I make no apology for the hyperproliferation of chapters dealing with recombinant retroviral methods and applications, since I believe this is clearly proportional to the recent expansion of interest in these techniques.

The user should note that not all basic molecular biology techniques are detailed in the methods described. Volumes 2 and 4 of this series provide excellent reference works for these more fundamental techniques.

In sum, *Practical Molecular Virology* is intended to aid research workers in the selection of appropriate viral vectors for their applications, and to provide basic protocols for their construction.

Mary Collins

Contents

Contributors

SALLY E. ADAMS • *British Bio-technology Ltd., Oxford, UK*
JEFFREY W. ALMOND • *Department of Microbiology, University of Reading, Reading, UK*
MARK BOYD • *Chester Beatty Laboratories, Institute of Cancer Research, London, UK*
PAUL M. BRICKELL • *Medical Molecular Biology Unit, University College and Middlesex School of Medicine, London, UK*
KAREN L. BURKE • *Institut für Mikrobiologie der Universitat Ulm, Abteilung Virologie, Ulm, FRG*
NIGEL R. BURNS • *British Bio-technology Ltd., Oxford, UK*
PAUL R. CLAPHAM • *Chester Beatty Laboratories, Institute of Cancer Research, London, UK*
OLIVIER DANOS • *Institut Pasteur, Paris, France*
VINCENT C. EMERY • *Department of Virology, Royal Free Hospital School of Medicine, Hampstead, London, UK*
DAVID J. EVANS • *Department of Microbiology, University of Reading, Reading, UK*
FARZIN FARZANEH • *Molecular Genetics Unit, King's College School of Medicine and Dentistry, London, UK*
JOOP GÄKEN • *Molecular Genetics Unit, King's College School of Medicine and Dentistry, London, UK*
JACQUELINE E. M. GILMOUR • *British Bio-technology Ltd., Oxford, UK*
JOHN F. HANCOCK • *Department of Haematology, Royal Free Hospital School of Medicine, Hampstead, London, UK*
JULIAN G. HOAD • *Chester Beatty Laboratories, Institute of Cancer Research, London, UK*
HELEN C. HURST • *ICRF Oncology Group, Hammersmith Hospital, London, UK*
LYNN M. KEMP • *Medical Molecular Biology Unit, University College and Middlesex School of Medicine, London, UK*
ALAN J. KINGSMAN • *British Bio-technology Ltd., Oxford, UK*
SUSAN M. KINGSMAN • *Department of Biochemistry, University of Oxford, Oxford, UK*
DAVID S. LATCHMAN • *Medical Molecular Biology Unit, University College and Middlesex School of Medicine, London, UK*
S. MARK • *British Bio-technology Ltd., Oxford, UK*
MYRA O. MCCLURE • *Chester Beatty Laboratories, Institute of Cancer Research, London, UK*
JANE MCKEATING • *Chester Beatty Laboratories, Institute of Cancer Research, London, UK*
PHILIP D. MINOR • *National Institute for Biological Standards and Control, Hertfordshire, UK*

John Moore • *Chester Beatty Laboratories, Institute of Cancer Research, London, UK*
Mukesh Patel • *Medical Molecular Biology Unit, University College and Middlesex School of Medicine, London, UK*
Clive Patience • *Chester Beatty Laboratories, Institute of Cancer Research, London, UK*
Colin D. Porter • *Department of Immunology, Institute of Child Health, London, UK*
Jack Price • *The National Institute of Medical Research, Mill Hill, London, UK*
H. Richardson • *British Bio-technology Ltd., Oxford, UK*
Javier M. Rodriguez • *Centro de Biologia Molecular (CSIS-UAM), Madrid, Spain*
Stephen J. Russell • *Chester Beatty Laboratories, Institute of Cancer Research, London, UK*
Thomas F. Schulz • *Chester Beatty Laboratories, Institute of Cancer Research, London, UK*
Antonio Talavera • *Centro Nacional de Biotechnologia, Madrid, Spain*
Richard Vile • *Chester Beatty Laboratories, Institute of Cancer Research, London, UK*
Karen H. Vousden • *Ludwig Institute, St. Mary's Hospital, London, UK*

CHAPTER 1

The Retroviral Life Cycle and the Molecular Construction of Retrovirus Vectors

Richard Vile

1. Introduction

The discovery of a filterable agent that allowed the transmission of cancers in chickens *(1)* was the first identification of the viruses now known as retroviruses. Subsequently, genes transmitted by some retroviruses were identified as transforming oncogenes. These findings suggested that retroviruses may be used as genetic vectors, since retroviral oncogenes (v-*onc*) are altered forms of "high-jacked" normal cellular genes *(2)*, and the retroviruses that transform cells in culture are often defective for replication because the v-*onc* genes have been substituted in place of one or more of the essential replicative genes *(3)*. Such defective oncogenic retroviruses can be propagated only in the presence of a wild-type "helper" virus, which supplies the functional gene products of the virus. Retroviruses can now be modified to become vehicles for the delivery and expression of cloned genes into a wide variety of cells, for both experimental and therapeutic purposes.

2. Retroviruses as Vectors

The motive behind gene delivery may be simply to mark the target cell population (elsewhere in this volume). Alternatively, it may lead to the gen-

From: *Methods in Molecular Biology, Vol. 8:*
Practical Molecular Virology: Viral Vectors for Gene Expression
Edited by: M. Collins © 1991 The Humana Press Inc., Clifton, NJ

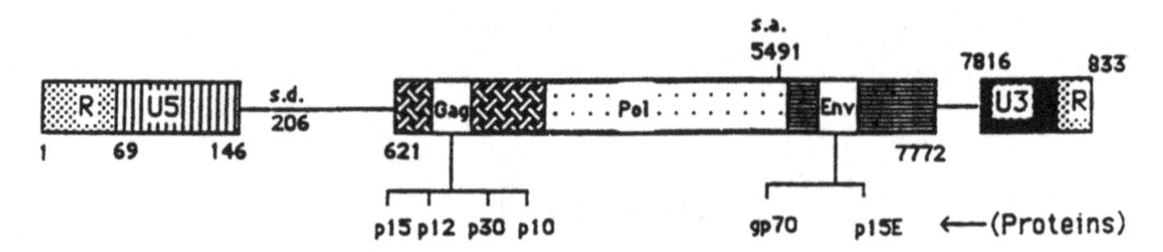

Fig. 1. RNA genome of MoLV.

eration of novel mutations that can be recognized by virtue of their linkage to the retroviral genome. More usually a gene is delivered to a specific cell type, where its expression is driven from sequence elements included in the vector and the effects of expression can be monitored. If the expressed gene encodes a correct version of a defective gene, its transfer into diseased cells may even allow the correcting of the genetic deficiency, either on a cellular level or in the animal as a whole. Such approaches may eventually lead to gene therapy for specific monogenic disorders (for review, *see* ref. *4*).

Advantages of retroviral vectors over other mechanisms of delivery of DNA to cells include the potential of infecting nearly every cell in the target population *(5)*. Physical methods can be used to induce cells to take up DNA, such as coprecipitation with calcium phosphate, forming complexes of DNA with lipids, electroporation, direct microinjection of DNA, or the encapsidation of DNA into liposomes. However, infection by high-titer retroviruses can usually exceed the efficiency (about 1% transfer) of these methods. In addition, infection leads to integration of a single copy of the vector genome in a predictable, stable fashion, which is rarely the case using nonviral transfer techniques.

3. The Retroviral Life Cycle

3.1. The Retroviral Particle

The retroviral genome within the virus particle is composed of two identical RNA molecules. Retroviral vectors are often derived from the murine virus, Moloney leukemia virus (MoMLV), the genome of which contains three open reading frames called *gag, pol,* and *env.* These reading frames are bound by regions that contain the signals essential for expression of the viral genes. At the 5' end of the RNA genome is the R and U5 region, followed by a short stretch of noncoding DNA before the start of the *gag* region. *Gag* encodes the structural proteins of the viral capsid, the *pol* region encodes the enzymatic activities for genome processing, and the *env* region specifies the proteins of the viral envelope. To the 3' side of the coding sequences lie additional regulatory sequences, the U3 region, and a copy of the R region. A single copy of the RNA genome, as found in the virion particle, is shown in Fig. 1.

3.2. Infection of the Target Cell

The virus particle contains two RNA strands forming a complex with viral enzymes (*pol* products) and host tRNA molecules within a core structure of *gag* proteins. Surrounding this capsid is a lipid bilayer derived from the host-cell membrane, containing the viral *env* proteins. These *env* proteins bind to the cellular receptor for the virus, and the particle enters the host cell, probably via a process of receptor-mediated endocytosis.

3.3. Reverse Transcription of the Viral RNA into DNA

Following shedding of the outer envelope, the viral RNA is copied into DNA by the process of reverse transcription. This takes place within the RNA/protein complex and is catalyzed by the reverse transcriptase enzyme of the *pol* region.

Reverse transcriptase uses as a primer a molecule of host-cell tRNA, which is carried into the infected cell in the particle and is homologous to a region just downstream of the 5' U5 region (–PBS). It commences its activity by copying the *donor* viral genome into DNA, moving into the U5 and R regions (toward the left end of the genome) (Fig. 2A). Since the copying extends to the extreme 5' end of the RNA template, a second enzymatic activity of reverse transcriptase digests away the portion of the RNA molecule that has already been copied (U5 region). This liberates the DNA copy (of U5-R) to base pair, via the R sequence, to another genomic RNA molecule, the *acceptor (6),* (Fig. 2B). Following this *inter*molecular "jump" of template by reverse transcriptase, the growing DNA molecule (minus strand) is extended by copying the acceptor RNA genome. It moves out of R, through the U3 region, and beyond, past a second Primer Binding Site (+PBS) that lies at the boundary of U3. The minus strand is elongated through the *env, pol,* and *gag* regions as far as the tRNA primer binding site of the acceptor RNA genome.

Meanwhile, a second molecule of reverse transcriptase becomes primed, and it commences DNA synthesis of a plus strand of DNA at the +PBS site on the minus strand (Fig. 2C). The plus strand is elongated to copy the U3-R-U5-(–PBS) cassette from the minus strand. Now, an *intramolecular* template-jump occurs, by which the plus-strand U3-R-U5-(–PBS) cassette is transposed to the 3' end of the minus strand in such a way that the –PBS on the plus strand is matched with the –PBS of the minus strand (Fig. 2D). The generation of the linear duplex DNA molecule is now relatively straightforward. The minus strand is completed by extension into the U5-R-U3 region of the plus strand; the plus strand is completed by copying the *gag-pol-env*-U3-R-U5 region of the minus strand. The final products of these complex molecular maneuvers are linear duplex DNA molecules with identical boundary regions

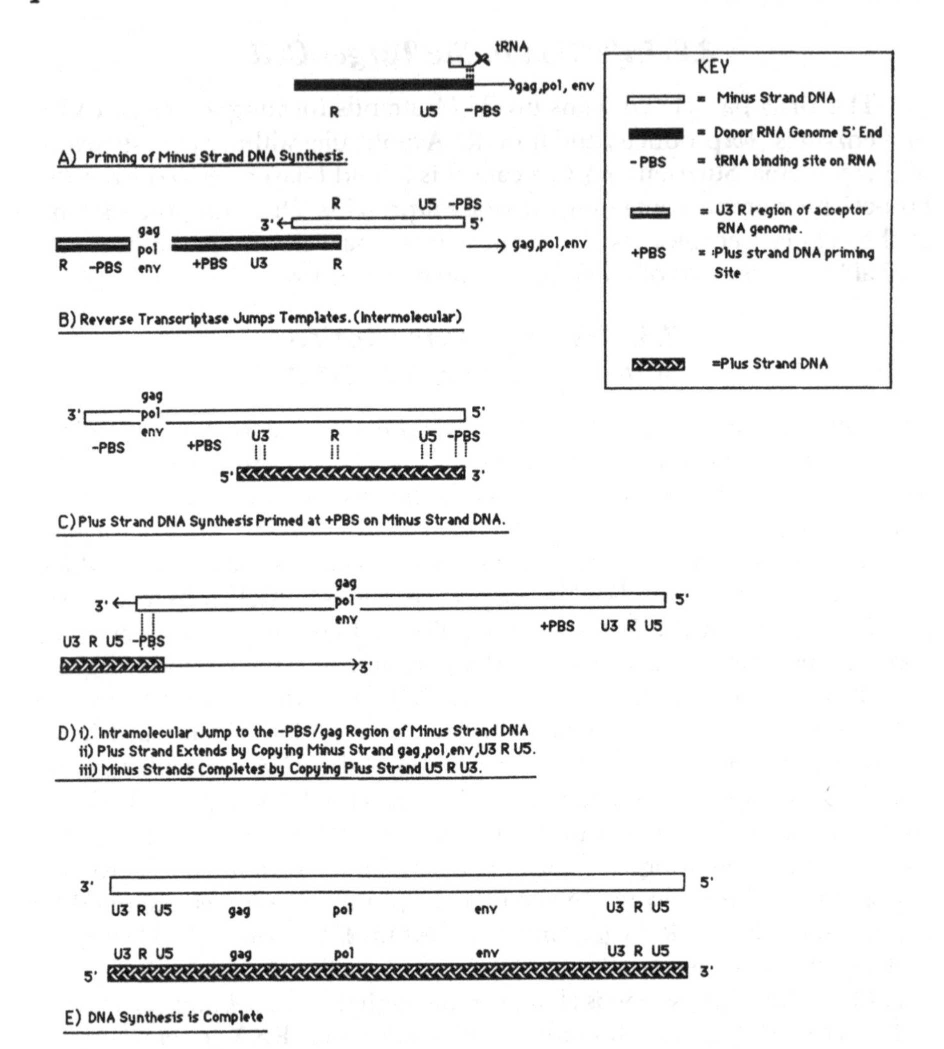

Fig. 2. Schematic representation of the molecular events of reverse transcription.

of U3, R, and U5 (Fig. 2E). Crucially, *the U3 region at both ends of the resulting DNA duplex can be seen to be derived from the 3' end of the RNA genome.* The reader is referred to refs. *6* and *7* for further details of the sequence of reactions.

3.4. Integration of the Viral Genome into Cellular DNA

The double-stranded linear DNA molecule produced by reverse transcription is made circular and inserted in random sites in the cell's genome, using a *pol* product, integrase protein. This protein recognizes sequences at

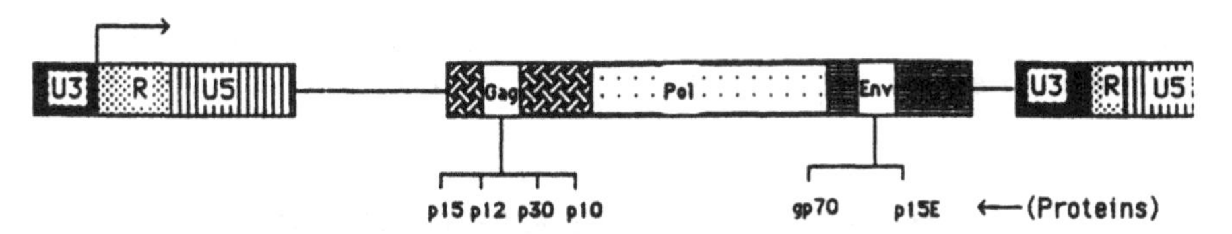

Fig. 3. DNA genome of MoMLV as integrated into the host-cell genome.

the ends of the DNA of the long terminal repeats (LTRs) to direct linear integration of the DNA provirus, in such a way that the provirus is always joined to host DNA 2 bp from the ends of the LTRs. The DNA copy of the genome now has two identical repeats at either end of the *gag-pol-env* regions, known as the LTRs. Each LTR consists of a U3-R-U5 cassette and represents the intact regulatory region, including all the transcriptional sequences necessary to allow expression of the viral DNA genome (Fig. 3).

3.5. Expression of the Provirus from the Host-Cell Genome

Host-cell enzymes (such as RNA polymerase II) are now used by the integrated virus to express the *gag, pol,* and *env* proteins from spliced RNA transcripts. A splice-donor signal (AGGT) at nucleotides 203–206 downstream of the 5' LTR is used in conjunction with splice-acceptor (sa) sites in the genome, principally a major 3' splice site at about 6 kbp into the MoMLV genome, upstream of the *env* sequences. Full-length transcripts of viral RNA are also produced, beginning at the start of the R region in the 5' LTR and terminating at the end of the R region in the 3' LTR.

3.6. Packaging of Viral Components

These full-length transcripts (*see* Fig. 1) are packaged into viral particles at the membrane of the cell, following recruitment of the proteins synthesized earlier in the cycle. Packaging of the genomic transcripts into virion particles requires an RNA signal known as the Ψ sequence that lies 3' of the 5' LTR *(8,9)* and, therefore, downstream of the splice-donor site. This ensures that any spliced messages from the viral genome will not include the Ψ site and, so, will not be packaged. Hence, only full-length genomic transcripts become packaged into virions. If any vector DNA is to be recognized as a retroviral genome, this region must be included. Further definition of the Ψ sequence has shown that higher-titer packaging of genomes is possible by inclusion of a small region of the *gag* gene (from bases 215 to 1039 in the MLV genome, as numbered in ref. *10*). This revised Ψ+ packaging sequence is included in the most recent generation of high-titer retroviral vectors *(11) (see below)*. A viral protease encoded in the *pol* region cleaves the envelope polyproteins into smaller components, and virus particles "bud off" from the cell surface to continue the next cycle of infection (for review, *see* ref. *12*).

4. The Principles of Retroviral Vector Construction

To achieve the aims of *carriage* and *expression* of nonviral genes, the construction strategy must encompass two considerations:

1. The vector must be able to behave as a viral genome to allow it to pass as a virus from the producer cell line. Hence, its DNA must contain the regions of the wild-type retroviral genome required *in cis* for incorporation in a retroviral particle.
2. The vector must contain regulatory signals that lead to the optimal expression of the cloned gene once the vector is integrated in the target cell as a provirus. These may or may not be provided by viral DNA sequences.

4.1. Manipulation of the Vector as DNA

Vectors are constructed as part of bacterial plasmid DNA molecules (e.g., pBR322), which can be grown in bacteria in the usual way. As such, they can be manipulated using standard techniques for recombinant DNA. The vector-containing plasmids were synthesized initially from a proviral clone of MoMLV, by using, or creating, suitable restriction sites to remove or insert relevant DNA sequences. The DNA sequences surrounding the virus-specific DNA contain as few restriction sites as possible, to simplify the insertion of foreign sequences into unique restriction sites in the constructs. The following discussions will focus only on the viral sequences within the plasmid.

4.2. Propagation of the Vector as a Virus

The finding that retroviruses can transduce cellular genes suggested that the LTRs do not discriminate which genes they express. In fact, it has been shown that all the viral genes can be discarded and replaced by exogenous coding sequences. However, consideration of the life cycle highlights those signals that must be incorporated into the vector DNA. These essential sequence elements are:

1. The Ψ sequence that ensures the packaging of the vector DNA into virions. This sequence corresponds to nucleotides 215–565 in the MoMLV sequence *(10)*. More recent vector constructs contain the extended Ψ sequence, which extends up to nucleotide 1039, incorporating the start

of the *gag* gene (Ψ+), but with the AUG start codon of the viral gene mutated;
2. The tRNA binding site that is necessary to prime reverse transcription of the RNA form of the vector (in the virion) into DNA within the target cell (–PBS);
3. The sequences in the LTRs that permit the "jumping" of the reverse transcriptase between RNA strands during DNA synthesis;
4. Specific sequences near the ends of the LTRs that are necessary for the integration of the vector DNA into the host-cell chromosome in the ordered and reproducible manner characteristic of retroviruses *(13)*;
5. The sequences adjoining the 3' LTR that serve as the priming site for synthesis of the plus-strand DNA molecule (+PBS).

The inclusion of these elements into the retroviral vector N2 *(14)* is shown in Fig. 4A.

4.3. Vector Size

A major consideration is that the overall length of the construct cannot exceed 10 kbp, since packaging efficiency declines with increasing length of RNA. This restriction is a major hindrance to the choice of genes—and accompanying expression-regulatory signals—that can be incorporated into vectors *(see below)*. An operational limit of about 8 kbp of added sequence limits the use of retroviral constructs.

As a consequence of the size limitations (as well as to ensure that the genome becomes replication-defective), the functions of the *gag*, *pol*, and *env* regions must be removed to make way for the gene of interest in the vector. These can be provided *in trans* in cells in which the recombinant genome is expressed, known as packaging lines (*see* Chapter 2). Therefore, it becomes possible to express different coding regions between the LTRs, which can be packaged into infectious viral particles.

4.4. Expression of the Coding Cassette After-Infection

Once the vector DNA has been passed, as a virus, into a target cell, it will be reverse-transcribed and integrated into the host-cell genome. Following integration, the LTR will direct transcription of the gene cloned into the construct. The transcription signals in the LTR reside in the U3 and R regions, and include recognition sequences for RNA polymerase II and a strong, largely cell-nonspecific enhancer. In addition, a polyadenylation site exists in the R region and is recognized in the 3' LTR. The life cycle of the retrovirus involves differential expression of genomic transcripts and the use of splicing

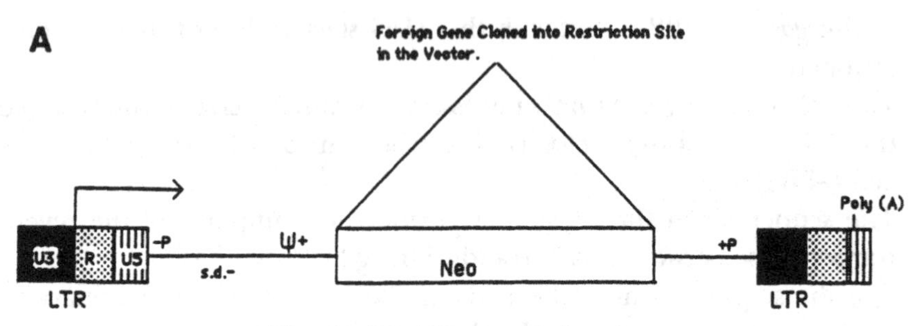

Fig. 4A. The N2 retroviral vector.

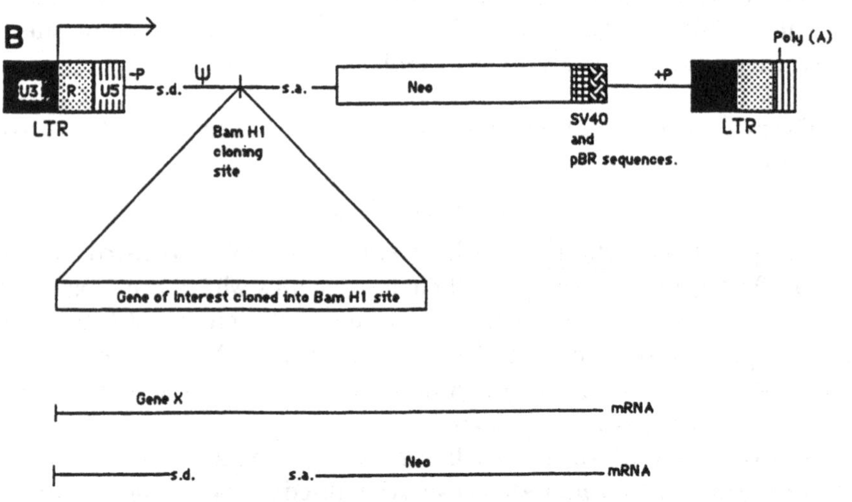

Fig. 4B. pZip Neo SV(X): A prototype double-expression vector.

signals to achieve expression of the *env* RNA. Inclusion of these splice-donor (sd) and splice-acceptor (sa) sites was initially thought to be necessary for efficient expression of cloned genes within a vector. However, in such constructs as the N2 vector (*see* Fig. 4A), the sd site of the MLV genome has been mutated to abolish the splicing signal. This feature removes the chance of aberrant splicing occurring from this signal to any cryptic sa sites that may exist in the inserted genes of the vector. Such unforeseen splicing in the producer line would reduce the levels of genomic transcripts and would lower the titer of any virus released.

If the expression of only a single function is required (i.e., a marker gene), the type of construct shown in Fig. 4A is adequate—provided that the LTR is known to function in the cell type of choice. In this N2 vector *(14)* (Fig. 4A), the *neo* gene has been inserted into a proviral clone of MoMLV from positions 1039 to 7673. When this vector is delivered to a cell by

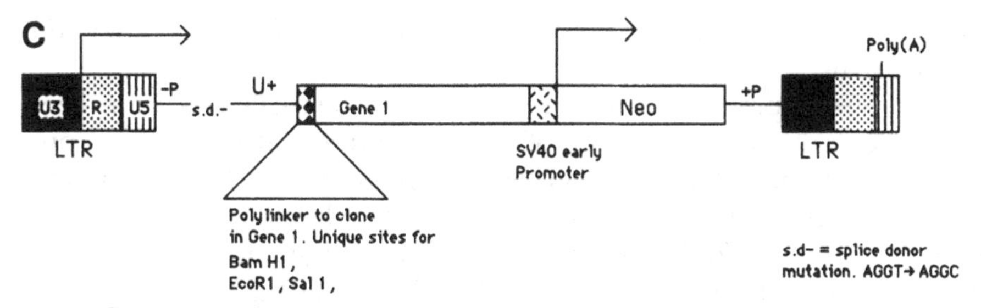

Fig. 4C. A double-expression vector with an internal promoter *(15).*

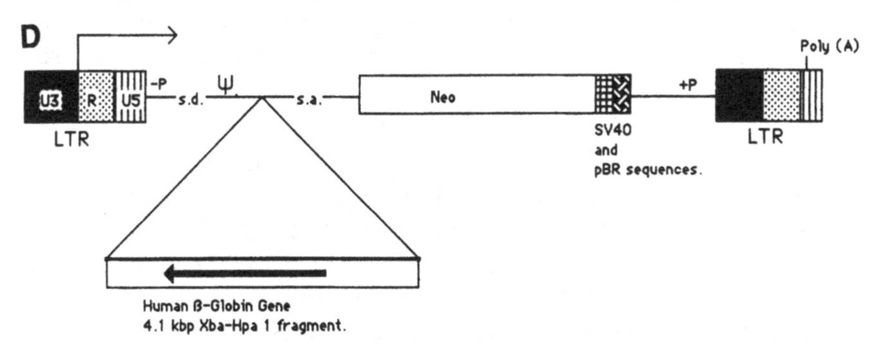

Fig. 4D. The HX19 double-expression vector *(17).*

viral infection, the LTR promoter elements drive expression of the *neo* gene, and infected cells can be recovered by their ability to grow under selection in G418.

However, it is frequently desirable to express more than one function from the vector. For instance, infection of a cell by a vector carrying a gene of biological interest may not be readily detectable. In this case, coexpression of a selectable marker will often save much effort by allowing ready identification of infected cells.

4.4.1. Double-Expression Vectors

The expression of the wild-type retroviral genome uses an sd signal close to the Ψ region at position 206 in the MLV genome (*see* Fig. 1), and an sa site located to the 5' side of the *env* region, to express the *env*-gene products. Early vectors attempted to use this mechanism by placing one gene in the *gag, pol* equivalent position and the second in the *env* equivalent position in the vector genome. An example of this type of double-expression vector is shown in Fig. 4B. The pZip Neo SV(X) vector allows the cloning of a second gene into a Bam H1 site that lies downstream of the 5' sd site, but upstream of the 3' sa site. The LTR transcription signals drive expression of both genes, relying on the utilization of the natural splicing mechanisms inherent to the wild-type virus.

However, expression of both functions from the LTR has often been shown to give disappointing levels of coexpression, and low titers from producer lines. This may be, in part, because the retroviral (*gag, pol*) message and the *env* message are naturally generated at different levels during expression of the provirus. In addition, the inclusion of a splicing requirement means that a lower proportion of RNA transcripts will be full length and, therefore, not packaged in the producer line. A solution to this problem has been the introduction of an *internal promoter* to drive expression of one of the foreign genes in a double-expression vector, or of just a single gene in a single-expression vector.

4.4.2. Internal Promoters

An example of a double-expression vector with an internal promoter is shown in Fig. 4C *(15)*. In this case, a polylinker region is present downstream from the Ψ+ signal. This allows a gene of choice to be inserted into the region that will be driven from the LTR. (The cloned gene should contain its own ATG for translation). The SV40 early promoter controls expression of the second gene. Since no splicing is required to express any messenger RNA species, the vector also contains the sd mutation described earlier.

Inclusion of internal promoter signals to drive expression of a gene makes it possible to remove dependence on the properties of the LTR as a promoter/enhancer cassette. The MoMLV LTR contains a largely cell-type-nonspecific, constitutive enhancer. This makes it suitable to express a selectable marker gene in a nonspecific way. However, the use of retroviral vectors to study more subtle phenotypic effects in specialized cell types demands that the expression of any introduced gene is correctly regulated within the target cell. Hence, it may be possible to include inducible, cell-type–specific, strong or weak promoters in vectors.

However, in vectors with two transcription units, in which one is expressed using the LTR-based promoter and the other is driven from an internal promoter, problems can arise, both in virus titers and in the relative levels of expression of the two functions. This arises from an effect known as "promoter selection" *(16)*. It has been noted that an LTR can exert a negative effect on other promoters directly downstream of it. Double-expression vectors are now being constructed with the internal transcription unit being inserted *in the orientation opposite* to the direction of transcription out of the LTR. Such reverse-orientation vectors appear to overcome some of the interference effects of the interaction of the promoters, although this is not a universal finding. In addition, reverse orientation constructs are especially useful in avoiding incorrect splicing of genomic transcripts included in the vector. Indeed, a globin gene, placed under its own promoter control but inserted in reverse orientation to the LTR, has been shown to be expressed at good levels and in the correct cell type (erythroid cells) *(17,18)*. (A theoretical objection to such vectors is that expression out of the LTR of the "viral" genome RNA in the producer cell line will be in the direction opposite to expression of the internal-construct RNA. This could lead to antisense blocking effects occurring between the two sets of transcripts, which may lead to a reduction in titer [i.e., levels of packageable genomic transcripts].) An example of a reverse orientation vector is shown in Fig. 4D.

4.4.3. Self-Inactivating Vectors

The retroviral DNA will integrate into the genome of the target cell largely at random (but *see* ref. *19*), but in a predictable, structurally and functionally intact fashion—a fact that is an attraction of the system. However, if a retrovirus integrates within a critical cellular gene, it may disrupt the coding sequence; alternatively integration near such a cellular gene may lead to the phenomenon of insertional mutagenesis. In these instances, the retroviral regulatory sequences can influence expression of the nearby cellular gene. This can lead to the inappropriate (over)expression of cellular sequences, primarily as a result of the influence of the strong retroviral-enhancer elements in both the 5' and 3' LTRs. If retroviral vectors are ever to be used in human therapies, it is important to minimize the possibility of insertional mutagenesis effects, which may cause transformation of the target cell. To avoid such potential dangers, vectors have been designed in which the viral LTR becomes transcriptionally inactive *after* infecting a cell. (This has the added advantage of removing any positive [or negative] effects that an active LTR may exert on the activity of any internal promoter in the construct.) As described in the life cycle, reverse transcription of the viral RNA dictates that the 3'-LTR U3 region is the heritable transcriptional unit of the retrovirus. That is, the 3' LTR in the virion (U3, R) is the template from which the 5' LTR of the provirus (U3, R, U5) is derived. Within the U3 region lie the promoter elements recognized by cellular RNA polymerase II, as well as a strong viral enhancer. Several "enhancer-deleted" (enh–) vectors with a deletion in the U3 region of the 3' LTR have been reported. The original such vector contained a 299-bp deletion that removed the 72-bp repeats (with enhancer and promoter activity) *(20)*, but others have included the entire promoter region in U3 *(21)* as well as the TATA box *(22)*. Viruses have been shown to be viable carrying as much as 327 bp that have been deleted from the 3' LTR *(23)*. On transfection into the packaging line, this enh– genome is transcribed, packaged, and subsequently copied into DNA. Upon reverse transcription, the enhancer deletion will be transferred into the 5' LTR, as well as being retained in the 3' LTR. This means that the vector now lacks its own promoter and enhancer sequences and, so, is transcriptionally crippled *(20,22)*. The gene

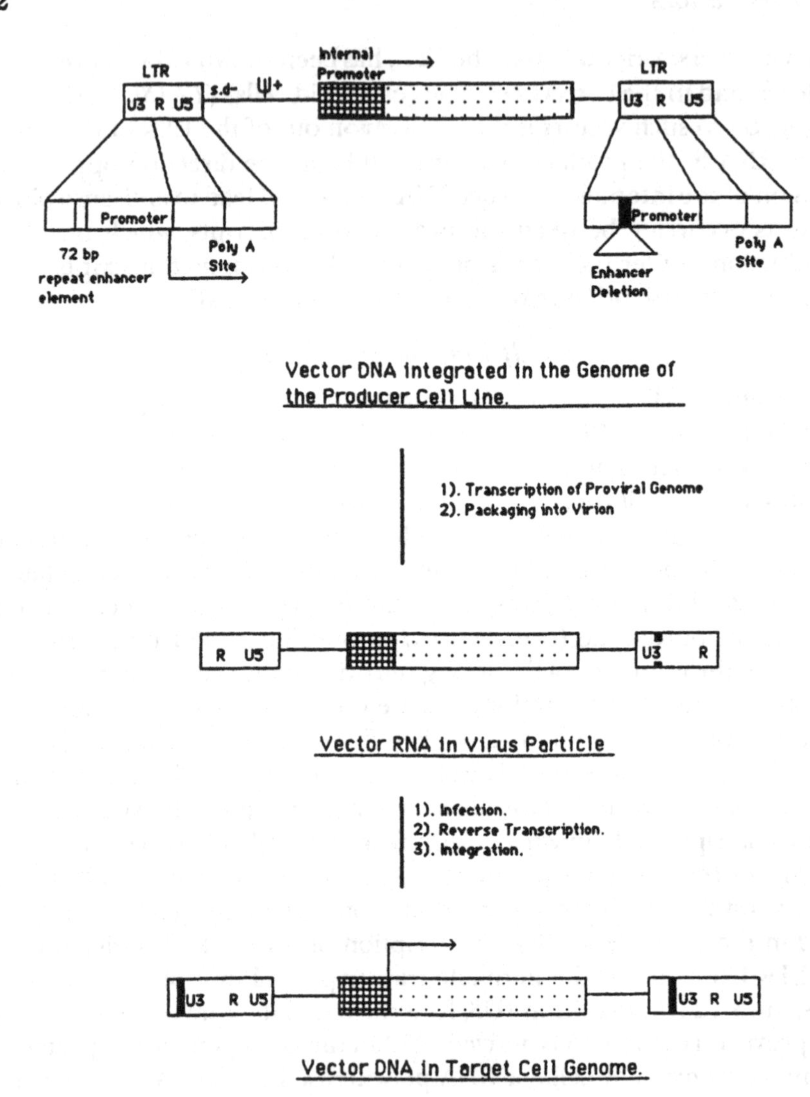

Fig. 5. A self-inactiving vector, showing inheritance of the enhancer deletion.

carried within enh– vectors must necessarily be accompanied by its own promoter, since no functions can be expressed from the LTR.

An added advantage of the enhancer deletion is that, with the LTR disabled, it is not possible to produce full-length transcripts of the integrated vector genome. This greatly reduces the chances of producing any wild-type virus from recombinations within the target cell. An example of an enh– vector currently in use is shown in Fig. 5.

Enhancer deletion has also allowed the use of a more unconventional vector design in an assay to identify cellular promoters *(24)*. In this case, a reporter gene was placed *within* the U3 region of the 3' LTR of an enh– vector, apparently with little effect on LTR function or packaging efficiency. Upon infection, the gene was placed only 50 nucleotides from flanking cellular DNA in the 5' LTR, and the LTR was transcriptionally silent. Only cellular promoters located near the position of integration could lead to expression of the reporter gene. Using cells expressing this gene, the vector sequences could easily be used to "tag" the integration site and clone the surrounding promoter elements.

5. Limitations of the System

Retroviral vectors can be used to transfer a wide variety of genes into different mammalian cell lines; expression of the encoded sequences can often confer phenotypic and genotypic changes that are stable. However, the functional and structural stability of the integrated proviral form of the vector is by no means absolute. Proviruses have been shown to exhibit instability, both of structure as a provirus and of expression of encoded functions. Viral sequences may be lost from infected cells, and viral gene expression can suddenly be shut off because of either genetic events within the cell or epigenetic events acting on the cell. For long-term usefulness, especially in clinical applications, the stability of gene expression from a vector must be assured, or at least predictable. The factors that play a role in determining the long-term stability of retroviral vectors are poorly characterized. Continuous growth of the infected cells in medium that selects for retention of the vector-encoded gene(s) helps to maintain the stability of integrated vectors, but other factors are also important. These include the target cell itself, the type of vector in use (including the nature of the promoter elements), and the foreign gene carried by the vector *(25)*.

6. Conclusion

The final choice of vector design depends on the requirements of each individual system. A balance must be struck between inclusion of the features that will lead to high levels of virus production in the producer cell line and those that will lead to good gene expression in the target cell line. The former relies on the vector carrying the smallest possible genetic elements that mark it as a retrovirus; the latter relies on the nature of the introduced genes and the information being sought. Often these considerations may oppose each other, and a compromise must be struck. Accurate prediction of how a vector will behave in any situation is difficult, and investigators should always try different constructs to optimize the chances of meeting their particular requirements. Nonetheless, the proven ability of retroviral vectors to permit highly efficient gene transfer at relatively high viral titers, with little or no contaminating wild-type virus present, makes them very attractive tools in an increasingly wide range of research fields.

References

1. Rous, P. (1911) A sarcoma of the fowl transmissible by an agent separable from the tumour cells. *J. Exp. Med.* **13,** 397–411.
2. Stehelin, D., Varmus, H. E., Bishop, J. M., and Vogt, P. K. (1976) DNA related to the transforming gene(s) of avian sarcoma viruses is present in normal avian DNA. *Nature* **260,** 170–173.
3. Neil, J. C., Hughes, D., McFarlane, R., Wilkie, N. M., Oinions, D. E., Lees, G., and Jarrett, O. (1984) Transduction and rearrangement of the *myc* gene by feline leukaemia virus in naturally occurring T-cell leukaemias. *Nature* **308,** 814–820.
4. Friedmann, T. (1989) Progress toward human gene therapy. *Science* **244,** 1275–1281.
5. Gluzman, Y. and Hughes, S. H. (1988) *Viral Vectors* (Cold Spring Harbor Laboratory, Cold Spring Harbor, NY).
6. Panganiban, A. T. and Fiore, D. (1988) Ordered interstrand and intrastrand DNA transfer during reverse transcription. *Science* **241,** 1064–1069.
7. Varmus, H. E. (1983)Retroviruses, in *Mobile Genetic Elements.* (Shapiro, J., ed.), Academic, NY, pp. 411–503.
8. Mann, R., Mulligan, R. C., and Baltimore, D. (1983) Construction of a retrovirus packaging mutant and its use to produce helper-free defective retrovirus. *Cell* **33,** 153–159.
9. Mann, R. and Baltimore, D. (1985) Varying the position of a retrovirus packaging sequence results in the encapsidation of both unspliced and spliced RNAs. *J. Virol.* **54,** 401–407.
10. Weiss, R. A., Teich, N., Varmus, J., and Coffin, J., eds. (1982,1985) *Molecular Biology of Tumor Viruses, RNA Tumor Viruses* (Cold Spring Harbor Laboratory, Cold Spring Harbor, NY), vols. 1,2.
11. Bender, M. A., Palmer, T. D., Gelinas, R. E., and Miller, A. D. (1987) Evidence that the packaging signal of Moloney murine leukaemia virus extends into the *gag* region. *J. Virol.* **61,** 1639–1646.
12. Varmus, H. E. (1988) Retroviruses. *Science* **240,** 1427–1435.
13. Panganiban, A. T. and Varmus, H. M. (1983) The terminal nucleotides of retrovirus DNA are required for integration but not virus production. *Nature* **306,** 155–160.
14. Keller, G., Paige, P., Gilboa, E., and Wagner, E. F. (1985) Expression of a foreign gene in myeloid and lymphoid cells derived from multipotent haematopoietic precursors. *Nature* **318,** 149–154.
15. Korman, A. J., Frantz, J. D., Strominger, J. L., and Mulligan, R. C. (1987) Expression of human class II major histocompatibility complex antigens using retrovirus vectors. *Proc. Natl. Acad. Sci. USA* **84,** 2150–2154.
16. Emerman, M. and Temin, H. M. (1984) Genes with promoters in retrovirus vectors can be independently suppressed by an epigenic mechanism. *Cell* **39,** 459–467.

17. Dzierzak, E. A., Papayannopoulou, T., and Mulligan, R. C. (1988) Lineage-specific expression of a human β-globin gene in murine bone marrow transplant recipients reconstituted with retrovirus-transduced stem cells. *Nature* **331,** 35–41.
18. Grosveld, F., van Assendelft, G. B., Greaves, D. R., and Kollias, G. (1987) Position-independent, high-level expression of the human β-globin gene in transgenic mice. *Cell* **51,** 975–985.
19. Shih, C.-C., Stoye, J. P., and Coffin, J. M. (1988) Highly preferred targets for retrovirus integration. *Cell* **53,** 531–537.
20. Yu, S.-F., von Ruden, T., Kantoff, P. W., Garber, C., Seiberg, M., Ruther, U., Anderson, W. F., Wagner, E. F., and Gilboa, E. (1986) Self-inactivating retroviral vectors designed for transfer of whole genes into mammalian cells. *Proc. Natl. Acad. Sci. USA* **83,** 3194–3198.
21. Cone, R. D., Weber-Benarous, A., Baorto, D., and Mulligan, R. C. (1987) Regulated expression of a complete human beta globin gene encoded by a transmissible retrovirus vector. *Mol. Cell. Biol.* **7,** 887–897.
22. Yee, J. K., Moores, J. C., Jolly, D. J., Wolff, J. A., Respess, J. G., and Friedmann, T. (1987) Gene expression from transcriptionally disabled retroviral vectors. *Proc. Natl. Acad. Sci. USA* **84,** 5197–5201.
23. Hawley, R. G., Covarrubias, L., Hawley, T., and Mintz, B. (1987) Handicapped retroviral vectors efficiently transduce foreign genes into haematopoietic stem cells. *Proc. Natl. Acad. Sci. USA* **84,** 2406–2410.
24. Von Melchner, H. and Ruley, H. E. (1989) Identification of cellular promoters by using a retrovirus promoter trap. *J. Virol.* **63,** 3227–3233.
25. Xu, Li., Yee, J.-K., Wolff, J. A., and Friedmann, T. (1989) Factors affecting long-term stability of Moloney murine leukaemia virus-based vectors. *Virology* **171,** 331–341.

CHAPTER 2

Construction of Retroviral Packaging Cell Lines

Olivier Danos

1. Introduction

Most of the time, retrovirus vectors retain only *cis*-acting sequences from the original viral genome. These sequences allow the recombinant structure to be transcribed (LTR promoter/enhancer) and the RNA to be processed (splicing and polyadenylation signals), packaged into a virion particle (packaging sequences), and replicated by the reverse transcriptase (tRNA binding site, R region, and polypurine track). The other viral functions have to be provided *in trans* for the assembly of recombinant viral particles to take place. This can be simply achieved by using a replication-competent helper virus, leading to the production of a mixed population. Nevertheless, helper-free stocks are desirable for most applications, since

1. The high frequency of recombination in a mixed virus stock is likely to lead to the appearance of recombinants with unknown structure and activity. These new chimeras, either spread by the helper virus or replication-competent themselves, create a potential safety problem.
2. Cell lineage analysis using retroviral marking can only be performed and interpreted in a helper-free context.
3. In vivo gene transfer experiments can be jeopardized by disease(s) associated with helper-virus infection *(1)*.

From: *Methods in Molecular Biology, Vol. 8: Practical Molecular Virology: Viral Vectors for Gene Expression*
Edited by: M. Collins

For these reasons, packaging cell lines have been developed to produce helper-free and replication-defective recombinant retroviruses. The viral *gag, pol,* and *env* gene products are stably expressed in these cell lines and can complement a defective construct introduced by transfection or infection. The packaging functions are usually expressed from a modified genome, the most important characteristic of which is a deletion of a *cis*-acting region required for packaging (Ψ, or E in the avian systems). As a consequence of this deletion, the helper structure provides only the packaging functions and, in first approximation, is not itself encapsidated. Table 1 summarizes the features and properties of helper cell lines that have been developed from both avian (reticuloendotheliosis virus, REV and Rous sarcoma virus, RSV) and murine (murine leukemia virus, MLV) viruses. This chapter focuses on MLV-based packaging systems.

Recently, we and others sought to solve the problem of helper-virus formation in producer clones isolated from packaging lines *(2–5)*. This problem is believed to originate in the leakiness of the Ψ^- deletion: Indeed, it is possible to show that Ψ^- genomes can be encapsidated at a 1000-fold-reduced rate *(6,7)*. Therefore, within the oversimplifying hypothesis by which the presence of Ψ is the only determinant of the encapsidation rate, one can expect to find among the particles released by a producer clone 0.1% of "heterozygous virions" containing the Ψ^- helper genome together with a Ψ^+ vector. During reverse transcription, recombination between the two RNA genomes is known to occur at a high frequency *(8)* and, in this way, rescue of the Ψ^- deletion (or of other *cis*-acting defects in the helper genome) can take place.

Other recombination events, such as gene conversion in the producer cell, have been proposed as an explanation for the emergence of replication-competent virus *(9)*. However, the frequency of such rearrangements is very low, and taking simple precautions for producer isolation and maintenance is enough to avoid them. The "helper-virus problem" has been solved by expressing the *gag, pol,* or *env* gene products in the packaging line, as complementing mutants on different DNA constructs. Even if copackaged with a vector genome, these mutants cannot be rescued.

Future improvements of packaging lines will have to deal with two main points. First, even though the safety level of packaging lines has been substantially improved, the encapsidation of endogenous genomes (or genome-like structures, such as VL30 in murine cells) and their transmission to the target cells are still potential problems *(10,11)*. A packaging cell line to be used in human gene therapy protocols would better be derived from one of the primate lines commonly used in vaccine production.

Second, one still needs to try to increase the average titer of recombinant virus obtained from a producer clone. This is a critical step in achieving

Table 1
Characteristics of Current Murine and Avian Retroviral Packaging Cell Lines

Packaging cell line	Ref.	Construction[a]	Cotransfected drug resistance[b]	Host range	Helper virus[c]
MURINE					
Ψ2	*14*	Ψ⁻ Mo-MuLV	gpt	Murine ecotropic	*2, 13, 22, 23*
Ψam	*18*	(Ψ⁻, 4070 A *env*)	gpt	Murine amphotropic	*18*
PA12	*19*	(Ψ⁻, 4070 A *env*)	HSV TK	Murine amphotropic	*24, 26*
PA 317	*2*	(Ψ⁻, 4070 A *env*, ppt⁻, SV40 polyA)	HSV TK	Murine amphotropic	*3*
Clone 32	*13*	(mMTI – Ψ⁻, SD⁻, *gag pol env*) + (mMTI – Ψ⁻, SD⁻, Δ *gag pol*), cotransfected	neo	Murine ecotropic	*13*
GP+E-86	*4*	(Ψ⁻, Δ *env*) + (Ψ⁻, Δ *gag pol*), cotransfected	gpt	Murine ecotropic	NR
GP+*env*Am12	*5*	(Ψ⁻, Δ *env*) + (Ψ⁻, Δ *gag pol*, 4070 A *env*), separately transfected	gpt + hph	Murine amphotropic	NR
ΨCRE	*3*	(Ψ⁻, *env*⁻, SV polyA) + (Ψ⁻, *gag pol*⁻, SV polyA), separately transfected	hph + gpt	Murine ecotropic	NR
ΨCRIP	*3*	Same as ΨCRE, with 4070 A *env*[d]	hph + gpt	Murine amphotropic	NR
AVIAN					
C3	*20*	(E⁻, SD⁻, REV-A *gag pol env*) + (E⁻, SD⁻, Δ *gag pol*, REV-A *env*), cotransfected	neo	Amphotropic	*9, 25*
DSN	*9*	(CMV IE – SNV *gag pol*) + (RSV LTR – SNV *env*), cotransfected	neo	Amphotropic	NR
DAN	*9*	(CMV IE promoter – SNV *gag pol*)+ (CMV IE promoter – MLV 4070 A *env*), cotransfected	neo	Murine amphotropic	NR
Q4dh	*21*	Ψ⁻ Rous associated virus (RAV 1) genome	hph	Chicken	*21*
pHF-g	*27*	Ψ⁻ Rous associated virus (RAV 1) genome ppt⁻, HSVTKpolyA	hph	Chicken	NR

[a]Abbreviations: ppt, poly purine track; SVpolyA, simian virus 40 polyadenylation signal; mMTI, mouse metallothionein I promoter; SD, splice donor; Δ, deletion; REV-A, reticuloendotheliosis virus subgroup A; CMV IE, Cytomegalovirus immediate early promoter; SNV, spleen necrosis virus; RSV, Rous sarcoma virus.

[b]Abbreviations: gpt, *E. coli* xanthine–guanine phosphoribosyltransferase; HSV TK, herpes simplex virus thymidine kinase; neo, *E. coli* neomycin phosphotransferase; hph, *E. coli* hygromycin B phosphotransferase; his, *S. thyphimurium* histidinol dehydrogenase.

[c]References of reports describing helper virus in producer clones derived from packaging lines are given; NR: none reported.

[d]Other cell lines expressing envelopes from murine xenotropic and polytropic viruses, as well as from the Sarma feline leukemia virus, have also been obtained (O. Danos, unpublished).

efficient and reproducible in vivo gene transfer to somatic tissue in animals larger than the mouse, including humans. Regarding this, one might need to produce packaging lines expressing a modified or substituted envelope gene, seeking increased stability of the protein and a tighter, more specific interaction with its receptor.

In this chapter, procedures are described and guidelines suggested that can be generally applied for the design, isolation, and maintenance of retrovirus packaging lines. Briefly, two defective helper genomes carrying complementary mutations in their coding sequences are engineered, starting from a cloned proviral DNA. These constructs are stably introduced into the parental cell line in two rounds of transfection and selection. The resulting cellular clones are then assayed for their ability to package a defective recombinant retrovirus.

An easy and sensitive assay for the presence of replication-competent helper virus is also described. Such an assay is necessary to show that cells infected with amphotropic viral stocks under P3 biological containment level are helper-free and can be handled according to less stringent safety guidelines. It is also advisable to check for helper–virus serum or tissues from animals engrafted with infected cells. Classical assays like XC plaque fprmation *(12)* are sensitive, but rather time-consuming and sometimes difficult to read out. Furthermore, they detect only actively replicating virus (i.e., able to spread from an infectious center and form a conspicuous syncitium or focus). Transfer of the packaging functions to the infected cells, potentially creating a secondary producer, is not detected. For these reasons, we prefer to use a "mobilization assay" by which a cell line harboring a defective recombinant provirus is challenged by the fluid to be tested. If helper virus is present, recombinant virus will be produced by (mobilized from) the challenged cells in the culture media. The mobilized provirus carries a selectable marker (*his*) or a reporter gene (*lac Z*) and can be easily detected.

1.1. Design of the Packaging Construct(s)

Most packaging constructs have been derived from infectious cloned provirus and retain a genome-like structure. This is based on the idea that, by minimizing the changes in the provirus structure, one would retain optimal expression and processing of the viral RNAs. However, packaging lines have been constructed with the viral functions expressed from heterologous promoters (Table 1). In these lines (which give titers equivalent to or lower than the others) recombination leading to helper-virus formation still occurs, although it was first thought that this problem would be avoided, since the regions of homology between the packaging constructs and the vector were minimal *(13)*. Therefore, there is no clear advantage to this strategy.

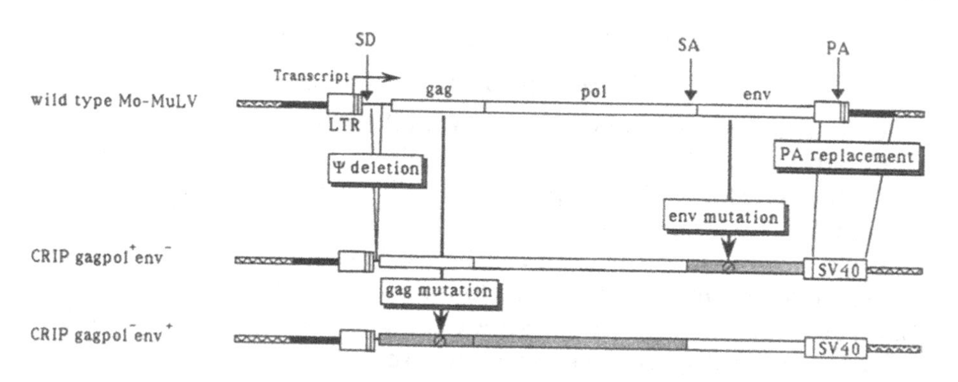

Fig. 1. General structure of the packaging constructs. The structure of a cloned infectious Mo-MuLV provirus with its two derived crippled structures is shown. The retroviral sequences are flanked by cellular (black boxes) and bacterial plasmid (decorated boxes) sequences. Active (open boxes) and inactive (shaded boxes) genes and control regions are indicated. Detailed structures of the CRIP constructs may be found in ref. *3*. Abbreviations: SD, splice donor; SA, splice acceptor; PA, polyadenylylation signal. Mutations are displayed as Ø.

Figure 1 shows the structure of a typical packaging construct. Its features are the following:

- *Ψ⁻ deletion:* The original deletion in Mo-MLV is 350 bp long and spans a region between the *env* splice donor and the *gag* ATG *(14)*. It has been shown that a shorter deletion can be sufficient *(4, 7)*, but considering the leakiness of the phenotype, it is probably advisable to remove as much as possible from this region.
- *Removal of the 3′ LTR:* Miller and Buttimore *(2)* have shown that a total deletion of the LTR and its replacement with SV40 late polyadenylation sequences still resulted in efficient expression of the packaging functions. The polypurine track used for second-strand synthesis during reverse transcription was also included in this deletion.
- *Mutations in the coding sequences:* It is preferable to work with mutations of known phenotype, already characterized and rescuable in the context of a cloned infectious provirus. By doing so, one will avoid introducing unwanted secondary effects on RNA transcription and processing. Short deletions or insertions are better than point mutations, since they are less prone to reversion during reverse transcription, if the packaging construct happens to be encapsidated.
- *Promoter elements:* If nonmurine cells are to be used as recipients, it might be useful to use different promoter elements to drive the packaging function. In human cells, for instance, the MLV LTR is inefficient and should be replaced by a stronger one.

1.2. Primary Transfections and Screening

1.2.1. Choice of the Parental Cell Line

Only attached cell lines have been used to derive packaging lines. All the MLV packaging lines shown in Table 1 are derived from mouse NIH 3T3 cells. Dog D17 and quail QT6 cells have been used to isolate REV and ASLV packaging lines, respectively.

Cells growing in suspension can be used, although the isolation of producer clones requires single-cell cloning and therefore is tedious. The advantage is that the cells can be grown at higher densities and, possibly, more concentrated virus stock can be recovered.

The recipient cell line should be easily transfectable and have a good cloning efficiency. It should be free of endogenous reverse transcriptase activity and contain as few active (or activatable) endogenous proviruses as possible. Finally, it must be nontransformed and nontumorigenic in vivo.

1.2.2. Choice of the Cotransfected Selectable Markers

Two different selectable markers will be necessary to isolate the packaging line. Among the positive selection systems, one should avoid those commonly used in retroviral vectors, such as G418 resistance (neo). The marker can be expressed on a plasmid of the pSV2 type *(15)*.

1.2.3. First Transfection and Cloning

The (*env*$^-$, *gagpol*$^+$) construct is transfected first, together with one of the selectable markers, with a molar ratio of 10:1. Any transfection protocol yielding stable transformants (calcium phosphate coprecipitate, electroporation, or lipofection) is suitable. Following transfection and selection, individual colonies are isolated using cloning cylinders and taking great care of crosscontamination (colonies should be sparse enough on the plate). About 30 colonies are picked and transferred to 24-well Falcon plates, and the 20 that grow best are tested for reverse transcriptase activity (*see* Methods section). Keep the 2 or 3 best clones, expand them, and keep several stock vials frozen in liquid nitrogen. Grow the best clone and submit it to a second round of transfection.

1.3. Second Transfection

The second construct (*gagpol*$^-$, *env*$^+$) is transfected with a second selectable marker and clones are selected as before. Since the envelope glycoprotein is being expressed on the cell membrane, it should be possible to select clones expressing high levels by FACS directly. This eliminates the need for a second marker and, through several sortings, more stable clones can be isolated. The clones are then tested for their ability to package a recombinant construct with the packaging assay (*see* Methods section).

2. Materials

2.1. Reverse Transcriptase Assay 2× Cocktail

In bidistilled water: 100 m*M* Tris-HCl, pH 8.3, 25 m*M* DTT, 1.2 m*M* $MnCl_2$, 120 m*M* NaCl, 0.1% NP40, 20 µ*M* dTTP, 0.33 µ*M* dTTP $\alpha^{32}P$ (3000 Ci/mmol), 10 µg/mL Oligo dT *(12–18)* (Pharmacia), 20 µg/mL Poly rA *(12–18)* (Pharmacia).

2.2. Cell Lines

NIH 3T3 cells were obtained, at an early passage, from G. Cooper (Dana Farber Cancer Institute, Boston, MA). The 3T3-116 and 3T3-BAG cells were isolated after infection of NIH 3T3 cells with the MSVhis *(3)* and the BAG *(16)* recombinant-defective viruses, respectively. These cell lines are available from the author.

2.3. BAG Virus Stock

A viral stock is prepared by transfecting the relevant packaging line with the DNA construct, as described elsewhere in this volume; isolating a high-titer producer; and harvesting virus from this producer. The stock is then kept at –70°C. The structure of the BAG virus is detailed in ref. *16*, and the construct is available from the authors.

3. Methods

3.1. Reverse Transcriptase Assay (Adapted from Ref. 17*)*

In order to select the clone with the highest RT activity, it is important to perform the test on exponentially growing cells having the same density.

1. Seed all clones to be tested at the same, subconfluent density (for 3T3-derived cells, use 3.10^4 cells/cm^2 in Falcon 6-well plates).
2. Change the medium after 8 h (2 mL/well).
3. After 16 h, take 500 µL of culture medium and spin for 1 min in an Eppendorf tube to remove detached cells and debris. The supernatants can be stored at –80°C and assayed later.
4. Prepare the 2× cocktail described in the Materials section.
5. Mix 50 µL of culture supernatant with 50 µL of 2× cocktail. If many samples are tested, it is convenient to use 96-well microtiter plates.
6. Incubate for 3 h at 37°C.
7. Filter on DE.81 paper (Whatman), using a dot blot apparatus, with 2 sheets of 3MM paper under the filter.
8. Wash 3 times with 2× SSC, and then wash the filter itself on a shaker, in 2× SSC, 3 times, 10 min each.

9. Wash the filter twice (1 min each) in 95% ethanol and dry quickly.
10. Expose with screen. There should be a strong signal after 3 h.

3.2. Packaging Assay

1. Prepare a high-titer, helper-free stock of a recombinant retrovirus containing the β-galactosidase gene and a selectable marker not already in the packaging line (i.e., BAG; *see* ref. *16*, and Note 1).
2. After selection and isolation, grow to confluence the clones to be tested and split them (1 in 20) in a Falcon 6-well plate for infection the next day.
3. Infect the clones with 10^3 cfu in 0.5 mL, with 2 µg/mL polybrene, following the procedure described in Chapter 3.
4. After 48 h, split the cells into selection (1 well into two 10-cm dishes) (*see* Note 2).
5. When individual colonies become apparent (8 d for NIH 3T3 in G418 or histidinol), pool them from the two plates and grow the population to confluence.
6. Split the infected population (1 in 10) and grow to subconfluence.
7. Change the medium and harvest the virus stock after 16 h.
8. Infect suitable target cells and score β-galactosidase-positive clusters after 48 h.

3.3. Growth and Storage of the Packaging Lines (see Note 3)

Immediately after isolation, the cell line is grown in sufficient quantity for storage in 40–50 vials in liquid nitrogen (from about 20 dishes [10 cm], 2–3 × 10^6 cells/mL in culture medium that contains 10% DMSO). If the cell line is carried continuously in the laboratory, management of the stock is as follows:

1. Thaw one stock vial, grow, and check packaging efficiency in a transient assay (*see* Section 3.2.).
2. Freeze 10 vials in liquid nitrogen, as described above, for a working stock.
3. Grow the cells, passaging them every 3–4 d, never allowing them to remain overconfluent.
4. Use a new vial from the working stock after 6 wk of continuous passage.

3.4. Mobilization Assay for Helper Virus (see Note 4)

1. Seed the indicator cell line at 2–3 × 10^5 cells/10-cm dish.
2. After 8–16 h, infect with 2-mL serial dilutions of the fluid to be assayed.
3. Split the cells (1 in 20), grow to confluence, and change the medium (10 mL/10-cm dish) (*see* Note 5).

4. After 16 h, harvest the medium from the cells.
5. Infect NIH 3T3 cells (or any suitable cell line) with 2 mL/10-cm dish of this culture supernatant, both undiluted and at dilutions of 10-, 100-, and 1000-fold.
6. After 48 h, split the infected NIH 3T3 (1 in 10) in selective media and grow colonies, or stain for β-galactosidase (β-gal) activity if the 3T3-BAG cell line is used.

4. Notes

1. Make sure that the host range of this test virus will allow it to infect the clones where you want to assay the packaging functions. Because of receptor interference, a cell expressing a retroviral envelope glycoprotein cannot be infected by the corresponding virus. Therefore, to test the packaging ability of clones expressing the ecotropic envelope, an amphotropic packaging line should be used to produce the β-gal virus.
2. At this stage, if many clones are tested and to reduce the amount of tissue-culture work, one can perform a transient assay. Change the media 16 h before splitting into selective media, take a transient harvest, and go to Step 7. Once the 4 or 5 best clones have been scored, one can discard the others and perform Steps 5–8 with only the best clones.
3. Some instability of the packaging lines has been observed (ref. 2; O. Danos, unpublished). The problem essentially concerns the lines expressing the murine 4070A amphotropic envelope, in which progressive loss of packaging functions sometimes occurs. Periodical reselection and recloning can be a solution, but it is time-consuming and tedious, since the new clones have to be tested again for packaging. By growing and storing the cells properly, one can reduce the instability problem to a tolerable level. The cells should always be grown in optimal conditions. The serum used, for instance, can be very important. In the case of Ψ CRIP and Ψ CRE cells, calf serum is much better than fetal calf serum, in which instability is more frequently observed. As a general rule, cells that have been submitted to harsh conditions (i.e., cell death >75%) during shipping, careless thawing, incubator failure, or massive mycoplasma contamination should be discarded, since surviving cells are the most likely to have lost the packaging functions.
4. As an indicator cell line, we use either 3T3-116 containing an his provirus or 3T3-BAG containing the BAG provirus. The assay with the 3T3-BAG cell line is quicker, since the read-out can be done 48 h after infection with the mobilized virus.
5. This step is optional, and the presence of the mobilized virus in the

culture media can be checked at this stage. However, splitting cells eventually allows reinfection by the helper virus to occur and significantly increases the level of detection. Using this procedure, the assay is 10- to 100-fold more sensitive than the XC assay.

References

1. Teich, N., Wyke, J., Mak, T., Bernstein, A., and Hardy, W. (1984) Pathogenesis of retrovirus-induced disease, in *RNA Tumor Viruses* (Weiss, R., Teich, N., Varmus, H., and Coffin, J., eds.), Cold Spring Harbor Laboratory, Cold Spring Harbor, NY, pp. 785–998.
2. Miller, A. D. and Buttimore, C. (1986) Redesign of retrovirus packaging cell line to avoid recombination leading to helper virus formation. *Mol. Cell. Biol.* **6,** 2895–2902.
3. Danos, O. and Mulligan, R. C. (1988) Safe and efficient generation of recombinant retroviruses with amphotropic and ecotropic host range. *Proc. Natl. Acad. Sci. USA* **85,** 6460–6464.
4. Markowitz, D., Goff, S., and Bank, A. (1988) A safe packaging line for gene transfer: Separating the viral genes on two different plasmids. *J. Virol.* **62,** 1120–1124.
5. Markowitz, D., Goff, S., and Bank, A. (1988) Construction and use of a safe and efficient amphotropic packaging cell line. *Virology* **167,** 400–406.
6. Danos, O., Mulligan, R. C., and Yaniv, M. (1986) Production of spliced DNA copies of the cottontail rabbit papillomavirus genome in a retroviral vector. *Ciba Found. Symp.* **120,** 68–77.
7. Mann, R. and Baltimore, D. (1985) Varying the position of a retrovirus packaging sequence results in the encapsidation of both unspliced and spliced RNAs. *J. Virol.* **54,** 401–407.
8. Linial, M. and Blair, D. (1984) Genetics of retroviruses, in *RNA Tumor Viruses* (Weiss, R., Teich, N., Varmus, H., and Coffin, J., eds.), Cold Spring Harbor Laboratory, Cold Spring Harbor, NY, pp. 649–783.
9. Dougherty, J. P., Wisniewsky, R., Yang, S., Rhode, B. W., and Temin, H. M. (1989) New retrovirus helper cells with almost no nucleotide sequence homology to retrovirus vector. *J. Virol.* **63,** 3209–3212.
10. Colicelli, J. and Goff, S. P. (1988) Isolation of an integrated provirus of Moloney murine leukemia virus with long terminal repeats in inverted orientation: Integration utilizing two U3 sequences. *J. Virol.* **62,** 633–636.
11. Scadden, D. T., Fuller, B., and Cunningham, J. M. (1990) Human cells infected with retrovirus vectors acquire an endogenous murine provirus. *J. Virol.* **64,** 424–427.
12. Rowe, W. P., Pugh, W. E., and Hartley, J. W. (1970) Plaque assay techniques for murine leukemia viruses. *Virology* **42,** 1136–1139.
13. Bosselman, R. A., Hsu, R.-Y., Bruszewski, J., Hu, S., Martin, F., and Nicolson, M. (1987) Replication-defective chimeric helper proviruses and factors affecting generation of competent virus: Expression of Moloney murine leukemia virus structural genes via the metallothionein promoter. *Mol. Cell. Biol.* **7,** 1797–1806.
14. Mann, R., Mulligan, R. C., and Baltimore, D. (1983) Construction of a retrovirus packaging mutant and its use to produce helper free defective retrovirus. *Cell* **33,** 149–153.
15. Mulligan, R. C. and Berg, P. (1980) Expression of a bacterial gene in mammalian cells. *Science* **209,** 1422–1427.

16. Price, J., Turner, D., and Cepko, C. (1987) Lineage analysis in the vertebrate nervous system by retrovirus-mediated gene transfer. *Proc. Natl. Acad. Sci. USA* **84,** 156–160.
17. Goff, S., Trackman, P., and Baltimore, D. (1981) Isolation and properties of Moloney murine leukemia virus mutants use of a rapid assay for release of virion reverse transcriptase. *J. Virol.* **38,** 239–248.
18. Cone, R. D. and Mulligan, R. C. (1984) High efficiency gene transfer into mammalian cells: Generation of helper-free recombinant retrovirus with broad mammalian host range. *Proc. Natl. Acad. Sci. USA* **81,** 6349–6353.
19. Miller, A. D., Law, M. F., and Verma, I. M. (1985) Generation of helper free amphotropic retroviruses that transduce a dominant-acting, methotrexate-resistant dihydrofolate reductase gene. *Mol. Cell. Biol.* **5,** 431–437.
20. Watanabe, S. and Temin, H. M. (1983) Construction of a packaging cell line for avian reticuloendotheliosis virus vectors. *Mol. Cell. Biol.* **3,** 2241–2249.
21. Stoker, A. W., and Bissel, M. J. (1988) Development of avian sarcoma and leukosis virus-based vector-packaging cell lines. *J. Virol.* **62,** 1008–1015.
22. Snodgrass R. and Keller, G. (1987) Clonal fluctuation within the haematopoietic system of mice reconstituted with retrovirus-infected stem cells. *EMBO J.* **6,** 3955–3960.
23. Belmont, J. W., MacGregor, G. R., Wager-Smith, K., Fletcher F. A., Moore, K. A., Hawkins, D., Villalon, D., Chang, S. M. -W., and Caskey, C. T. (1988) Expression of human adenosine deaminase in murine hematopoietic cells. *Mol. Cell. Biol.* **8,** 5116–5125.
24. Miller, A. D., Trauber, D. R., and Buttimore, C. (1986) Factors involved in the production of helper virus-free retrovirus vectors. *Somatic Cell Mol. Genet.* **12,** 175–183.
25. Hu, S., Bruszewski, J., Nicolson, M., Tseng, J., Hsu, R. Y., and Bosselman, R. (1987) Generation of competent virus in the REV helper cell line C3. *Virology* **159,** 446–449.
26. Wolf, J. A., Yee, J. K., Skelly, H. F., Moores, J. C., Respess, J. G., Friedmann, T., and Leffert, H. (1987) Expression of retrovirally transduced genes in primary cultures of adult rat hepatocytes. *Proc. Natl. Acad. Sci. USA* **84,** 3344–3348.
27. Savatier, P., Bagnis, C., Thoraval, P., Poncet, D., Belakebi, M., Mallet, F., Legras, C., Cosset, F.-L., Thomas, J. L, Chebloune, Y., Faure, C., Verdier, G., Samarut, J., and Nigon, V. (1989) Generation of a helper cell line for packaging avian leukosis virus-based vectors. *J. Virol.* **63,** 513–522.

CHAPTER 3

Making High Titer Retroviral Producer Cells

Stephen J. Russell

1. Introduction

Retroviral vectors are uniquely suitable for high-efficiency gene transfer to a large target cell population and this (among other things) has led to the exploration of previously impractical strategies for somatic gene therapy *(1,2)*. Thus, for example, cultured bone marrow can be used to recolonize mice with hematopoietic progenitors expressing a variety of genes after retroviral gene transfer *(3)*, and efforts to achieve the same with human bone marrow may eventually result in effective therapies for sickle cell anemia, the thalassemias, and a variety of other single-gene disorders. The major factors limiting the success of this type of experiment are the maximum achievable concentration of recombinant virus particles, the density of virus receptors on the target cell population (which must be actively proliferating, since the processes of reverse transcription and integration are S-phase-dependent) and the emergence of replication-competent "helper virus" in the system.

1.1. Virus Concentration

Overnight supernatant from a good retroviral packaging cell line contains about 10^8 empty virions/mL, and this represents the theoretical maximum achievable virus titer. Virion production by packaging cell lines may diminish with time, making the use of a frozen early passage advisable for

From: *Methods in Molecular Biology, Vol. 8: Practical Molecular Virology: Viral Vectors for Gene Expression*
Edited by: M. Collins © 1991 The Humana Press Inc., Clifton, NJ

each new experiment. Some ailing packaging cell lines may be revived by the use of appropriate selective media when the plasmids coding for viral structural proteins are linked to selectable marker genes. In practice, the maximum achievable recombinant viral titer is closer to 10^7/mL, and a more realistic goal is about 10^6/mL. Titers below 10^6/mL are likely to be inadequate for the type of murine bone marrow experiment discussed above.

Viral titer depends not only on the quality of the packaging cell line, but also on the quantity of packagable RNA per cell and the extent of packaging sequence included. Unspliced, single-gene, LTR-driven constructs with no internal promoter and extended packaging sequence (including the 5' end of the gag coding region) give the highest titers, particularly when the single gene is itself a selectable marker. The use of internal promoters, multiple genes, RNA splicing, reverse-orientation inserts (which generate antisense RNA), and shorter packaging sequences all serve to reduce attainable viral titers to a variable degree. The choice of construct (which may be dictated by other considerations—*see* elsewhere in this vol) therefore has a profound effect on viral titer.

1.2. Virus Receptors

Currently available retroviral packaging cell lines generate murine leukemia virus-based recombinant viruses with either ecotropic or amphotropic host range. Ecotropic virus receptors are expressed on murine and some other rodent cells, but not on the cells of unrelated species; amphotropic receptors are present on the cells of a wide range of animal species as diverse as mouse, dog, and human *(4)*. Amphotropic receptor density may vary considerably between species and between different cell populations from a single species. Thus a milliliter of viral supernatant from an amphotropic producer cell line capable of infecting 10^6 NIH3T3 may only infect 10^4 human epithelial cells. As a general rule, amphotropic viral titers are lower on human than on murine target cells. It is theoretically possible to construct packaging cells and gene transfer vectors based on virtually any retrovirus from any species (including human), which may give a better choice of vectors for specific target cell populations.

1.3. Helper Virus

A particularly troublesome aspect of retroviral gene transfer has been the emergence of helper virus, either in the packaging cell line or in the infected target cell population. This is a result of homologous recombination between plasmids in the packaging cell and recombination during reverse transcription between copackaged psi^+ and psi^- RNA transcripts (psi^- RNA transcripts coding for viral structural proteins are packaged at low frequency).

Newer packaging cell lines in which viral structural proteins are encoded by multiple plasmids have largely overcome this problem (*see* Chapter 2), but it is still advisable to check viral producer cell lines and infected target cells in a sensitive helper-virus assay.

The methods to be discussed include generation of cell lines that release recombinant retroviruses, techniques for assessing the titer of these viruses, and, finally, some tricks to increase both the titer and the efficiency of target cell infection.

2. Materials

1. Retroviral plasmids, transfection solutions, a suitable retroviral packaging cell line, NIH3T3 fibroblasts, selective and nonselective tissue-culture media, and tissue-culture plates (*see* elsewhere in this vol).
2. Sterile cloning rings.
3. A tube of sterile high-vacuum grease. Autoclaving is not recommended, since it may considerably reduce viscosity.
4. Trypsin, 0.05%, in versene, 0.02%.
5. Nunclon® multidish 24-well tissue-culture plates.
6. Sterile disposable 0.45-μm syringe filters (e.g., Acrodisc®, Gelman Sciences, Ann Arbor, MI, product no. 6224184).
7. Disposable 1 and 5-mL plastic syringes.
8. Polybrene. Make a sterile stock solution of 8-mg/mL (1000×) in water.
9. Cell-colony stain. Make a solution of 50% methanol and 5% Giemsa in water.
10. DNA lysis buffer: 1% SDS, 10 m*M* Tris-HCl (pH 8.0), 1 m*M* EDTA (pH 8.0), 100 m*M* NaCl.
11. Proteinase K. Make a stock solution of 10 mg/mL (100×) in water and store 1-mL aliquots at –20°C.
12. Tris-equilibrated phenol. Phenol should be neutralized, preferably on the same day on which it is used. Thoroughly mix equal volumes of phenol and 1*M* Tris-HCl (pH 8.0) by shaking in a glass or polypropylene container (wear gloves and do not use plastic containers). Allow the phenol to settle (30 min) and discard the upper (aqueous) layer. Repeat with 100 m*M* Tris-HCl (pH 8.0), and keep the phenol overlaid with 100 m*M* Tris in a foil-wrapped container.
13. Salt/ethanol: 25 vol ethanol:1 vol 3MK/5MAc (pH 4.8). To make 100 mL of 3MK/5MAc, add 11.5 mL of glacial acetic acid and 28.5 mL of H_2O to 60 mL of 5MKAc.
14. TE: 10 m*M* Tris-HCl (pH 8.0),1 m*M* EDTA (pH 8.0).

15. Denaturing solution: 0.4*M* NaOH, 0.6*M* NaCl.
16. Neutralizing solution: 1.5*M* NaCl, 0.5*M* Tris-HCl (pH 7.5).
17. GeneScreenPlus™ hybridization transfer membrane (NEN Research Products, Boston, MA, product no. NEF-976).
18. 20 × SSC: 3*M* NaCl, 0.3*M* sodium citrate (adjust pH to 7.0).
19. Deionized formamide: 100 mL of formamide is mixed with 5 g of AG 501-X8(D) mixed-bed resin (Bio-Rad Laboratories, Richmond, CA), stirred for 30 min at room temperature, and filtered to remove resin. Aliquots may be frozen, but freshly prepared formamide gives better results.
20. Prehybridization solution: 50% deionized formamide, 1% SDS, 1*M* NaCl, and 10% dextran sulfate. This solution should be freshly prepared from stocks of 10% SDS, 5*M* NaCl, and 50% dextran sulfate.
21. Salmon-sperm DNA. This should be sheared either by sonication or by repeatedly syringing through a 21-gage needle. Prepare a 10 mg/mL stock in water and store at –20°C.
22. Oligo-labeling buffer containing random hexanucleotides, dATP, dGTP, and dTTP is available commercially, but can easily be made up from the following components:

    - Solution O: 1.25*M* Tris-HCl and 0.125*M* $MgCl_2$, at pH 8.0. Store at 4°C.
    - Solution A: 1 mL solution O + 18 µL 2-mercaptoethanol + 5 µL dATP, 5 µL dTTP, and 5 µL dGTP (each triphosphate previously dissolved in TE [3 m*M* Tris-HCl, 0.2 m*M* EDTA, pH 7.0] at a concentration of 0.1*M*). Store at –20°C.
    - Solution B: 2*M* HEPES titrated to pH 6.6 with 4*M* NaOH. Store at 4°C.
    - Solution C: Hexadeoxyribonucleotides evenly suspended in TE at 90 OD_{260} U/mL. Store at –20°C.

    Mix solution A, B, and C in a ratio of 100:250:150 to make OLB. Store at –20°C. Repeated freeze-thawing of OLB over a period of at least 3 mo does not adversely affect it.
23. TNE: 10 m*M* Tris-HCl (pH 8.0),1 m*M* EDTA (pH 8.0),100 m*M* NaCl.
24. Alpha^{32}dCTP. Stabilized aqueous solution (containing 5 m*M* β-mercaptoethanol) at 370 MBq/mL, 10mCi/mL. Available from Amersham International, product no. PB 10205.
25. Beckman Ultra-Clear™ centrifuge tubes (25 × 89 mm), product no. 344058.

3. Methods

3.1. Transfection

The procedure is as described in Chapter 21. For constructs encoding a selectable marker, 10 µg of plasmid should give a transfection efficiency between 0.1 and 1.0%. Thus, if the plates are split 1:10 into selective medium 48 h after transfection, this should give upwards of 20 colonies/plate.

For constructs lacking a selectable marker, cotransfection of a G418 resistance plasmid such as pSV2*neo,* will allow selection for uptake of the retroviral plasmid. A tenfold molar excess of the retroviral plasmid should be used to ensure that stable pSV2*neo* transfectants have taken up both plasmids. This usually works out at about 0.5 µg of pSV2*neo,* which gives about 10 G418-resistant colonies/plate when the cells are split as above.

Transfected packaging cells should not normally be transferred into appropriate selective media (described in Chapter 5) until 48 h after selection to allow time for uptake, integration, and expression of the selectable marker gene.

3.2. Transfer and Expansion of Colonies

Colonies should be easily visible and of a suitable size for transfer (2–5 mm diameter) after 10–20 d in selection.

1. Use a marker pen to mark the positions of viable colonies on the underside of the 90-mm tissue-culture dish.
2. In the tissue-culture hood remove the medium from the cells, briefly (5 s) rinse the dish with trypsin/versene, and drain it thoroughly.
3. Sparingly apply vacuum grease around the entire rim of several sterile cloning rings (up to 10) and press each ring firmly onto the tissue-culture dish at the site of a marked colony so that the vacuum grease forms a watertight seal with the base of the dish. This creates a small well over each colony (*see* Note 1).
4. Drop 100 µL of trypsin/versene into each well and incubate at 37°C for 5 min.
5. Using a p200 Gilson pipetman, pipet each well up and down to detach all the cells from the dish. Transfer the cell suspension into 1 mL of medium in a 24-well plate. Replace the medium after overnight incubation at 37°C. Individual clones grow at variable rates and reach confluence 1–4 d after transfer, at which point they should be trypsinized into 200 µL of trypsin/versene and replated in 90-mm tissue-culture dishes.

3.3. Determination of Viral Titer

Recombinant retroviral titer is a measure of the rate of release from the producer cell of infectious enveloped retroviral particles containing two strands of the expected full-length recombinant messenger RNA. Unfortunately, it is not possible to measure titer directly, but there are a variety of techniques available that give a more or less accurate approximation.

Dot blots of RNA extracted from virus-containing supernatants give an indication of the order of magnitude of virus titers above 10^5/mL. For a more accurate determination of titer even when very low, viral transfer of a selectable marker gene to NIH3T3 or other receptive cells can be quantitated. In the absence of a selectable marker, Southern hybridization analysis of proviral copy number in DNA from infected target cells is useful for detection of titers in excess of 10^4/mL. Alternatively, titer may be determined by counting the number of cells expressing the transferred gene. This can be straightforward, for example, when the gene encodes a cell-surface protein to which monoclonal antibodies are available, allowing FACS analysis of large numbers of infected cells. However, for secreted proteins or intracellular proteins detectable only by bioassay or some other means, it may be necessary to analyze multiple subclones derived from a bulk-infected cell population.

Whatever the titer, it is absolutely essential, before proceeding further, to ensure that the transferred gene is also expressed and that the producer cell is not also producing helper virus. Titrating by nonselectable gene expression is not discussed further, since the method needs to be tailored to suit the gene concerned.

3.3.1. Harvesting Producer Supernatants

Test supernatants should be harvested from confluent monolayers of producer cells in 90-mm tissue-culture dishes after overnight incubation in 8 mL of new medium (*see* Note 2). The supernatants should be passed through 0.45-μm filters prior to use. The filter removes viable producer cells from the test supernatant. A 10-min centrifugation run at 5000*g* is an alternative to the use of 0.45-μm filters, since this will pellet the cells but not the virus. Retroviral particles will not pass through a 0.2-μm filter.

3.3.2. RNA Dot Blot of Producer-Cell Supernatant

This technique (described in Chapter 4) gives a very rough approximation of viral titer and is recommended only for rapid initial screening of large numbers of colonies, prior to more accurate titration of clones that give a strong signal.

3.3.3. Determination of Titer on NIH3T3 Fibroblasts

3.3.3.1. Infection (*see* Note 3)

1. Trypsinize NIH3T3 fibroblasts and replate 10^5 cells/90-mm dish. After overnight incubation at 37°C, infect the cell monolayer (now approx 2×10^5 cells) with test supernatant according to the following protocol:
   a. Remove the medium and add dilutions of each test supernatant to the cells in a final vol of 2.5 mL containing 2.5 µL of 1000× polybrene. Polybrene is a polycation that enhances viral attachment to target cells, probably by reducing charge repulsion. In the absence of polybrene, viral titers may fall by one or more orders of magnitude.
   b. Gently rock the dishes for 2 min, incubate at 37°C for 2 h, and then add a further 7.5 mL of medium to each dish.
   c. Incubate at 37°C for a further 48 h and proceed to either of the protocols below.

3.3.3.2. Titer by Expression of a Selectable Marker Gene

1. Trypsinize the confluent monolayer of NIH3T3 fibroblasts 48 h after infection and split 1:10 into the appropriate selective medium.
2. Refeed the cells with selective medium twice per week until resistant colonies with diameters of 2–5 mm are visible. This should take 10–20 d.
3. Remove the medium from each dish and stain for 30 min with 2.5 mL of 50% methanol/5% Giemsa.
4. Wash the plates with phosphate-buffered saline and count the stained, viable colonies.
5. Viral titer (infectious particles/mL) is calculated as

$$[\text{colony no./volume of supernatant (mL)}] \times (10/4)$$

   This formula corrects for the two cell divisions that occur prior to the 1:10 split into selection and for the initial virus dilution.

3.3.3.3. Titration by Southern Hybridization Analysis of Proviral Copy Number

3.3.3.3.1. Preparation of High-Molecular-Weight DNA

1. Grow the monolayer of infected NIH3T3 to confluence (48–72 h), remove the medium from the cells, add 2 mL of DNA lysis buffer containing 100 µg of proteinase K, and incubate at 37°C for 5 min.

2. Swirl the cell lysate around the dish and pour into a 10-mL polypropylene tube, which should then be firmly capped.
3. Incubate the lysate in a 37°C water bath overnight while the proteinase digests cellular nucleases. The samples may be stored frozen at –20°C after this stage.
4. Add 2 mL of phenol to each sample and mix gently for 15 min on a rotor wheel and spin for 15 min at 1500*g*.
5. With a pair of scissors, cut the end off a pipet tip and transfer the top layer of each sample to a new polypropylene tube, taking care not to disturb the proteinaceous debris at the interface. The high-mol-wt DNA solution is highly viscous, and removing the end of the pipet tip helps to prevent shearing of the DNA. It is important to continue using cut-off pipet tips at every stage from here until the DNA has been digested.
6. Following the same procedure, gently extract the DNA solution with a 1:1 phenol:chloroform mixture and, finally, with chloroform.
7. Transfer the high-mol-wt DNA solution to a clear plastic conical-bottomed sterilin tube.
8. Add 2.5 vol of salt/ethanol, firmly cap the tube, turn it on its side, and gently rock to and fro until the DNA has formed a contracted, cotton-wool-like precipitate.
9. Form the melted tip of a glass pipet into the shape of a blunt hook, hook the DNA out of the sterilin tube, allow the DNA to drain for a few seconds, and then immerse the glass hook in 200 µL of sterile TE solution in an Eppendorf tube, moving it gently until the DNA floats off into the TE.
10. Leave the Eppendorf tube at room temperature overnight while the DNA dissolves.
11. Calculate the concentration of DNA from the od 260 (should be approx 100 µg total).

3.3.3.3.2. Digestion and Electrophoresis of DNA

1. Digest 10 µg of each sample of DNA to completion with approx 50 U of an appropriate restriction enzyme (*see* Note 4).
2. Run the samples overnight on a 0.8% agarose gel. On the same gel, run DNA size markers and one-tenth-, single-, and five-copy controls of the appropriately digested retroviral plasmid (*see* Note 5). Photograph the gel alongside a ruler.
3. Denature the gel by gentle agitation in denaturing solution for 30 min at room temperature.
4. Neutralize the gel by agitation in neutralizing solution for 30 min at room temperature.

3.3.3.3.3. TRANSFER OF DNA

A number of membranes are available for transfer and hybridization of DNA. The protocol described below gives good results with GeneScreenPlus™ (NEN Research Products) and is largely based on one of the protocols supplied by the manufacturers.

1. GeneScreenPlus™ membrane has convex and concave sides when dry. It is important to mark the convex side with a pencil, since the DNA will be transferred onto this side of the membrane.
2. Cut a sheet of GeneScreenPlus™ between liner sheets to the exact size of the gel. Wear gloves.
3. Lay the membrane on the surface of a tray of deionized water and allow it to become wet by capillary action.
4. Soak the membrane for about 15 min in 10× SSC solution.
5. Place a rectangular sheet of glass across a tray filled with 500 mL of 10× SSC.
6. Cut two sheets of Whatman 3MM filter paper and place them over the sheet of glass with their sides immersed in the 10× SSC.
7. Wet the 3MM paper with 10× SSC and, using a disposable 10-mL pipet, roll it flat onto the glass plate, making sure to exclude all air bubbles.
8. Place the gel face down onto the paper-covered glass plate and remove air bubbles by rolling with the 10-mL pipet.
9. Place gel spacers or Saran Wrap™ along each side of the gel.
10. Place the hybridization membrane on the gel so that the concave side is in contact with the gel, and remove any air bubbles.
11. Place 6 gel-sized pieces of dry 3MM paper on top of the membrane.
12. Place a 6-inch stack of absorbent paper towels (cut to the size of the gel) on top of the 3MM paper.
13. Place a suitable weight on top of the paper towels. A medium-sized book is usually sufficient.
14. Allow the transfer to continue for 6–24 h. Ideally, the transfer should be overnight, but most of the DNA will be on the membrane after 6 h.
15. Remove the paper towels without disturbing the membrane.
16. Wear gloves. Carefully lift and overturn the gel, membrane, and remaining sheets of filter paper as a unit.
17. Use a pencil to mark the positions of the gel slots by gently pushing the tip through this part of the gel onto the membrane.
18. Carefully lift the membrane away from the gel.
19. Immerse the membrane in an excess of 0.4*M* NaOH for 30–60 s to ensure complete denaturation of immobilized DNA.

20. Remove the membrane from the NaOH solution and immerse in an excess of 0.2*M* Tris-HCl, pH 7.5/2× SSC.
21. Place the membrane, with transferred DNA face up, on a piece of filter paper, and allow it to dry at room temperature. As the membrane dries, it will assume its characteristic curl.

3.3.3.3.4. Hybridization of DNA (Note 6)

The following is a formamide procedure for hybridization at 42°C, which works well for GeneScreenPlus™.

1. In a strong, sealable plastic bag, prehybridize the membrane with 10–20 mL (depending on the size of the membrane) of prehybridization solution. There is no need to prewet the membrane, but ensure that there are no air bubbles in the bag when it is sealed. Incubate the bag with constant agitation for at least 15 min at 42°C.
2. Add 300 µL of 10 mg/mL sonicated salmon-sperm DNA to about 50 ng of ^{32}P oligo-labeled probe DNA (*see below* for oligo-labeling protocol) in a capped Eppendorf tube.
3. Use a needle to puncture the cap of the Eppendorf tube and denature the probe plus salmon-sperm DNA by immersing the bottom of the Eppendorf tube in boiling water for 10 min.
4. Cut open a corner of the hybridization bag, add the contents of the Eppendorf tube, reseal the bag, and incubate with constant agitation overnight at 42°C. The bag should lie flat so that all parts of the membrane are constantly exposed to the hybridization mix.
5. Remove the membrane from the hybridization bag and wash as follows: 2 × 100 mL of 2× SSC at room temperature for 5 min with constant agitation, then 2 × 200 mL of a solution containing 2× SSC and 1% SDS at 65°C for 30 min with constant agitation, and finally 2 × 100 mL of 0.1× SSC at room temperature for 30 min with constant agitation.
6. Place the membrane face up on a piece of filter paper and allow surface water to drain away. Complete drying of the membrane leads to irreversible binding of the probe, precluding the possibility of reprobing at a later date.
7. Seal the damp membrane between sheets of Saran Wrap™, expose to X-ray film (initially overnight, and then for longer or shorter periods of time as necessary) and develop the autoradiograph.
8. Assess proviral copy number by comparing the test sample with copy number control band intensities.

$$\text{Virus titer (virus/mL)} = [\text{No. cells infected } (2 \times 10^5) \times \text{copy no./volume of supernatant used (mL)}]$$

3.3.3.3.5. PREPARATION OF ^{32}P OLIGO-LABELED DNA PROBE

This procedure utilizes random hexanucleotides to prime synthesis of ^{32}P-labeled complementary strands of DNA from single-stranded unlabeled template. Klenow polymerase is used to catalyze the chain elongation reaction in the presence of unlabeled (dATP, dGTP, dTTP) and alpha ^{32}P-labeled (dCTP) deoxynucleotides.

1. Dilute 10–50 ng of the (gel-purified) DNA fragment to be used as a probe in 30 µL of sterile H_2O in a microfuge tube, pierce the cap with a needle, and immerse in boiling water for 10 min to denature the DNA.
2. Cool on ice and add 10 µL of OLB, 2 µL of bovine serum albumin (1 mg/mL), and 5 µL of α ^{32}PdCTP.
3. Add 1 µL(1 U) Klenow polymerase, mix thoroughly, and incubate at 37°C for 2 h or at room temperature overnight.
4. To separate labeled probe from unincorporated α^{32}PdCTP, the sample can now be passed through a Sephadex G-25 or G-50 spun column. Discard the plunger of a disposable plastic 1-mL syringe, push a thin layer of siliconized glass wool to the bottom of the barrel, and fill the syringe with TNE-equilibrated Sephadex G-25, taking care to exclude air bubbles.
5. Place the syringe in a sterilin tube (it will sit across the rim of the tube) and allow excess TNE to drain through the glass wool.
6. Overlay the Sephadex with 100 µL TNE and spin for 4 min at 200*g*.
7. Add 50 µL of TNE to the labeled probe and overlay the Sephadex with the probe/TNE mixture.
8. Remove the cap from a microfuge tube and place it under the spun Sephadex column in the bottom of the sterilin tube. Spin for 4 min at 200*g* and retrieve the microfuge tube, which now contains the labeled probe. Check the incorporation of radioactivity. If necessary, store at –20°C.

3.4. Aftercare of Producer Cells

1. Freeze putative virus-producing cell clones as soon as supernatants have been harvested for titrating.
2. Thaw, expand, and freeze several vials of those clones subsequently shown to have a high, helper-free titer of virus that can transfer and express the relevant gene.
3. High-level virus production is not always sustained during long-term

passage of producer cells, and recombinations can lead to emergence of helper virus, so use a new vial of early passage cells for each new gene transfer experiment.

4. Recombinant retroviruses are stable for long periods of time when frozen at–70°C, but repeated freeze-thawing will considerably reduce the titer. Therefore, freeze not only cells, but also aliquots of 0.45-µm-filtered supernatant from an early passage of the best producer cell lines.

3.5. Boosting Virus Titer

Frequently the titer of recombinant virus in supernatant from the best producer cell line is insufficient for the intended application. If this is the case, a variety of strategies can be adopted to increase the frequency of successful gene transfer to the target cell population. Broadly, these can be classified as virus concentration, amplification of virus production, producer plus target cell cocultivation, and target cell priming.

3.5.1. Virus Concentration

3.5.1.1. Method A

Harvest virus from almost-confluent producer-cell monolayers into the smallest volume of medium possible. Use roller bottles rather than flat-bottomed tissue-culture plates or flasks. This will give an increase in virus titer of up to tenfold. Harvesting virus for longer than 16–24 h gives diminishing returns, because the infectivity of recombinant retroviruses at 37°C decays with a half-life of 8 h.

3.5.1.2. Method B

For low-titer (10^3) ecotropic virus, we have found that concentrating particles 1000-fold by this method raised the infectious titer by 100- to 500-fold. High-titer (10^5 plus) ecotropic virus and amphotrophic virus do not concentrate well. Centrifuging through a sucrose cushion gives a cleaner virus pellet (less proteinaceous debris), but seriously impairs virus infectivity.

1. Place 40 mL of 0.45-µm-filtered viral supernatant in each of six 25- × 89-mm Beckman Ultra-Clear™ centrifuge tubes.
2. Ultracentrifuge at 4°C, 10,000 rpm for 16 h (overnight) using the SW28 rotor.
3. Discard the supernatant, gently resuspend the (visible) pellets in a total volume of 250 µL of neat fetal calf serum and store at –70°C.
4. Check the titer of concentrated virus on NIH3T3 fibroblasts.

3.5.2. Amplification of Virus Production

A major determinant of virus titer is the number of packagable RNA transcripts per producer cell, which is dependent on the integrated proviral

DNA copy number. Retroviral packaging cell lines are coated with viral envelope glycoproteins and are therefore resistant to superinfection by virus of the same, but not of a different, host range (interference). Consequently, recombinant retroviruses can shuttle back and forth between amphotropic and ecotropic packaging cell lines in a mixed culture ("ping pong") leading to amplification of proviral DNA copy number and virus titer *(7)*. Transfer of packaging functions between ecotropic and amphotropic lines leads eventually to generation of replication-competent helper virus. Increasing numbers of cells express both ecotropic and amphotropic envelope proteins and are therefore resistant to further infection. Moreover, cells with large numbers of proviruses are unhealthy.

There is thus a period during the "ping-pong" process when virus titer is high and helper virus is absent. The time of appearance of helper virus is relatively constant for a given ecotropic plus amphotropic packaging line combination and varies with the number of recombination events necessary to generate replication-competent virus from the retroviral sequences present in the packaging lines concerned *(8)*. Single-cell cloning of "ping-pong" cultures before the generation and spread of helper virus can give producer cell clones with titres in excess of 10^{10}/mL *(9)*.

1. Plate a 1:1mixture of ecotropic and amphotropic packaging cells (2×10^5 total) in 90-mm tissue-culture plates, and grow them overnight.
2. Transfect the cells with a retroviral plasmid or infect them with the recombinant virus to be amplified (which may be of ecotropic or amphotropic host range). As an alternative to steps 1 and 2, ecotropic or amphotropic producer cell lines (or both) can be used in place of their respective parental packaging cell lines.
3. Trypsinize and replate the cells at several densities when they reach confluence.
4. Harvest and store supernatant when the cells again reach confluence.
5. Trypsinize and replate the cells at several densities. Harvest and freeze supernatant, as before, until you have frozen supernatants from days 4 to 20.
6. Analyze the supernatants for titer and for helper virus. Use the supernatant with optimal characteristics for subsequent infections.

3.5.3. Co-Cultivation of Virus Producer and Target Cells

Often the target cells for gene transfer are either nonadherent or weakly adherent, and can be grown in suspension culture. Lymphocytes and bone marrow cultures, for example, fall into this category. Such cells are usually more resistant to infection than those that grow as adherent monolayers, but their nonadherence makes them suitable for infection by cocultivation with producer cells.

Cocultivation brings the virus-producing cells into very close contact with the target cell population for long periods of time, increasing the chance of virus–host-cell interaction before the virus loses its infectivity or is diluted into the culture supernatant.

1. Plate the virus-producer cells in tissue culture flasks and grow until they are semiconfluent. Use media and culture conditions in which the target cells are normally grown. In our experience, NIH3T3-derived producer cells can be grown in a variety of media without loss of titer.
2. Add target cells to the culture at a 1:1 ratio to producer cells in new medium containing 2.5 µg/mL polybrene. Polybrene should be excluded if previously shown to be toxic to the target cells.
3. Incubate the culture for 48 h and remove the supernatant that contains infected target cells. Gently wash the producer-cell monolayer with medium to dislodge any weakly adherent target cells and add the washings to the culture supernatant.
4. Transfer the culture supernatant to a large flask and lie it flat in the tissue-culture incubator for a further 4 h. This separates most of the remaining producer cells (which adhere) from the infected target cells (which do not).
5. Gently harvest the culture supernatant, and pellet and resuspend the cells, which can now be selected for expression of a marker gene or used immediately for reconstitution experiments.

3.5.4. Target-Cell Priming

Successful retroviral gene transfer requires that the target cells should enter the S-phase shortly after infection to allow proviral integration. Culture conditions and growth factor supplements that increase the likelihood of cell division will therefore enhance the efficiency of gene transfer. Thus, in one recent experiment, primate bone marrow was infected by cocultivation with an amphotropic producer-cell clone with a very high titer (10^{10}/mL), which had previously been transfected with gibbon IL3 and human IL6 expression plasmids and was derived from a "ping-pong" culture *(9)*. The gene transferred encoded G418 resistance and the producer cell-derived IL3 and IL6 acted synergistically to stimulate proliferation of bone marrow progenitor cells.

4. Notes

1. *Colony transfer:* Speed is essential at this stage to avoid drying of exposed colonies, which causes loss of viability. In practice, it is not possible to

transfer much more than 10 colonies/plate. To improve the survival of transferred colonies, it may be advisable to use nonselective medium. Long-term culture in the absence of selection, however, may lead to some loss of viral titer. The chance of identifying a high-titer producer increases with the total number of colonies analyzed. In one representative experiment using a two-gene "Zip" vector, of 80 PA317 colonies analyzed most gave titers less than 10^3/mL, seven gave titers of approx 10^4/mL, and one gave a titer of 10^6/mL. For single-gene constructs, high-titer producers arise at much higher frequencies.

Colony transfer may be achieved more rapidly without the use of cloning rings. Simply scrape the colony with the tip of a Gilson p20 pipetman while pipeting 10 µL of trypsin/versene up and down. Transfer the resulting cell suspension to 1 mL of medium in a 24-well plate. The cells will grow erratically in clumps; overall, results are better with the use of cloning rings, which give evenly distributed cell monolayers.

2. *Harvesting producer supernatants:* Retroviral titers are conventionally determined on overnight (16 h) supernatants from confluent producer-cell monolayers. The number of viral particles containing two copies of the desired full-length RNA transcript cannot be directly determined. Titers are therefore estimated by a number of less direct techniques, involving analysis of RNA in a producer-cell supernatant or proviral DNA copy number and viral gene expression in infected target cells. Titers are initially determined on NIH3T3 fibroblasts and later on more relevant (usually less infectable) target cells.
3. *Choosing supernatant dilutions for titrating:* The dilutions should be chosen based on consideration of the likely titer and the sensitivity of the technique being used. For viruses carrying a selectable marker, the sensitivity of the assay is high and titer can be most accurately assessed when about 100 recombinant viruses are added to each dish. Thus, 10 µL of test supernatant with a viral titer of 10^4/mL, or 1 µL with a titer of 10^5/mL would be ideal. When there is no selectable marker and titration is by Southern analysis of DNA copy number, the assay will not detect $<2 \times 10^4$ infected cells (one-tenth copy). In this case, it is often better to use 2.5 mL of test supernatant initially and to repeat the assay using appropriate dilutions of those test supernatants that give a high proviral copy number.
4. *Choice of restriction enzymes when titrating by Southern blot:* Each infected target cell will most likely have a unique proviral integration site. To generate a single band on the Southern blot it is therefore necessary to use a restriction enzyme that cuts at two sites within the provirus that

spans the gene of interest. Enzymes that cut once within each proviral LTR fulfill this criterion and give additional information of the expected size of the full length recombinant provirus. The LTR cutters Nhe I, Xba I, Sac I, and Kpn I all fall into this category. Choose one that does not cut the cDNA insert, and remember that Kpn I is inhibited by salt and Nhe I is expensive.

5. *Copy number controls when titrating by Southern blot:* The limit of detection for most Southern blots is one-tenth copy when 10 µg cellular DNA is loaded. It is advisable to run several copy-number controls between one-tenth and 10-copy equivalents of appropriately digested retroviral plasmid. Assuming 3×10^9 bp/haploid genome, the amount of cut-plasmid insert to use for a single-copy-number control (µg) when loading 10 µg of cellular DNA/track can be calculated as $[(\text{Size of insert (bp)} \times 10)/(3 \times 10^9)]$.
6. *Choice of DNA probe when titrating by Southern blot:* The choice of probe will be guided by the retroviral construct under study. It is usually convenient to use whichever gene was initially inserted into the vector. For optimal signal-to-background ratio, the final concentration of probe in the bag should be <10 ng/mL ($1–4 \times 10^5$ dpm/mL).

References

1. Friedmann, T. (1989) Progress toward human gene therapy. *Science* **244,** 1275–1281.
2. Temin, H. M. (1989) Retrovirus vectors: Promise and reality. *Science* **246,** 983.
3. Dzierzak, E. A., Papayannopoulou, T., and Mulligan, R. C. (1988) Lineage-specific expression of a human β-globin gene in murine bone marrow transplant recipients reconstituted with retrovirus-transduced stem cells. *Nature* **331,** 35–41.
4. Rein, A. and Schultz, A. M. (1984) Different recombinant murine leukemia viruses use different surface receptors. *Virology* **136,** 144–152.
5. Southern, E. M. (1975) Detection of specific sequences among DNA fragments separated by gel electrophoresis. *J. Mol. Biol.* **98,** 503.
6. Feinberg, A. P. and Vogelstein, B. (1983) A technique for radiolabelling DNA restriction endonuclease fragments to high specific activity. *Anal. Biochem.* **132,** 6–13, and addendum **133,** 266–267.
7. Bestwick, R. K., Kozak, S. L., and Kabat, D. (1988) Overcoming interference to retroviral superinfection results in amplified expression and transmission of cloned genes. *Proc. Natl. Acad. Sci. USA* **85,** 5404–5408.
8. Muenchau, D. D., Freeman, S. M., Cornetta, K., Zwiebel, J. A., and French Anderson W. (1990) Analysis of retroviral packaging lines for generation of replication-competent virus. *Virology* **176,** 262–265.
9. Bodine, D. M., McDonagh, K. T., Brandt, S. J., Ney, P. A., Agricola, B., Byrne, E., and Nienhuis, A. W. (1990) Development of a high-titer retrovirus producer cell line capable of gene transfer into rhesus monkey haematopoietic cells. *Proc. Natl. Acad. Sci. USA* **87,** 3738–3742.

CHAPTER 4

Methods of Titrating Nonselectable Recombinant Retroviruses

Colin D. Porter

1. Introduction

When using recombinant retroviruses for gene transduction, it is necessary to have a measurement of the concentration of virus particles in the medium conditioned by virus-producing cells (i.e., the viral titer). In the case of viruses encoding selectable markers, this can readily be determined by infecting a susceptible cell line (usually 3T3 fibroblasts), culturing in selection medium, and scoring colonies of resistant cells (Chapter 3, this volume), which gives a concentration of virus particles in the preparation as a number of colony-forming units/mL (CFU/mL). In cases in which the gene of interest is not selectable, quantitation can still be achieved by the use of recombinant viruses additionally encoding a selectable marker under the control of a separate promoter (Chapter 1). In some cases a selectable gene may be desirable, but there is evidence indicating that in viruses possessing two transcriptional units expression from each is depressed *(1,2)*. Thus, for maximum viral titer and maximum gene expression in the target cell, the virus construct of choice will encode a single gene. Other means of estimating viral titers are then required.

Titers can be determined by using Southern blotting to estimate the relative gene copy number in DNA extracted from a population of infected cells (Chapter 3). Knowing the number of target cells and having suitable

From: *Methods in Molecular Biology, Vol. 8: Practical Molecular Virology: Viral Vectors for Gene Expression*
Edited by: M. Collins © 1991 The Humana Press Inc., Clifton, NJ

DNA controls representing gene copy-number equivalents, the number of cells in which integration of the construct has occurred can be calculated. Although this method is accurate, it is insensitive (copy-number equivalents <0.1 are not detectable). It is also time-consuming and not suited to screening the large numbers of producer clones resulting from the transfection of packaging cells.

Often, the gene to be transduced encodes a cell-surface antigen. This can be used as a marker of infection to estimate titer by subsequently staining a population of infected cells with an antibody tagged with a fluorescent label and using a fluorescence-activated cell sorter (FACS) to quantitate the proportion of antigen-positive cells (personal observations). The sorting facility of the FACS can be used to select positive cells, and analysis of virus-producing cells has shown that their level of cell-surface-antigen expression correlates with their production of virus *(3)*.

A quick and simple method of estimating titers, suited to screening large numbers of producer clones, is based on dot-blot hybridization of radiolabeled probes to viral genomic RNA prepared from the medium conditioned by these cells *(4)*. The strength of the signal following autoradiography is a measure of the concentration of virus particles in the medium. The main value of this technique is to determine qualitatively those clones producing the highest viral titers prior to their quantitation by Southern blotting. However, this technique can be used quantitatively if virus preparations of known titer are available for use as hybridization standards *(4)*. A variation of the method, making use of a Schliecher and Schuell (Keene, NH) dot-blot manifold apparatus, is given below. It gives a reliable measure of relative titers, as determined by the study of several clones producing varying titers (CFU/mL) of viruses transducing neomycin resistance (personal observations).

2. Materials

1. 10% SDS, treated with diethyl pyrocarbonate (DEPC).
2. 0.25*M* EDTA, treated with DEPC.
3. Yeast tRNA, 10 mg/mL dissolved in DEPC-treated water.
4. Proteinase K, 10 mg/mL.
5. "Phenol/chloroform:" Phenol, chloroform, isoamyl alcohol (25:24:1, v/v).
6. "Chloroform:" Chloroform, isoamyl alcohol (24:1, v/v).
7. 3*M* sodium acetate, pH 5.2.
8. Blotting membrane, e.g., Hybond-N™ (Amersham International).
9. 20× SSC: 3*M* NaCl; 0.3*M* sodium acetate, pH 7.0.
10. 20× SSPE: 3.6*M* NaCl, 0.2*M* sodium phosphate, pH 7.7, 0.02*M* EDTA.

11. Formamide, deionized.
12. Denhardt's solution: 0.02% BSA, 0.02% Ficoll, 0.02% polyvinylpyrollidone.
13. Salmon-sperm DNA, sonicated and denatured.
14. [α^{32}P] dCTP.
15. X-ray film.

3. Methods

3.1. Preparation of Viral RNA

1. Grow a plate of virus-producing cells to confluence. Drain the plate and add 5–10 mL of fresh growth medium. After overnight incubation, harvest the conditioned medium and pass it through a 0.45-µm filter. This supernatant, containing the virus to be titrated, may be stored at –70°C until required. Thus, if a large number of clones are to be screened, supernatants may be harvested and stored until all have been collected.
2. Combine the following: 5 mL of 10% SDS, 2 mL of 0.25*M* EDTA, 1 mL of 10 mg/mL yeast tRNA, and 5 mL of 10 mg/mL proteinase K. Add 260 µL of this mixture to 1.74 mL of virus supernatant (or pro rata) and incubate at 37°C for 45 min.
3. Extract with an equal vol of phenol/chloroform followed by chloroform. Precipitate the RNA by the addition of 0.1 vol 3*M* sodium acetate, pH 5.2, and 2.5 vol of ethanol. Leave overnight at –70°C. Pellet the RNA in a bench centrifuge, at maximum speed, for 20 min. The pellet is easily visible by virtue of the carrier tRNA. Pour off the supernatant and air-dry. Dissolve the RNA in 50 µL of water.

3.2. RNA Dot-Blotting and Hybridization

1. Assemble the dot-blot apparatus, using a membrane of Hybond-N or its equivalent, prewetted in 20× SSC. Apply suction to the manifold and flush the wells with 20× SSC. Load 30 µL of each viral RNA preparation (equivalent to 1 mL of the original conditioned medium), as well as a tenfold dilution of each sample. After the samples have blotted, flush the wells with water to minimize losses.
2. Disassemble the dot-blot apparatus and air-dry the membrane. Fix the RNA to the membrane by UV transillumination, in the case of Hybond-N, or by baking at 80°C.
3. Prepare prehybridization solution consisting of: 5× SSPE, 50% formamide, 5× Denhardt's solution, and 0.5% SDS. Prehybridize the filter for 2 h at 42°C. Add a suitable radiolabeled probe—e.g., the cDNA fragment present in the viral construct, labeled by incorporation of

$[\alpha^{32}P]$dCTP by random priming of DNA polymerase "Klenow" fragment *(5)* —and denatured sonicated salmon-sperm DNA to give a concentration of 20 µg/mL. Hybridize overnight at 42°C.

4. Wash the filter (a) briefly in 2× SSC and 0.1% SDS, (b) for 30 min in 2× SSC and 0.1% SDS at 42°C, and (c) for 30 min in 1× SSC and 0.1% SDS at 42°C. Autoradiograph using preflashed X-ray film.

4. Notes

1. If RNA is available from a virus of known titer encoding the same cDNA, then aliquots corresponding to a standard series may be included.
2. A radiolabeled probe specific for the virus, rather than the cDNA insert (e.g., an LTR probe), could be used, in which case quantitation can be achieved using a standard series of any similar virus of known titer.
3. Accurate quantitation of autoradiographic signals (by scanning densitometry) requires preflashing of X-ray film to ensure linearity of film response.

References

1. Emerman, M. and Temin, H. M. (1986) Quantitative analysis of gene suppression in integrated retrovirus vectors. *Mol. Cell. Biol.* **6,** 792–800.
2. Bowtell, D. D. L., Cory, S., Johnson, G. R., and Gonda, T. J. (1988) Comparison of expression in haematopoietic cells by retroviral vectors carrying two genes. *J. Virol.* **62,** 2464–2473.
3. Strair, R. K., Towle, M. J., and Smith, B. R. (1988) Recombinant retroviruses encoding cell surface antigens as selectable markers. *J. Virol.* **62,** 4756–4759.
4. Huszar, D., Ellis, J., and Bernstein, A. (1989) A rapid and accurate method for determining the titer of retrovirus vectors lacking a selectable marker. *Technique* **1,** 28–32.
5. Feinberg, A. P. and Vogelstein, B. (1984) A technique for radiolabelling DNA restriction endonuclease fragments to high specific activity. *Anal. Biochem.* **137,** 266,267.

CHAPTER 5

Selectable Markers for Eukaryotic Cells

Richard Vile

1. Introduction

The transfer of DNA sequences into a population of cells can rarely, if ever, be achieved with 100% efficiency. Typically, transfection of cells with the calcium phosphate method will transduce only between 0.1 and 1% of the cells with the sequences of interest *(1)*, although some workers have achieved higher efficiencies *(2)*. If the transferred sequences do not confer a selective growth advantage, it is essential to use selection for transduced cells. The marker gene itself may be the gene of interest, e.g., to label a certain cellular population; alternatively, expression of the marker may merely be a convenient selection for the cellular population expressing another gene that has been cotransfected with the marker. Critically, selectable markers permit *positive* selection (i.e., the cells of interest are not killed); this is in contrast to systems in which demonstration of infection with a virus leads to death of the recipient cell (such as VSV pseudotypes, *see* Chapter 9).

1.1. Types of Selectable Markers

1.1.1. Dominant Selectable Markers

Many dominant selectable markers are bacterial genes that can be expressed in eukaryotic cells when placed under the control of suitable promoter elements. The recipient cells do not have to posses specific genotypes.

From: *Methods in Molecular Biology, Vol. 8: Practical Molecular Virology: Viral Vectors for Gene Expression*
Edited by: M. Collins © 1991 The Humana Press Inc., Clifton, NJ

Frequently, two or more of these selection systems can be used together to introduce sequentially different sequences into the same cells. In such cases, it is necessary to titrate the selective conditions in tandem, rather than relying on the optimal conditions worked out for each selection separately.

1.1.1.1. Neomycin Phosphotransferase *(NEO)*

One of the most frequently used dominant selectable markers has been the gene that confers resistance selectively to a group of antibiotics, including Geneticin® (G418-sulfate, or G418). Geneticin® is an aminoglycoside that is toxic to both prokaryotic and eukaryotic cells (obtained from GIBCO). However, certain bacterial strains contain mobile genetic elements (transposons) that encode dominantly acting genes with protein products that convert the toxic drug into tolerable byproducts. Two separate resistance genes are located on bacterial transposons Tn601 and Tn5. These genes are called aminoglycoside phosphotransferase 3' I and II [APH(3')I and APH(3')II], respectively, or neomycin phosphotransferase, *neo,* for short. Their functional transfer into eukaryotic cells permits those cells to grow in media containing G418.

1.1.1.2. Histidinol Dehydrogenase

The prokaryotic protein histidinol dehydrogenase catalyzes the NAD-linked oxidation of L-histidinol to histidine. L-Histidinol is normally toxic to eukaryotic cells, but cells stably expressing the histidinol dehydrogenase *(His D)* gene can survive in the presence of otherwise lethal doses of this drug *(3)*. Therefore, in medium lacking the essential amino acid histidine, but containing histidinol, expression of *His D* has the dual effect of removing the inhibitor and providing a required nutrient from its breakdown. Cells expressing the *His D* gene (following transfection or infection) can be grown in DMEM medium *lacking histidine* with added L-histidinol (Sigma Chemical Company), which has the advantage of being cheaper than G418 over long periods of selection.

1.1.1.3. Hygromycin Phosphotransferase

Hygromycin B is an aminocyclitol antibiotic produced by *Streptomyces hygroscopius (4)*. It is toxic to prokaryotic and eukaryotic cells, since it inhibits protein synthesis at the level of ribosomal translocation and aminoacyl-tRNA recognition. A plasmid-coded gene, hygromycin B phosphotransferase, has been isolated from *Escherichia coli* and has been shown to permit direct selection for hygromycin B resistance following transfection of eukaryotic cell lines *(5)*.

1.1.1.4. Xanthine-Guanine Phosphoribosyltransferase (GPT)

The bacterial enzyme xanthine-guanine phosphoribosyltransferase (GPT) *(6)* can provide, when expressed in mammalian cells, a purine salvage function that is normally provided by expression of the mammalian enzyme hypoxanthine phosphoribosyltransferase (HPRT). However, unlike the mammalian enzyme, which cannot use xanthine for purine nucleotide synthesis, expression of *gpt* permits the synthesis of GMP from xanthine, via XMP. Therefore, it is possible to select for cells that are expressing *gpt* by supplying xanthine as the sole source of precursor for guanine nucleotide synthesis. Cells must also be grown in the presence of aminopterin, to prevent synthesis of inosine monophosphate (IMP) and, simultaneously, mycophenolic acid is added to inhibit IMP dehydrogenase, thus preventing the formation of guanosine monophosphate. Therefore, if cells are grown in hypoxanthine and xanthine, only cells transformed with the *gpt* gene will be capable of generating the guanine that they require to grow.

1.1.1.5. Other Dominant Selectable Markers

Several other dominant-selectable-marker systems exist that rely on the expression of prokaryotic genes in eukaryotic cells. The *neo* marker is very versatile, since it allows selection in both prokaryotes and eukaryotes, and is widely used. However, it is very expensive to use in the relatively large quantities required to keep cell lines under continual selection. As well as those discussed above, other marker genes are available, including the dihydrofolate reductase gene (DHFR) *(7,8)*, with selection using methotrexate in the range of 0.1–10 μ*M*; the *E. coli* tryptophan synthetase B gene (*trp B*), with selection in medium containing indole in place of tryptophan *(3)*, and, more recently, the puromycin resistance gene *(9)*, with subsequent selection in puromycin in the range of 1.0–10 μg/mL. The particular choice of system depends on the availability of the respective antibiotics/marker plasmids, the preexisting phenotype of the target cell types, and, increasingly, on budgetary constraints.

1.1.2. Selection by Exploitation of Cellular Mutations

1.1.2.1. Thymidine Kinase Selection

This selection relies on the existence of cells deficient in the expression of the *TK* gene *(10)*. Such cells are unable to make IMP from thymidine and can be grown in aminopterin to prevent other pathways being used for IMP synthesis. Such conditions will kill tk^- cells. If the *TK* gene is now delivered by infection or transfection, along with thymidine and hypoxanthine added to the medium, transfected cells will survive. The only recipient cells that can be used for this type of selection are, necessarily, tk^- (e.g., mouse L tk^- cells), so such a selection system is not as broadly useful as the dominantly acting ones described earlier.

1.1.2.2. Adenine Phosphoribosyl Transferase *(APRT)*

Cells that lack the *APRT* gene will die when grown in HAT medium as described in Section 2.6. If the *APRT* gene is transferred to such mutants, cells will survive *(11)*. As for tk selection, only a severely limited range of cells can be used if this selection is to be used.

1.1.3. Amplifiable Selectable Markers

When cells are exposed to the appropriate concentration of a toxic drug, it is sometimes possible to isolate clones that show an increased tolerance of the drug. This effect should not be confused with an inadequate concentration of the drug to kill normal cells, but can represent the overproduction of the enzyme that is *inhibited* by the drug's action. This effect has been observed to be attributable to an *increase in the number of copies* of the structural gene for the enzyme, and has been suggested to be as a result of errors in DNA replication *(12,13)*. The increased levels of protein production from the amplified gene leads to survival in an otherwise normally toxic concentration of the inhibitor of that protein. The region of chromosomal DNA that is amplified is not restricted to the enzyme gene alone, and may extend over 1000 kbp. Significantly, if a transfected gene is amplified, the sequences near it will also become amplified in the cell. Therefore, if an amplifiable marker is introduced into a cell that also acts as a selectable marker, then genes cointroduced on the same plasmid will be coamplified on selection for the marker.

Dominant selectable marker genes can be used only if an inhibitor exists that will form complexes with the marker gene product, thus allowing cell survival only following selective amplification of the gene. (Hence, enzymes like *HisD* or *neo* cannot be amplified by selection in antibiotic.) Alternatively, the amplifiable gene can be used as a selectable marker if a recipient cell line that lacks the enzyme that is the marker is used, for example, DHFR–Chinese hamster ovary (CHO) cells *(7)*. In these cells, DHFR can be used as an amplifiable selectable marker to supply the missing endogenous enzyme activity when cells are grown in the absence of nucleosides and glycine.

2. Materials

The majority of the antibiotics that are used for selection of cells are commercially available. The sources of the major ones, as well as a guide to the usual effective concentration ranges over which they can be used, are given here.

2.1. Geneticin—To Select for the Neomycin Phosphotransferase (neo) *Gene*

Geneticin®, from GIBCO/BRL, can be added directly to DMEM medium from a stock solution of 50 mg/mL in water, and its effects should be titrated on test cells over the range 200 μg–2 mg/mL. G418 selection is also heavily dependent on cell density; nonresistant cells can survive in otherwise toxic concentrations if they are crowded on a tissue-culture dish. The liquid stock solution should be stored below –20°C, but as a solid should be stored at room temperature (15–30°C).

2.2. L-Histidinol—To Select for the Histidinol Dehydrogenase Gene

L-Histidinol is supplied by the Sigma Chemical Company. The concentration range should fall between 0.1 and 10 m*M*. Interestingly, NIH3T3 cells *select* best at a concentration of 0.125 m*M*, but once selected, grow better at a higher concentration of 0.5 m*M*. Presumably, the higher concentration provides a better supply of the essential amino acid histidine. Histidinol can be made up as a stock solution of 20 m*M* and stored frozen (–20°C).

2.3. Hygromycin B—To Select for the Hygromycin Phosphotransferase Gene

Hygromycin B can be obtained from Calbiochem Corporation and should be tested at concentrations in the range of 0.1–10 mg/mL to select a range of mammalian cells. We have used hygromycin B from a stock solution of 100 mg/mL stored at 4°C.

2.4. Selection Medium for Xanthine-Guanine Phosphoribosyl Transferase (gpt)

The selection medium for use of the *gpt* gene is more complex than for other dominant marker genes *(14)* and is described below:

To 100 mL of DMEM. with dialysed calf serum are added

1. Xanthine stock solution (5 mL): 5 mg/mL in 0.1*M* NaOH, pH 10.5; filter-sterilized; stored at room temperature.
2. Thymidine (2 mL): 59 μg /100 mL of water; filter-sterilized; stored at 4°C.
3. Hypoxanthine (1 mL): 13.6 mg/mL in 9.5 mL of water; 0.5 mL of 0.1*M* NaOH, pH 9.5; filter-sterilized; stored at room temperature.
4. Aminopterin (0.225 mL): 8.8 mg in 10 mL of PBS; filter-sterilized; stored frozen, wrapped in aluminum foil to protect it from the light.
5. Mycophenolic acid (0.1 mL): 25 mg/mL in absolute ethanol in a glass tube.
6. Glycine (0.2 mL): 5 mg/mL in water; filter-sterilized; stored at 4°C.

2.5. Puromycin—To Select for the Puromycin Resistance Gene

Puromycin can be obtained from the Sigma Chemical Company. Cells carrying the selectable marker gene should be assayed for toxicity in the range

Table 1
Inhibitors for Amplification of Four Genes

Introduced amplifiable gene	Selective inhibitor	Range of inhibitor concentration for amplification
Dihydrofolate reductase	Methotrexate	10 – 1000 n*M* *(17)*
Glutamine synthetase	Methionine sulphoximine	15 – 100 µ*M* *(15)*
Adenosine deaminase	Deoxycoformycin	3 – 30 n*M* *(18)*
Metallothionein-1	Heavy metal ions (Cd^{2+})	3 – 10 µ*M* *(19)*

of 0.5–10 µg/mL puromycin. The stock solution can be made up in water to 100× and should be stored below 0°C.

2.6. HAT Medium—To Select for the Thymidine Kinase (tk) Gene and the Adenine Phosphoribosyl Transferase (APRT) Gene

This selection relies on the existence of cells deficient in the expression of the *tk* gene *(10)*. The medium used is called HAT medium and contains final concentrations of aminopterin at $2 \times 10^{-5} M$; thymidine at $10^{-4} M$, and hypoxanthine at $10^{-4} M$. These components can be added to DMEM medium from diluted stocks of $10^{-3} M$, $4 \times 10^{-2} M$, and $10^{-2} M$, respectively (*see also* Section 2.4.).

2.7. Amplifiable Selectable Markers

Four genes that can be used as dominant amplifiable markers are listed in Table 1.

The medium requirements for each specific system are dependent on the selection regime used. For example, selection of cells using the glutamine synthetase (GS) system requires the growth of cells in medium with no glutamine, but with high levels of glutamate. Cells can synthesize the essential amino acid glutamine from glutamate and ammonia using the GS enzyme. Therefore, the selective medium includes the inhibitor methionine sulfoximine *(15)*, so only cells that show amplified levels of GS will contain sufficient enzyme to overcome the inhibitor concentration and allow synthesis of glutamine for the cell.

For selection of GS-expressing vectors, cells must first be grown in the glutamate-rich, glutamine-deficient GMEM-S medium *(16)* to ensure that they can survive. GMEM-S can be constituted, per liter, with the following additions (filter-sterilized through 0.2-µm filters):

1. Water, 800 mL.
2. 10× Glasgow MEM medium lacking glutamine (from GIBCO) (100 mL):

should be stored at 4°C.

3. 7.5% Sodium bicarbonate (GIBCO) (37 mL): should be stored at 4°C.
4. 100 × Nonessential amino acids solution (10 mL): 10 m*M* alanine, aspartate, glycine, proline, and serine; 50 m*M* asparagine and glutamate; 3 m*M* guanosine, cytidine, adenosine, and uridine; 1 m*M* thymidine, stored at 4°C.
5. 100 × Glutamate/asparagine mixture (10 mL): 600 mg of glutamic acid and 600 mg of asparagine in 100 mL of distilled water, stored at 4°C.
6. 100 × Sodium pyruvate (GIBCO) (10 mL).
7. 50 × Nucleoside solution (20 mL): 35 mg each of cytidine, uridine, guanosine, and adenosine; 12 mg of thymidine; in 100 mL of distilled water, stored at –20°C.
8. Dialyzed fetal calf serum (100 mL): heat-inactivated at 56°C for 30 min, stored at –20°C (GIBCO).
9. Penicillin/streptomycin (10 mL): 5000 U/mL (GIBCO).

Twenty-four hours after transfection of the target cells with the GS-containing plasmid, the MSX inhibitor is added, at varying concentrations (15–100 μ*M*), to the GMEM-S growth medium, as described in Section 3.3. MSX can be obtained from Sigma, and should be made up to 18 mg/mL in tissue-culture medium, filtered, and stored at –20°C.

If the marker gene is being used to replace a missing endogenous function in these selection systems, then highly specific media compositions are often required to permit effective selection. For a detailed review and specific experimental details of different systems, *see* ref. *16*.

3. Methods

3.1. Determination of Optimal Drug-Selection Concentrations

The amount of any selective antibiotic required to select a cell population is dependent on many factors, including the strength of the promoter driving expression of the marker gene and, critically, on the cell type. It is essential to test every separate cell line to find the optimal concentration of any given selecting drug. This will depend on

1. The lowest concentration of drug required to kill *all* the cells not carrying the marker gene; and
2. The lowest concentration (above that in 1) at which the marker gene construct can be expressed from the promoter elements of choice to confer resistance to the cell type.

The sensitivity of each relevant cell line to a particular drug can be assayed by the ability of 10^4 adherent cells, plated in individual wells of a

24-multiwell plate, to grow in a range of concentrations of the drug. (For cells in suspension, 10^4 cells in 100 μL should be plated in 96-well plates.) The optimal concentration to be used for selection is generally determined to be the lowest concentration that kills all the cells within 10–14 days, although some antibiotics can kill cells faster than this, so the dying cells should be continually monitored. The selective medium should be changed every 2–3 days during this period to remove dead and dying cells. As stated earlier, it is not possible to quote definitive concentrations for any selection system, since this depends on the cell type used, the construct from which the marker is expressed, and the growth conditions.

3.2. Time-Course for DNA Transfer and Subsequent Selection

A typical experimental protocol for the transformation of adherent test cells to antibiotic resistance is described below.

1. Plate 5×10^5 cells in a 90-mm plate.
2. After 16–24 h, transfer DNA into the cells using the method of choice (calcium phosphate precipitation, lipofection, electroporation, DEAE-dextran, or viral infection). The marker gene is included either by itself (~15 μg), as cotransfected plasmid (~0.5 μg), or as an integral component of the test DNA sequences.
3. Leave the transformed cells at least 36–48 h to grow in normal medium. This allows the integration and expression of the transferred sequences, and the accumulation of the enzymes required to degrade the antibiotic.
4. Split the cells into the appropriate selection medium. The strength of the split depends on the efficiency of the transfer process and, also on the optimal cell density requirements for the selection. Typically, 5×10^4 cells are plated into selection medium per 90-mm plate.
5. Grow the selecting cells for up to 14 d by removing dying cells every 3–4 d and renewing the selection medium. Separate colonies should become visible on the bottom of the plate, which can be cloned using cloning rings and expanded for further study.

3.3. Selection for Gene Amplification

Selection for gene amplification frequently requires specific growth media for the recipient cells, especially if the amplification is sought after introduction of the marker gene into a cell deficient in endogenous activity of the gene. Details of these requirements are given in ref. *14*. Otherwise, transfection and selection are carried out largely as described in Section 3.2.

1. Seed, for example, 8 plates each at 5×10^5 cells/90-mm plate.
2. After 16–24 h, transfer plasmid DNA (~10 μg) containing the amplifiable marker gene into the cells of four plates, using the method of choice. It is particularly important to leave four plates into which no DNA is added, to monitor any amplification of endogenous sequences that may occur.
3. One day after transfection, add the inhibitor at four different concentrations, both to the transfected plates and to the corresponding control plates.
4. After 3–4 d, remove dead cells and add fresh medium with the appropriate concentration of inhibitor.
5. After about 2 wk in selection, the number of colonies are counted and compared to the control plates to confirm that surviving colonies represent selection by amplification of the marker gene. Genuine colonies can then be screened for expression of the gene of interest, which will have been coamplified.

4. Using Selectable Marker Genes

4.1. Cotransfection of the Gene of Interest in Large Excess over the Marker Gene

The source of the selectable marker gene in the laboratory is most often cloned, synthetic plasmids that can be grown to high concentrations in bacterial host cells. For example, the *neo* or *gpt* DNA coding sequence can readily be used from such plasmids as pSV2-*neo* *(20)* or pSV2-*gpt* *(21)*. Such plasmids are frequently used to coselect cells that have acquired sequences of DNA that themselves do not confer a selective growth advantage.

Several methods exist to deliver DNA into cells, including calcium phosphate transfection *(1,2)*, DEAE-dextran *(22)*, electroporation *(23)*, protoplast fusion *(24)*, and lipofection *(25)*. In each case, a large excess of the test plasmid can be cotransferred with a selectable marker plasmid, and cells that take up the rare marker sequences will also be likely to be transfected with the vastly more abundant test sequences (e.g., *see* ref. *26*). Typically, cotransfection of 0.5 μg of pSV2-*neo* and 15 μg of a test plasmid (of about 5–10 kbp), or 1 μg pSV2-*neo* and 50 μg of total genomic DNA, yield sufficient colonies to be confident of inclusion of test sequences in many of the resulting G418-resistant clones *(20)*.

4.2. Inclusion of the Selectable Marker Gene on the Plasmid Containing the Gene of Interest

DNA sequences introduced by transfection or infection are notably unstable *(27)*, so an improvement over cotransfection has been the inclusion of marker genes directly in plasmids that contain the relevant test sequences.

For example, the plasmid pSV2-*neo* contains a unique Pvu 11 restriction site that permits the introduction and subsequent expression of cloned DNA fragments downstream of an SV40 late promoter. Any cells transfected by such a construct, that survived in G418 would also be expected to contain part, or all, of these additional sequences in the SV40 late promoter–DNA transcription unit. It is still highly advisable to clone many individual resistant cell clones in order to be sure that at least one will contain an intact unit of the gene of interest and its promoter regions, inserted correctly into the target cell genome.

An example of the use of this strategy is in the construction of retroviral packaging cell lines based on avian leukosis virus *(9)*. Continued expression of the packaging functions in transfected cells was assured by linking the (*gag-pol*) genes to the hygromycin B gene, and the *env* gene, on a separate plasmid, to a selectable phleomycin gene.

4.3. Inclusion of the Marker Gene into Infectious Viral Genomes

Selectable marker genes can be engineered into recombinant viral genomes, as described in Chapter 1, for instance. The marker can then be used directly to indicate infection of the target cell population. The first approaches to gene transfer into humans have used a *neo* gene in a retroviral vector genome to mark the fate of tumor-infiltrating lymphocytes reintroduced into cancer patients *(28)*. Simple use of a marker gene alone can also be used in probing cellular processes, such as the identification of cellular promoter or enhancer elements *(29)*. Alternatively, the marker gene can be incorporated into a viral genome designed to coexpress a sequence of functional interest. In the case of recombinant retroviruses, genomes have been generated with a selectable marker on its own, in series with a second gene, or even in triple-gene constructs *(30)*. Because of the precise nature of the viral infection and integration process, any infected cell should express both marker and test genes. Therefore, this provides the most efficient way of introducing a functional gene into a selected cell population; theoretically, 100% of all cells expressing resistance to the selection will also contain the test sequences as part of the integrated viral genome structure, which is not true with the more random processes governing integration of transfected plasmid sequences.

References

1. Graham, F. L. and Van der Eb, A. J. (1973)A new technique for the assay of human adenovirus 5 DNA. *Virology* **52,** 456–467.
2. Chen, C. and Okayama, H. (1987) High efficiency transformation of mammalian cells by plasmid DNA. *Mol. Cell. Biol.* **7,** 2745–2752.
3. Hartman, S. C. and Mulligan, R. C. (1988) Two dominant acting selectable markers for gene transfer studies in mammalian cells. *Proc. Natl. Acad. Sci. USA* **85,** 8047–8051.
4. Pettinger, R. C., Wolfe, R. N., Hoehn, M. M., Marks, P. N., Dailey, W. A., and McGuire, J. M. (1953) Hygromycin. 1. Preliminary studies on the production and biological activity of a new antibiotic. *Antibiot. Chemother.* **3,** 1286–1278.
5. Blochlinger, K. and Diggelmann, H. (1984) Hygromycin B phosphotransferase as a selectable marker for DNA transfer experiments with higher eukaryotic cells. *Mol. Cell. Biol.* **4,** 2929–2931.
6. Mulligan, R. and Berg, P. (1981) Selection for animal cells that express the escherichia coli gene coding for xanthine-guanine phosphoribosyltransferase. *Proc. Natl. Acad. Sci. USA* **78,** 2072–2076.
7. Urlaub, G. and Chasin, L. A. (1980) Isolation of Chinese hamster cell mutants deficient in dihydrofolate reductase activity. *Proc. Natl. Acad. Sci. USA* **77,** 4216–4220.
8. O'Hare, K., Benoist, C., and Breathnach, R. (1981) Transformation of mouse fibroblasts to methotrexate resistance by a recombinant plasmid expressing a prokaryotic dihydrofolate reductase. *Proc. Natl. Acad. Sci. USA* **78,** 1527–1531.
9. Cosset, F. L., Legras, C., Chebloune, Y., Savatier, P., Thoraval, P., Thomas, J. L., Samarut, J., Nigon, V. M., and Verdier, G. (1990)A new avian leukosis-based packaging cell line that uses two separate transcomplementing helper genomes. *J. Virol.* **64,** 1070–1078.
10. Wigler, M., Silverstein, S., Lee, L., Pellicer, A., Cheng, V., and Axel, R. (1977) Transfer of purified herpes virus thymidine kinase gene to cultured mouse cells. *Cell* **11,** 223–232.
11. Lowy, I., Pellicer, A., Jackson, J. F., Sim, G.-K., Silverstein, S., and Axel, R. (1980) Isolation of transforming DNA: Cloning the hamster aprt gene. *Cell* **22,** 817–823.
12. Stark, G. R. (1986)DNA amplification in drug resistant cells and in tumors. *Cancer Surveys* **5,** 1–23.
13. Stark, G. R. and Wahl, G. M. (1984) Gene amplification. *Annu. Rev. Biochem.* **53,** 447–491.
14. Gorman, C. (1985) High efficiency gene transfer into mammalian cells, in *DNA Cloning: A Practical Approach* vol. 11 (Glover, D. M. ed.), IRL, Oxford, pp. 143–190.
15. Hayward, B. E., Hussain, A., Wilson, R. H., Lyons, A., Woodcock, V., McIntosh, B., and Harris, T. J. R. (1986) The cloning and nucleotide sequence of cDNA for an amplified glutamine synthetase gene from the Chinese hamster. *Nucleic Acids Res.* **14,** 999–1008.
16. Bebbington, C. R. and Hentschel, C. C. G. (1987) The use of vectors based on gene amplification for the expression of cloned genes in mammalian cells, in *DNA Cloning: A Practical Approch* vol. 111 (Glover, D. M., ed.), IRL, Oxford, pp. 163–188.
17. Subramani, S., Mulligan, R., and Berg, P. (1981) Expression of the mouse dihydrofolate reductase complementary deoxyribonucleic acid in simian virus 40 vectors. *Mol. Cell. Biol.* **1,** 854–864.
18. Kaufman, R. J., Murtha, P., Ingolia, D. E., Yeung, C.-Y., and Kellems, R. E. (1986)

Selection and amplification of heterologous genes encoding adenosine deaminase in mammalian cells. *Proc. Natl. Acad. Sci. USA* **83**, 3136–3140.

19. Hamer, D. H. and Walling, M. J. (1982) Regulation in vivo of a cloned mammalian gene: Cadmium induces the transcription of a mouse metallothionen gene in SV40 vectors. *J. Mol. Appl. Genet.* **1**, 273–288.
20. Southern, P. J. and Berg, P. (1982) Transformation of mammalian cells to antibiotic resistance with a bacterial gene under control of the SV40 early region promoter. *J. Mol. Appl. Genet.* **1**, 327–341.
21. Mulligan, R. and Berg, P. (1980) Expression of a bacterial gene in mammalian cells. *Science* **209**, 1422–1427.
22. Sompayrac, L. and Danna, K. (1981) Efficient infection of monkey cells with DNA of simian virus 40. *Proc. Natl. Acad. Sci. USA* **12**, 7575–7578.
23. Potter, H., Weir, L., and Leder, P. (1984) Enhancer-dependent expression of human k immunoglobulin genes introduced into mouse pre-B lymphocytes by electroporation. *Proc. Natl. Acad. Sci. USA* **81**, 7161–7165.
24. Schaffner, W. (1980) Direct tranfer of cloned genes from bacteria to mammalian cells. *Proc. Natl. Acad. Sci. USA* **77**, 2163–2167.
25. Felgner, P. L., Gadek, T. R., Holm, M., Roman, R., Chan, H. W., Wenz, M., Northrop, J. P., Ringold, G. M., and Danielsen, M. (1987) Lipofection: A highly efficient, lipid-mediated DNA-transfection procedure. *Proc. Natl. Acad. Sci. USA* **84**, 7413–7417.
26. Danos, O. and Mulligan, R. C. (1988) Safe and efficient generation of recombinant retroviruses with amphotropic and ecotropic host range. *Proc. Natl. Acad. Sci. USA* **85**, 6460–6464.
27. Xu, L., Yee, J.-K., Wolff, J. A., and Friedmann, T. (1989) Factors affecting long-term stability of moloney murine leukemia virus-based vectors. *Virology* **171**, 331–341.
28. Culliton, B. J. (1989) Fighting cancer with designer cells. *Science* **244**, 1430–1433.
29. Von Melchner, H. and Ruley, H. E. (1989) Identification of cellular promoters by using a retrovirus promoter trap. *J. Virol.* **63**, 3227–3233.
30. Overell, R. W., Weisser, K. E., and Cosman, D. (1988) Stably transmitted triple promoter retroviral vectors and their use in transformation of primary mammalian cells. *Mol. Cell. Biol.* **8**, 1803–1808.

Chapter 6

Virus Genome Detection by the Polymerase Chain Reaction (PCR)

Julian G. Hoad and Thomas F. Schulz

1. Introduction

Since the polymerase chain reaction (PCR) technique was developed in 1986 *(1)*, it has found wide application in molecular biology and is now regarded as an irreplaceable tool in many laboratories. This chapter will deal with its use for the detection of virus genomes for diagnostic and experimental purposes.

1.1. Theory of PCR

PCR consists of three steps—melting of double-stranded template DNA, annealing of the oligonucleotide primers, and synthesis of new DNA—that are performed repeatedly.

The first step of PCR involves the denaturation, at 93–96°C, of the double-stranded DNA into single-stranded DNA to provide the templates for the amplification procedure. Next, two short oligonucleotides (13–30 bp) flanking the sequence to be amplified, each complementary to a different strand of the target DNA, are annealed to the template DNA at a temperature dependent on the melting temperature of these oligonucleotides. The third step is an elongation step in which both strands of template are duplicated between the primers, resulting in a doubling of the amount of DNA of interest. The heteroduplex is then heated again to denature it; the DNA is

From: *Methods in Molecular Biology, Vol. 8: Practical Molecular Virology: Viral Vectors for Gene Expression*
Edited by: M. Collins

reannealed to the primers, since they are present in excess; and the process is repeated 25–30 times, resulting in a theoretical total amplification factor of 2^{25}–2^{30}. However, in practice, the amplification achieved is somewhat lower.

Section 3.1. shows an example for a standard PCR from DNA. A number of points in this protocol merit comment:

The correct magnesium ion concentration is critical for the efficient amplification of DNA. Taq1 DNA polymerase is sensitive to the concentration of Mg^{2+} ions and functions optimally in the range 0.5–1.0 m*M*. A concentration >10 m*M* Mg^{2+} will inhibit Taq1 polymerase. Since magnesium ions bind to dNTPs, a $MgCl_2$ concentration of between 1.5 and 4.0 m*M* is usually chosen. In our experience, a 10× PCR buffer containing 20 m*M* $MgCl_2$, giving a final concentration in the PCR mix of 2 m*M*, works in most cases. There are various reports that an excess of magnesium ions may cause multiple bands; if this is a problem, it may be necessary to determine empirically the optimal $MgCl_2$ concentration, which may vary for individual primers.

The dNTPs should be of high quality, and stock solutions should be aliquoted to avoid repeated freeze-thawing and also to minimize the risk of cross contamination. The concentration in the PCR should lie between 40 μ*M* and 1.5 m*M*; an insufficient amount will lead to a reduction in sensitivity. High concentrations of dNTP (>4–6 m*M*) will inhibit Taq1 polymerase. The relative amounts of each deoxynucleotide should be equal, since an imbalance may induce mismatching of bases.

The addition of extra ingredients, such as gelatin, bovine serum albumin (BSA), and dimethylsulfoxide (DMSO), into the PCR buffer are, to a large extent, dependent on choice and previous success. Some workers add DMSO, to a final concentration of 10%, to improve PCR efficiency and reduce artifactual products by abolishing any strong secondary structure within the template DNA and by effectively reducing the melting temperature. BSA and gelatin act as carrier molecules, thus increasing the efficiency of the PCR. In our experience, however, DMSO has not usually been found to improve efficiency, although BSA is often added at 20 μg/mL in the final reaction mixture.

Taq1 polymerases from different commercial sources differ widely in their performance and may prefer slightly different buffer conditions. We normally use the Cetus cloned polymerase "Amplitaq™" (Perkin Elmer, Norwalk, CT). According to Cetus' specifications, Amplitaq has an enzymatic activity of approx 70 nucleotides/s at 75°C, with an optimum temperature range of 75–80°C. The activity decreases to 60 nucleotides/s at 72°C and only 24 dNTP/s at 55°C. At 37°C, the rate of addition is 1.5 dNTP/s. This may well influence the choice of extension time for a PCR dependent on the annealing temperature. For example, a PCR with an annealing temperature of about 60°C will require a shorter extension time (depending on the size of the product) than a reaction with an annealing temperature of about 37°C, since 60°C will support a higher rate of polymerization than 37°C. The error frequency of Taq1 polymerase (the rate at which wrong nucleotides are added to the extending DNA molecule) is about 1 in 75,000/cycle/base pair *(2)*. This is a point to be considered if sequencing from a PCR product. A simple way of avoiding this problem is to run a repeat PCR to compare sequences, since it is extremely unlikely that the same error will occur twice.

Template DNA or RNA can be prepared by a variety of standard methods, e.g., SDS/proteinase K lysis of cells followed by phenol/chloroform extractions and ethanol precipitations (details of such protocols can be found in ref. *3*). Although we usually try to use fairly "clean" nucleic acid preparations as templates, rather crude preparations are often good enough. For example, peripheral blood lymphocytes can be lysed at 6×10^5 cells/mL in 10 m*M* Tris-HCl (pH 8.3), 50 m*M* KCl, 2.5 m*M* $MgCl_2$, 1 mg/mL gelatin, 0.5% NP40, 0.5% Tween 20, and 60 µg/mL proteinase K at 55°C for 2 h and the proteinase K can then be inactivated at 95°C for 10 min. Between 10 and 15 µL of this extract can be used as a template in a 50-µL PCR reaction *(4)*. Nucleic acid template (DNA or RNA) can also be prepared from serum or plasma by treating 100 µL of serum or plasma with 100 µg/mL of proteinase K in 10 m*M* Tris-HCl (pH 7.5), 1 m*M* EDTA, and 0.2% SDS at 37°C for 4 h, followed by phenol/chloroform extraction and ethanol precipitation.

It is essential to obtain good-quality oligonucleotide primers. We usually use 20-mer oligonucleotides for a standard experiment, although 15-mers are sufficient in most instances. In choosing a sequence, it is normal to select a region with a GC content of 40–60%: Too high a GC content will lead to more mismatches and nonspecific amplification; too low a GC content implies a lower annealing temperature, which can also be accompanied by nonspecific reaction products. As a rough guide to determining the annealing temperature, we use the formula $4\times (G + C) + 2\times (A + T) - 5$, which usually results in the desired amplification. However, in some instances it may be better to raise the annealing temperature empirically by a few degrees above the calculated value to obtain a more specific amplification and fewer nonspecific bands (*see* Notes).

The oligonucleotide primers do not have to match the target sequence perfectly in order to produce the desired amplification. As a general rule, mismatches toward the 3' end of the oligonucleotide are more critical for a successful amplification than changes at the 5' end, although this does depend on the GC content of the region involved and on the length of the oligonucleotide. As a practical consequence it is possible to introduce mismatches at the 5' end—e.g., creation of artificial restriction sites to facilitate cloning of the amplified product, start or stop codons essential for the expression of a nucleotide sequence, and so on. In designing an oligonucleotide to incorporate an artificial restriction site into the amplified product, we usually choose a 28- to 30-mer that includes 18 bp of the sequence to be amplified, 4–6 bp of the desired restriction site, and 6 bp as a "stuffer fragment" to facilitate restriction-enzyme digestion of the amplified product. The annealing temperature for such a primer is usually higher than the calculated temperature for the 18 matching base pairs.

There are many variations on this standard PCR protocol. One very exciting approach is that of double or nested PCR. This is a two-stage amplification procedure designed to improve the sensitivity of the PCR by several orders of magnitude and cut down to some extent the problem of background. Basically, the technique relies on amplifying small amounts of DNA using primers to a specific sequence along the DNA, following a typical PCR protocol. A portion (5–10%) of the reaction mixture is then removed to a fresh reaction containing primers internal to the region already amplified. Most of the DNA carried over to the second reaction is the short product from the first reaction, amplified to a level many thousands of times higher than that of the nonspecific DNA. The sensitivity of this is such that as little as 80 ag (10^{-18} g) of DNA has been detected. It is thus possible to detect a single target molecule per tube, and this technique may therefore be an important means for detecting and confirming low levels of virus in clinical samples. In the case of HIV, this technique has been used to amplify and sequence genome fragments derived from individual HIV molecules *(5)*.

It is possible to perform PCR from RNA by incorporating a reverse transcription step into the process, as illustrated in Section 3.2. In this protocol, total cellular RNA prepared by standard methods is used as the starting material. (For a collection of protocols, *see*, for example, ref. *3*.)

1.2. Detection of PCR Products

A variety of methods are being employed to detect the amplified product of a PCR. The most straightforward method is to run an aliquot of the PCR mixture (e.g., 25 µL of a 50-µL reaction vol) on an agarose gel, the concentration of which will depend on the size of the amplified band. Since it is difficult to pour agarose gels of a concentration >1.5% agarose, it is advisable to use either special brands of agarose (e.g., Nusieve™) or a polyacrylamide gel for very small amplification products. In many instances it is sufficient to make the amplified band visible by staining the agarose gel with ethidium bromide. However, Southern blotting of the agarose gel, followed by hybridization with an end-labeled oligonucleotide or labeled restriction fragment, may detect bands that are not visible with ethidium bromide staining. Section 3.3. shows an example of this procedure.

Southern blotting and hybridization to an oligonucleotide can also be used to verify the specific nature of the amplified band. In this case it is advisable to use an internal oligonucleotide as a probe (i.e., an oligonucleotide located within the amplified sequence and distinct from the primers used for this amplification), since probing with one of the PCR primers may also detect spurious amplification products produced by that primer. Probing with a suitable restriction fragment can also be used to verify an amplified band.

Since Southern blotting and subsequent hybridization compose a somewhat lengthy procedure, quicker alternatives have been developed. One possibility is incorporating a labeled primer into the PCR *(6)*. This will result in a labeled PCR product that can be separated on a gel and made visible by autoradiography. Although much quicker than Southern blotting and subsequent probing, and more sensitive than detection by ethidium bromide staining, this method will also make visible spurious amplification products and is therefore not ideal for ascertaining the specific nature of the amplification product. However, if the specificity of the amplified band is beyond doubt, it is a very quick and sensitive way to detect the reaction product. Another alternative is the use of liquid hybridization to a labeled oligonucleotide, restriction enzyme digestion of the labeled product, and detection by gel electrophoresis and autoradiography *(6,7)*.

It is evident from this list of techniques that the choice of methods will depend on the individual problem that PCR is used to tackle.

1.3. Cloning and Sequencing PCR Products

Section 3.4. provides an example of this. As a general rule, PCR products clone less efficiently than classical restriction fragments; when trying to clone a PCR fragment "blunt-ended" into a suitable blunt-ended site, it is therefore advisable to "repair" the ends of the PCR band by treating with T4 polynucleotide kinase and Klenow fragment, as described in Section 3.4. This is not necessary when cloning PCR fragments that have been generated by primers that include "sticky-end" restriction sites. We usually sequence positive clones directly from "miniprep" DNA prepared by the alkaline lysis method *(8)*.

PCR fragments can be sequenced directly after separation on an agarose gel, using one or both of the PCR primers as sequencing primers. We usually use the protocol described by Winship *(9)*, as described in Section 3.5., but other methods have been described *(10)*.

1.4. PCR in Diagnostic Virology

PCR is beginning to enjoy great popularity in the detection of virus genomes in patient samples. However, because of some problems typically associated with PCR (in particular, the danger of false positive results), it is controversial whether it should, at present, be employed in a routine diagnostic setting. Whether or not it is advisable to use PCR as a routine diagnostic test certainly depends on the experience available in individual laboratories. However, as a clinical research tool, it has proved invaluable, although its use has given rise to a few controversial and probably erroneous reports regarding the association of some viruses with certain diseases *(11,12)*.

Table 1 (pp. 68–71) offers a choice of primers that have been successfully used for some viruses. This list is obviously not comprehensive and, given the wide application that PCR is enjoying at the moment, it is impossible to list every primer pair ever used. Nevertheless, Table 1 may be of some help to the newcomer to this field.

2. Materials

1. 10× Taq buffer: 100 m*M* Tris-HCl, pH 8.3, 500 m*M* KCl, 20 m*M* $MgCl_2$, 200 µg/mL BSA.
2. 40 m*M* dNTP mixture: 10 m*M* with respect to each dNTP, hence 40 m*M* total dNTP.
3. 100× Denhardt's solution: Ficoll, 1 g, polyvinylpyrrolidone, 1 g, BSA, 1 g, H_2O to 100 mL.
4. SSPE (20×): NaCl, 174.0 g, $NaH_2PO_4 \cdot H_2O$, 27.6 g, EDTA, 7.4 g, 10*N* NaOH to adjust pH to 7.4, H_2O to bring vol to 1 L.
5. T4 kinase buffer (10×): 0.5*M* Tris-HCl, pH 7.5, 0.1*M* $MgCl_2$, 50 m*M* dithiothreitol (DTT), 0.5 mg/mL BSA or gelatin.
6. Prehybridization buffer: 5× SSPE, 5× Denhardt's solution, 0.5% SDS, 50 µg/mL salmon-sperm DNA.
7. Annealing buffer: 250 m*M* KCl, 10 m*M* Tris-HCl, pH 8.3, 1 m*M* EDTA.
8. cDNA buffer: 24 m*M* Tris-HCl, pH 8.3, 16 m*M* $MgCl_2$, 8 m*M* DTT, 0.4 m*M* each of dATP, dGTP, dCTP, dTTP.
9. Taq buffer for PCR from RNA: 100 m*M* Tris-HCl, pH 8.8, 30 m*M* $(NH_4)_2SO_4$, 6 m*M* $MgCl_2$, 10 m*M* β-mercaptoethanol.

3. Methods

3.1. Standard PCR from DNA

1. Use autoclaved 500-µL Eppendorf tubes and autoclaved pipet tips.
2. Prepare "PCR premixture:" For a final reaction volume of 50 µL:

10× Taq buffer	5 µL
40 m*M* dNTP mix	1 µL of 50 µL
5' primer	120–250 ng (*see text*)
3' primer	120–150 ng (*see text*)

Taq1 polymerase	2.5 U
H_2O	30 µL

As a precaution against contamination, this "PCR premixture" can be irradiated for 5 min on a short-wavelength UV transilluminator *(13)*.

3. Add the DNA template (either plasmid or chromosomal) in a total of 20 µL. Use 200 ng–1 µg of genomic DNA, or 1 ng or less of plasmid DNA. Mix thoroughly, and overlayer with 70–100 µL of paraffin oil in order to prevent evaporation. Handle the positive control template last, after the other tubes have been capped.
4. The reaction mixture is then subjected to about 25 cycles in a PCR machine. Times and temperatures will vary depending on the primers and the size of the amplified product, but the following give a reasonable guideline for a reaction yielding a product of approx 200–1000 bp:

   denaturation temperature, 96°C for 1 min;
   annealing temperature, about T_m – 5°C *(see text)* for 1 min;
   extension temperature, 72°C for 2 min.
5. Then 25 µL of the reaction mixture is run on a 1–1.5% agarose gel and made visible by ethidium bromide staining.

3.2. PCR from RNA

1. Lyophilize 10 µg of total RNA in a 500-µL Eppendorf tube *(3)*.
2. Add 10 µL of annealing buffer followed by 1 µg of negative strand primer in 1 µL of H_2O. Heat to 85°C for 3 min, incubate for 15 min at 69°C, and then place tube(s) in a 200-mL beaker containing water at 69°C and allow to cool to 43°C. This will take roughly 15–20 min.
3. Add 15 µL of cDNA buffer and 5 U of reverse transcriptase; incubate at 43°C for 1 h.
4. Add 55 µL of RNA Taq buffer, 20 µL of 2.5 m*M* dNTP (0.75 m*M* of each dNTP), 10 µL of DMSO (optional), and 1 µg of the positive strand primer. Heat the mixture to 93°C for 7 min and anneal at about T_m– 5°C for 5 min. Add 2.5 U of Taq polymerase and cycle 25–30 times: 1 min at 93°C, 1 min at T_m– 5°C, and 2–4 min at 72°C.
5. Analyze 30 µL of the PCR mixture on an agarose gel.

3.3. Abbreviated Southern Transfer for PCR Products

1. Run PCR samples on a 1–1.5% agarose gel, depending on the expected size of the products.
2. Wash gel twice in 0.4*M* NaOH for 20 min and blot overnight in the same, onto a Genescreen nylon filter.

Table 1
Some Successfully Used PCR Primers

Virus	Primer name	Primer sequence	Region amplified		Ref.
HIV-1	M667 (496-526)	5'-GGCTAACTAGGGAACCCACTG-3'			18
	M668 (656-637)	5'-CGCGTCCCTGTTCGGGCGCC-3'	(with M667) 161bp	Env	
	LA8 (711-730)	5'-GCGCGCACAGCAAGAGGCGA-3'			
	LA9 (805-786)	5'-GACGCTCTCGCACCCATCTC-3'	(with LA8) 95bp	Env	
	LA45 (5964-5984)	5'-GGCTTAGGCATCTCCTATGGC-3'	(with LA41) 123bp	Tat/Rev	
	LA41 (8395-8374)	5'-TGTCGGGTCCCCTCGTTGCTGG-3'	(with LA8) 133bp	Rev	
HIV-1	GAG4 (730-748)	5'-AGGAAGCTGCAGAATGGG-3'			19
	GAG5 (951-977)	5'-GAGCATTCTGGACATAAGACAAGGACC-3	'(with GAG4) 247bp	Gag	
HIV-1	C011 (8462-8442)	5'-CGGGCCTGTCGGGTTCCCTC-3'			20
	C012 (5990-6009)	5'-CTTAGGCATCTCCTATGGCA-3'	(with C011) 127bp	Tat	
HIV-2	SK100 (1132-1150)	5'-ATCAAGCAGCCATGCAAAT-3'			21
	SK104 (1401-1422)	5'-CCTTTGGTCCTTGTCTTATGTC-3'	(with SK100) 290bp	Gag	
	SK89 (9432-9449)	5'-AGGAGCTGGTGGGGAACG-3'			
	SK90 (9577-9596)	5'-GTGCTGGTGAGAGTCTAGCA-3'	(with SK89) 146bp	LTR	
HIV-2	Tat/1 (5834-5854)	5'-CTATACTAGACATGGAGACA-3'			37
	Tat/2 (6082-6062)	3'-CCTTTCGTTCATAACATATC-5'	(with Tat/1) 249bp	Tat	
	TM/1 (8070-8090)	5'-GAACAGGCACAAATTCAGCAA-3'			
	TM/2 (8330-8310)	3'-TCCCCGGTCCTTGTGGATATG-S'	(with TM/1) 221bp	Transmembrane	
HSV-1	TK A	5'-ATACCGACGATATGCGACCT-3'			22
	TK B	5'-TTATTGCCGTCATAGCGCGG-3'	(with TK A) 110bp		
	ICPO A (IE110)	5'-GACCCTCCAGCCGCATACGA-3'			

	ICPO B (IE110)	5'-TTCGGTCTCCGCCTCAGAGT-3'	(with ICPO A)157bp	ICPO (IE110)	
	ICPO C (IE110)	5'-AACTCGTGGGTGCTGATTGA-3'	(with ICPO B)144bp	ICPO (IE110)	
HSV-1&2	1	5'-CATCACCGACCCGGAGAGGGAC-3'			23
	2	5'-GGGCCAGGCGCTTGTTGGTGTA-3'	(with 1)92bp	Polymerase	
HRV-14	OL26 (182-197)	5'-GCATTCTGTTTCCCC-3'		5'noncoding	24
	OL27 (547-562)	3'-GATGAAACCCACAGGC-S'	(with OL26)380bp	region	
EBV	TC60 (1396-1416)	5'-CCAGAGGTAAGTGGACTT-3'			25
	TC61 (1520-1503)	5'-GACCGGTGCCTTCTTAGG-3'	(with TC60)124bp	BamW	
	TC67(80220-80246)	5'-CAGGCTTCCCTGCAATTTTACAAGCGG-3'			
	TC69(80507-80482)	5'-CCCAGAAGTATACGTGGTGACGTAGA-3'	(with TC67)287bp	BMRF1	
	TC70(83520-83545)	5'-CTTGGAGACAGGCTTAACCAGACTCA-3'			
	TC72(83874-83850)	5'-CCATGGCTGCACCGATGAAAGTTAT-3'	(with TC70)354bp	BMLF1	
HBV	1 (844-824)	5'-TTAGGGTTTAAATGTATACCC-3			26
	2 (637-754)	5'-TTCCTATGGGAGTGGGCC-3'	(with 1)207bp	S	
	3 (429-450)	5'-CATCTTCTTGTTGGTTCTTCTG-3'	(with 1)394bp	S	
	4 (187-170)	5'-GTCCTAGGAATCCTGATG-3'			
	5 (2837-2818)	5'-GGGTCACCATATTCTTGGGA-3'	(with 4)2667bp	PreS1,PreS2	
HBV	1763 (1763-1793)	5'-GCTTTGGGGCATGGACATTGACCCGTATAA-3'			27
	2032R (2032-2002)	5'-CTGACTACTAATTCCCTGGATGCTGGGTCT-3'	(with 1763)270bp	Core	
HBV	I (1937-1960)	5'-CTGTGGAGTTACTCTCGTTTTTGC-3'			28
	II (2434-2460)	5'-CTAACATTGAGATTCCCGAGATTGAGA-3'	(with I524bp	Core	

Table 1 *(continued)*

Virus	Primer name	Primer sequence	Region amplified	Ref.
CMV	IE1	5'-CCACCCGTGGTGCCAGCTCC-3'		29
	IE2	5'-CCCGCTCCTCCTGAGCACCC-3'	(with IE1) 159bp IE	
	LA1	5'-CCGCAACCTGGTGCCCATGG-3'		
	LA2	5'-CGTTTGGGTTGCGCAGCGGG-3'	(with LA1) 139bp gp64	
CMV	MIE-4 (731-755)	5'-CCAAGCGGCCTCTGATAACCAAGCC-3'		30
	MIE-5 (1165-1150)	5'-CAGCACCATCCTCCTCTTCCTCTGG-3'	(with MIE-4) 435bp MIE antigen	
	LA-1 (2101-2125)	5'-CACCTGTCACCGCTGCTATATTTGC-3'		
	LA-6 (2500-2476)	5'-CACCACGCAGCGGCCCTTGATGTTT-3'	(with LA-1) 400bp Late Antigen	
HPV-11	HPV-11 (221-240)	5'-CGCAGAGATATATGCATATG-3'		31
	HPV-11'(291-301)	5'-AGTTCTAAGCAACAGGCACA-3'	(with HPV-11) 90bp E6 ORF	
-16	HPV-16 (421-440)	5'-TCAAAAGCCACTGTGTCCTG-3'		
	HPV-16'(521-540)	5'-CGTGTTCTTGATGATCTGCA-3'	(with HPV-16) 120bp E6 ORF	
-18	HPV-18 (371-390)	5'-ACCTTAATGAAAAACCACGA-3'		
	HPV-18'(451-470)	5'-CGTCGTTGGAGTCGTTCCTG-3'	(with HPV-18) 100bp E6 ORF	
HPV-16	1a (5551-5570)	5'-TTATTGCTGATGCAGGTGAC-3'		32
	1b (5704-5723)	5'-AGCACGGATGAATATGTTGC-3'	(with 1a) 173bp L1/L2 ORF	
HPV-16	H1 (320-339)	5'-ATTAGTGAGTATAGACATTA-3'		33
	H2 (410-429)	5'-GGCTTTTGACAGTTAATACA-3'	(with H1) 109bp E6 ORF	
-18	H3 (418-437)	5'-GGTTTCTGGCACCGCAGGCA-3'	(with H1) 117bp E6 ORF	
HTLV-I	RPX3 (5096-5115)	5'-ATCCCGTGGAGACTCCTCAA-3'		34
	RPX4 (7357-7338)	5'-AACACGTAGACTGGGTATCC-3'	(with RPX3) 145bp Tax_1/Rex_1	

HTLV-I	SG160 (51-74)	5'-CCCGGGGGCTTAGAGCCTCCCAGT-3'			35
	SG161 (768-745)	5'-GAATTCTCTCCTGAGAGTGCTATA-3'	(with SG160) 717bp	LTR	
	SG166 (1388-1411)	5'-CTGCAGTACCTTTGCTCCTCCCTC-3'			
	SG167 (1957-1934)	5'-CCCGGGGGGGGACGAGGCTGAGT-3'	(with SG166) 569bp	3'Gag	
	SG231 (2801-2820)	5'-CCCGGGCCCCCTGACTTGTC-3'			
	SG238 (3037-3018)	5'-CTGCAGGATATGGGCCAGCT-3'	(with SG231) 236bp	Pol	
	SG219 (5273-5292)	5'-CAGCTGCTGTACTCTCACAA-3'			
	SG220 (5804-5785)	5'-CTCGAGGATGTGGTCTAGGT-3'	(with SG219) 531bp	Env I	
	SG221 (5799-5818)	5'-CTCGAGCCCTCTATACCATG-3'			
	SG227 (6125-6106)	5'-GGATCCTAGGGTGGAACAG-3'	(with SG221) 326bp	Env II	
HTLV-I	Tax/3 (7339-7359)	5'-GGATACCCAGTCTACGTGTTT-3'			*
	Tax/4 (7531-7550)	5'-GTGGTTCTCTGGGTGGGGAA-3'	(with /3) 212bp	Tax	
HTLV-II	1 (7248-7265)	5'-CGGATACCCAGTCTACGT-3'			38
	2 (7386-7407)	5'-CGATGGACGCGTTATCGGCTC-3'	(with 1) 159bp	Tax	
BKV	PEP-1	5'-AGTCTTTAGGGTCTTCTACC-3'			36
(polyoma)	PEP-2	5'-GGTGCCAACCTATGGAACAG-3'	(with PEP-1) 176bp	Early Region	

* Schulz, unpublished observations.

3. Rinse nylon membrane in 2× SSC, expose to UV light for 3 min to crosslink the DNA, and bake for 90 min at 80°C.
4. Prehybridize for 1–3 h in prehybridization buffer.
5. For an oligonucleotide probe, end-label the oligonucleotide with T4-polynucleotide kinase:

100 ng	appropriate oligonucleotide
2 µL	10× T4 kinase buffer
5 µL	[-^{32}p]ATP
1 µL	T4 polynucleotide kinase (10 U/µL)
H_2O	to 20 µL total

Incubate at 37°C for 45 min; then separate the labeled oligonucleotide from the unincorporated label using a spun column containing Sepharose G-50 *(3)*.

6. Add the probe to 5 mL of prehybridization buffer and seal the filter in a bag (or hybridization bottle) together with this hybridization mixture. Incubate with shaking (or rotating) overnight at a temperature $T_m - 5°C$, where $T_m = [(G + C) \times 4 + (A + T) \times 2]$.
7. Wash the filter in

2× SSPE, 0.1%SDS	(2× 10 min, r.t.)
1× SSPE	(20 min, r.t.)

and, if the filter is still very hot, in

0.5× SSPE	(1h, $T_m - 5°C$)
0.25× SSPE	(2hr, $T_m - 5°C$)

8. Expose (a few hours to overnight) to X-ray film.

3.4. Cloning PCR Products

1. Separate PCR products on an agarose gel, cut out the band of interest, and elute from gel using Geneclean™, resuspending the DNA in 7 µL of water.
2. For blunt-ended cloning, the ends of the PCR band have to be "repaired" using T4 kinase and Klenow fragment, as follows:

Kinasing:

7 µL	eluted PCR band
1 µL	10× T_4 kinase buffer
2 µL	100 m*M* ATP
0.5 µL	T_4 kinase (10 U/µL)
	37°C, 30 min

Generation of blunt ends:

10 µL	kinased DNA
1.5 µL	100 m*M* Tris-HCl, pH 8.0, 50 m*M* $MgCl_2$
0.5 µL	2 m*M* dNTP
2.5 µL	H_2O
0.5 µL	Klenow fragment (5 U/µL)
	37°C, 20 min
	85°C, 5 min, to inactivate Klenow

3. Ligation into Sma, PvuII, EcoRV sites of pUC, bluescript, or the like:

7.5 µL	blunt-ended PCR-fragment
0.5 µL	digested plasmid (50 ng/µL)
1 µL	10× ligation buffer
1 µL	10 m*M* ATP
0.5 µL	T4 ligase (10 U/µL)
	16°C, 16 h

4. For PCR bands generated with primers that contain restriction sites, the eluted PCR band is digested with the appropriate restriction enzyme, phenol/chloroform is extracted, and remaining traces of phenol/chloroform are removed by ethanol precipitation or treatment with Geneclean™. Ligation is performed as for blunt ends, except that less enzyme can be used.
5. Transformation of 5 µL of the ligation reaction into competent bacteria is done as in a standard cloning experiment (for technical details, *see*, for example, ref. *3*).

3.5. Direct Sequencing of PCR Products

This protocol is based on a report by P. Winship *(10)*. Some of the ingredients used in this protocol are from the Sequenase™ kit of USB (Cleveland, OH).

1. It is necessary to start off with a "clean" PCR reaction (i.e., few or no nonspecific bands). The band of interest is isolated by agarose-gel electrophoresis, treatment of the gel fragment with Geneclean™, and elution in 5 µL of water.
2. Annealing of one of the PCR primers with the PCR band:

5 µL	PCR band (about 200 ng)
1 µL	primer (140 ng)
1 µL	DMSO

2 μL 5× Sequenase™ reaction buffer
1 μL H_2O

The mixture is boiled for 3 min, and then snap-cooled on dry ice for 20 s.

3. The labeling mixture is prepared as follows:

10 μL	primer–template mixture (as above)
2 μL	Sequenase™ dGTP labeling mixture (1:5 dilution)
1 μL	DTT
0.5 μL	^{35}SdATP (specific activity > 1000Ci/mmol)
2 μL	Sequenase™ (diluted 1:8 in ice-cold TE, pH 7.4)
0.5 μL	DMSO

   This is mixed well and incubated at room temperature for 10 min.
4. Next, 3.5 μL of the labeling mixture is removed and added to tubes G, A, T, and C, each containing 2.5 μL of a solution of one of the four dideoxynucleotides (ddCTP,ddATP,ddGTP, and ddTTP, respectively) and 0.25 μL of DMSO, the tubes having been prewarmed to 37°C for at least 2 min. This is incubated for 5 min at 37°C.
5. Then 4 μL of "stop solution" (Sequenase™) is added to the tubes, which are then spun down to ensure adequate mixing.
6. The samples are analyzed on an 8% sequencing gel containing 8*M* urea, as in a standard sequencing reaction.

4. Notes: Troubleshooting for Problems Encountered with PCR

1. Nonspecific amplification/spurious bands: One of the more common problems to occur in PCR is nonspecific amplification, leading to spurious bands. This can range from a few bands with the wrong molecular weight, or even one band with the correct or nearly correct molecular weight, to a massive smear of amplified products. A number of factors may contribute to this problem. If we are certain of the quality of our buffers, dNTPs, and the correct concentration of the oligonucleotide, we usually try to increase the annealing temperature until the result becomes more specific. Some oligonucleotide primers tend to produce more nonspecific bands than others, and varying the annealing temperature may take care of that. Sometimes varying the Mg^{2+} concentration in the buffer is worthwhile: As mentioned above, the correct Mg^{2+} concentration can be very critical and depend on both the concentration of dNTPs and the primer concentration used. When dealing with a new pair of primers, it is sometimes a good idea to titrate the amount of

primer used from approx 100 ng to 2 µg. Indeed, we have encountered examples in which 1 µg of primers was necessary to produce optimal results. This is, however, an extreme case; usually 0.2–1 µM/L of primer (roughly equivalent to 60–300 ng of a 20-mer in a 50-µL PCR reaction) is sufficient. Another cause for nonspecific amplification is the use of too much template DNA, especially when chromosomal DNA is used. Sometimes it is better to use only 200 ng of chromosomal DNA, instead of the standard amount of 1 µg. In our hands, more than 30 cycles of amplification increase background without producing noticeably more of the specific product.

2. No amplification: If a PCR experiment produces the wrong product, or none at all, a number of potential causes have to be considered. It is usually a good approach to determine the optimal conditions on a dilution series of a plasmid containing the target sequence of interest, if possible. As stated above, it is worthwhile to vary the annealing temperature, the Mg^{2+}, and the primer concentration, and perhaps to test other enzyme batches or sources. Using a dilution series of a plasmid enables one to calculate the level of sensitivity that can be achieved under the conditions chosen and, thus, to determine whether the original experiment had any chance of being successful. Detection limits of PCR vary with the primers and conditions used.

   An obvious reason for a failure to amplify is a critical mismatch between one of the primers and the target sequence. As mentioned above, mismatches toward the 3' end of the primer tend to be more critical in this respect. Sometimes a lowering of the annealing temperature can help to overcome this problem. Since the first few rounds of amplification produce newly synthesized strands, which contain the primer sequences, it is sometimes possible to run the initial five cycles with a low annealing temperature and then to increase the annealing temperature in the subsequent rounds of amplification.

3. Contamination of PCR: Contamination of buffers, tubes, and the like with nucleotide sequences not present in the sample to be tested is one of the major problems encountered with PCR, and may be one of the most difficult to deal with. Indeed, there have been a number of publications erroneously reporting the presence of viruses in patient material. In fact, having amplified a sequence from clinical and experimental samples, the burden of proof is often shifted toward demonstrating that this particular sequence was present in the original material in the first place. This problem is particularly acute with the nested or double PCR described above. The three most common sources of PCR contamina-

tion are plasmids containing the sequence in question that have been handled in the same laboratory, amplification products from previous PCR runs, and cross-contamination during the simultaneous preparation of several patient or experimental samples. It is therefore common in many laboratories to prepare the sample DNA or RNA in a room different from the one in which the PCR reaction is set up and run, and to ensure that no plasmids containing the sequence of interest are handled in the same room. Additional measures include the use of separate sets of pipets for sample preparation, the use of positive-displacement pipets, setting up the positive control tubes last (often after the other tubes have been physically removed), and frequent glove changes, especially after handling the positive control. One measure, which we have found particularly useful, has been to expose all PCR ingredients, except for the target DNA to be amplified, to short-wave (254-nm) UV light for 5 min, using a standard UV transilluminator of the type used for ethidium bromide stained agarose gels *(13)*. We now routinely treat the PCR buffer, dNTP stock, primers, water, and paraffin oil separately with UV light, and also treat the "PCR premixture" (*see* Section 3.1.). Since low-level contamination tends to be a random event, i.e., it does not occur in every tube, it is important to include more than one negative control tube (without target sequence to be amplified) if one is to have a reasonable chance of detecting it.

A variety of approaches can be used to demonstrate that a repeatedly amplifiable sequence is not a product of contamination. To rule out contamination with a plasmid handled in the same lab, one can amplify the same sample with a pair of primers derived from other regions in that plasmid. Furthermore, if the virus in question exhibits a sufficient degree of variability, it may be possible to sequence the amplified product and to compare the sequence to that of other viruses handled in the lab. If available, repeat samples from the same patient, processed in a different room, may need to be tested.

4. Variable results: It is not uncommon to observe a certain day-to-day variability in PCR, especially with regard to the level of sensitivity reached in individual experiments. Depending on the purpose of the experiment, it may therefore be valuable to include a titration series of a positive control in the PCR run. In setting up such a dilution series, it is important to dilute the target DNA into a constant amount of carrier DNA—otherwise the effective concentration of free plasmid will be difficult to control. A not-uncommon source of this kind of variability is the PCR machine, since its performance may vary from day to day.

References

1. Saiki, R. K., Gelfand, D. H., Stoffel, S., Scharf, S. J., Higuchi, R., Horn, G. T., Mullis, K. B., and Erlich, H. A. (1988) Primer-directed enzymatic amplification of DNA with a thermostable DNA polymerase. *Science* **239,** 487–491.
2. Goodenow, M., Huet, T., Saurin, W., Kwok, S., Sninsky, J., and Wain-Hobson, S. (1989) HIV-1 isolates are rapidly evolving quasispecies: Evidence for viral mixtures and preferred nucleotide substitutions. *J. Acqu. Immunodefic. Syndr.* **2,** 344–352.
3. Sambrook, J., Fritsch, E. F., and Maniatis, T. (1989) *Molecular Cloning: A Laboratory Manual,* 2nd Ed. (Cold Spring Harbor Laboratory, Cold Spring Harbor, NY).
4. Rodgers, M. F., Ou, C.-Y., Rayfield, M., Thomas, P. A., Schoenbaum, E. E., Abrams, E., Krasinski, K., Selwyn, P. A., Moore, J., Kaul, A., Grimm, K. T., Bamji, M., and Schochetman, G. (1989) Use of the polymerase chain reaction for early detection of the proviral sequences of human immunodeficiency virus in infants born to seropositive mothers. *N. Engl. J. Med.* **320,** 1649–1654.
5. Simmonds, P., Balfe, P., Peutherer, J. F., Ludlam, C. A., Bishop, J. O., and Brown, A. J. L. (1990) Human immunodeficiency virus-infected individuals contain provirus in small numbers of peripheral mononuclear cells and at low copy numbers. *J. Virol.* **64,** 864–872.
6. Abbott, M. A., Poiesz, B. J., Byrne, B. C., Kwok, S., Sninsky, J. J., and Ehrlich, G. D. (1988) Enzymatic gene amplification: Qualitative and quantitative methods for detecting proviral DNA amplified in vitro. *J. Infect. Dis.* **158,** 1158–1169.
7. Ou, C.-Y., Kwok, S., Mitchell, S. W., Mack, D., Sninsky, J. J., Krebs, J. W., Feorino, P., Warfield, D., and Schochetman, G. (1988) DNA amplification for direct detection of HIV-1 in DNA of peripheral blood mononuclear cells. *Science* **239,** 295–297.
8. Chen, E. Y. and Seeburg, P. H. (1985) Supercoil sequencing: A fast and simple method for sequencing plasmid DNA. *DNA* **4(2),** 165–170.
9. Winship, P. R. (1989) An improved method for directly sequencing PCR amplified material using dimethyl sulphoxide. *Nucleic Acids Res.* **17,** 1266.
10. Gyllensten, U. B. and Erlich, H. A. (1988) Generation of single stranded DNA by the polymerase chain reaction and its application to direct sequencing of the *HLA-DQA* locus. *Proc. Nat. Acad. Sci. USA* **85,** 7652–7656.
11. Reddy, P., Sandberg-Wollheim, M., and Mettus, R. (1989) Amplification and molecular cloning of HTLV-1 sequences from DNA of MS patients. *Science* **243,** 529–533.
12. Richardson, J., Wucherpfennig, K., Endo, N., Rudge, P., Dalgleish, A. G., and Hafler, D. A. (1989) PCR analysis of DNA from MS patients for the presence of HTLV-1. *Science* **246,** 821–823.
13. Sarkar, G. and Sommer, S. S. (1990) Shedding light on PCR contamination. *Nature* **343,** 27.
14. Bell, J. and Ratner, L. (1989) Specificity of polymerase chain amplification reactions for human immunodeficiency virus type 1 DNA sequences. *AIDS Res. Hum. Retroviruses* **5,** 87–95.
15. Laure, F., Courgnaud, V., Rouzioux, C., Blanche, S., Veber, F., Burgard, M., Jacomet, C., Griscelli, C., and Brechot, C. (1988) Detection of HIV-1 DNA in infants and children by means of the polymerase chain reaction. *Lancet* **2,** 538–541.
16. Loche, M. and Mach, B. (1988) Identification of HIV-infected seronegative individuals by a direct diagnostic test based on hybridisation to amplified viral DNA. *Lancet* **2,** 418–421.

17. Ou, C.-Y., Kwok, S., Mitchell, S. W., Mack, D. H., Sninsky, J. J., Krebs, J. W., Feorino, P., Warfield, D., and Schochetman, G. (1988) DNA amplification for direct detection of HIV-1 in DNA of peripheral blood mononuclear cells. *Science* **239,** 295–297.
18. Arrigo, S. J., Weitsman, S., Rosenblatt, J. D., and Chen, I. S. (1989) Analysis of rev gene function on human immunodeficiency virus type 1 replication in lymphoid cells by using a quantitative polymerase chain reaction method. *J. Virol.* **63,** 4875–4881.
19. Weber, J., Clapham, P., McKeating, J., Stratton, M., Robey, E., and Weiss, R. (1989) Infection of brain cells by diverse human immunodeficiency virus isolates: Role of CD4 as receptor. *J. Gen. Virol.* **70,** 2653–2660.
20. Hart, C., Schochetman, G., Spira, T., Lifson, A., Moore, J., Galphin, J., Sninsky, J., and OU, C.-Y. (1988) Direct detection of HIV RNA expression in seropositive subjects. *Lancet* **2,** 596–599.
21. Rayfield, M., De-Cock, K., Heyward, W., Goldstein, L., Krebs, J., Kwok, S., Lee, S., McCormick, J., Moreau, J. M., Odehouri, K., Schochetman, G., Sninski, J., and Ou, C.-Y. (1988) Mixed human immunodeficiency virus (HIV) infection in an individual: Demonstration of both HIV type 1 and type 2 proviral sequences by using polymerase chain reaction. *J. Infect. Dis.* **158,** 1170–1176.
22. Lynas, C., Cook, S. D., Laycock, K. A., Bradfield, J. W., and Maitland, N. J. (1989) Detection of latent virus mRNA in tissues using the polymerase chain reaction. *J. Pathol.* **157,** 285–289.
23. Cao, M., Xiao, X., Egbert, B., Darragh, T. M., and Yen, T. S. (1989) Rapid detection of cutaneous herpes simplex virus infection with the polymerase chain reaction. *J. Invest. Dermatol.* **92,** 391–392.
24. Gama, R. E., Horsnell, P. R., Hughes, P. J., North, C., Bruce, C. B., al-Nakib, W., and Stanway, G. (1989) Amplification of rhinovirus specific nucleic acids from clinical samples using the polymerase chain reaction. *J. Med. Virol.* **28,** 73–77.
25. Saito, I., Servenius, B., Compton, T., and Fox, R. I. (1989) Detection of Epstein-Barr virus DNA by polymerase chain reaction in blood and tissue biopsies from patients with Sjogren's syndrome. *J. Exp. Med.* **169,** 2191–2198.
26. Thiers, V., Nakajima, E., Kremsdorf, D., Mack, D., Schellekens, H., Driss, F., Goudeau, A., Wands, J., Sninsky, J., Tiollais, P., and Brechot, C. (1988) Transmission of hepatitis B from hepatitis-B seronegative subjects. *Lancet* **2,** 1273–1276.
27. Kaneko, S., Miller, R. H., Feinstone, S. M., Unoura, M., Kobayashi, K., Hattori, N., and Purcell, R. H. (1989) Detection of serum hepatitis B virus DNA in patients with chronic hepatitis using the polymerase chain reaction assay. *Proc. Natl. Acad. Sci. USA* **86,** 312–316.
28. Sumazaki, R., Motz, M., Wolf, H., Heinig, J., Jilg, W., and Deinhardt, F. (1989) Detection of hepatitis B virus in serum using amplification of viral DNA by means of the polymerase chain reaction. *J. Med. Virol.* **27,** 304–308.
29. Shibata, D., Martin, W. J., Appleman, M. D., Causey, D. M., Leedom, J. M., and Arnheim, N. (1988) Detection of cytomegalovirus DNA in peripheral blood of patients infected with human immunodeficiency virus. *J. Infect. Dis.* **158,** 1185–1192.
30. Demmler, G. J., Buffone, G. J., Schimbor, C. M., and May, R. A. (1988) Detection of cytomegalovirus in urine from newborns by using polymerase chain reaction DNA amplification. *J. Infect. Dis.* **158,** 1177–1184.
31. Young, L. S., Bevan, I. S., Johnson, M. A., Blomfield, P. I., Bromidge, T., Maitland, N. J., and Woodman, C. B. (1989) The polymerase chain reaction: A new epidemiologi-

cal tool for investigating cervical human papillomavirus infection. *Br. Med. J.* **298,** 14–18.

32. Maitland, N. J., Bromidge, T., Cox, M. F., Crane, I. J., Prime, S. S., and Scully, C. (1989) Detection of human papillomavirus genes in human oral tissue biopsies and cultures by polymerase chain reaction. *Br. J. Cancer* **59,** 698–703.
33. Shibata, D. K., Arnheim, N. , and Martin, W. J. (1988) Detection of human papilloma virus in paraffin-embedded tissue using the polymerase chain reaction. *J. Exp. Med.* **167,** 225–230.
34. Kinoshita, T., Shimoyama, M., Tobinai, K., Ito, M., Ito, S., Ikeda, S., Tajima, K., Shimotohno, K., and Sugimura, T. (1989) Detection of mRNA for the *tax1/rex1* gene of human T-cell leukemia virus type I in fresh peripheral blood mononuclear cells of adult T-cell leukemia patients and viral carriers by using the polymerase chain reaction. *Proc. Natl. Acad. Sci. USA* **86,** 5620–5624.
35. Greenberg, S. J., Ehrlich, G. D., Abbott, M. A., Hurwitz, B. J., Waldman, T. A., and Poiesz, B. J. (1989) Detection of sequences homologous to human retroviral DNA in multiple sclerosis by gene amplification. *Proc. Natl. Acad. Sci. USA* **86,** 2878–2882.
36. Arthur, R. R., Dagostin, S., and Shah, K. V. (1989) Detection of BK virus and JC virus in urine and brain tissue by the polymerase chain reaction. *J. Clin. Microbiol.* **27,** 1174–1179.
37. Schulz, T. F, Whitby, D., Hoad, J. G., Corrah, T., Whittle, H., and Weiss, R. A. Biological and molecular variability of HIV-2 isolates from the Gambia. *J. Virol.* **64 (10),** 5177–5182.
38. Lee, H., Swanson, P., Shorty, V. S., Zack, J. A., Rosenblatt, J. D., and Chen, I. S. Y. (1989) High rate of HTLV-II infection in seropositive IV drug abusers in New Orleans. *Science* **244,** 471–475.

CHAPTER 7

Syncytial Assays

Myra O. McClure

1. Introduction

Investigation of the biological activity of viruses in vitro necessitates some means of identifying their presence within cells and assaying their activity. Since virus particles themselves are metabolically inert, their detection and quantitation is dependent on their cellular effects or on synthesis of viral antigens, enzymes, and the like. Virus-induced cell fusion is a cytopathic cellular response easily recognizable by microscopic examination, which has been exploited to considerable advantage in virology, as will be briefly discussed.

Both DNA- and RNA-enveloped viruses are capable of inducing cell fusion. In this respect the best characterized family of viruses is the paramyxovirus *(1–3)*. In this chapter, however, discussion of cell fusion will be confined to that induced by retroviruses.

Retroviruses are RNA viruses that replicate in infected cells via a DNA intermediate. The existence of a specific viral enzyme, reverse transcriptase *(4,5)*, allows DNA to be transcribed from RNA. It is this distinction that classified the Retroviridae, a diverse family of viruses consisting of three subfamilies. The oncoviruses, or cancer-causing viruses (formerly called RNA tumor viruses), constitute the first group. Not all the member viruses of this group carry an oncogene in their genome or are associated with malignancy. Nonpathogenic endogenous viruses and viruses causing aplasia and immu-

From: *Methods in Molecular Biology, Vol. 8: Practical Molecular Virology: Viral Vectors for Gene Expression*
Edited by: M. Collins © 1991 The Humana Press Inc., Clifton, NJ

nodeficiency (D-type) or neuropathy (such as HTLV, the human T-cell leukemia virus) are included. Oncoviruses are further classified into B-type (murine mammary tumor), C-type (the animal leukemia viruses), and D-type (the simian retroviruses found in macaques) on the basis of their morphology.

The second retroviral subfamily are the foamy viruses, which have not yet been associated with specific diseases. The third subfamily are the lentiviruses, so called because of the long latency period between infection and disease in the prototype, visna virus of sheep. The human immunodeficiency viruses (HIVs) belong to this subfamily. Although the HIVs and related viruses (SIV, FIV, BIV) are currently the most intensely studied of this group, they are recent recruits, the founding members being the visna/maedi virus of sheep and the equine infectious-anemia virus.

All retroviruses from all host species share common structural characteristics. The retroviral envelope consists of a lipid bilayer into which is embedded a single glycoprotein (the "spike") that mediates virus binding to, and fusion with, host cells. The spike glycoprotein is initially expressed as a precursor polyprotein that lacks fusion activity. The newly synthesized glycoprotein is proteolytically cleaved by a cellular protease as it is transported to the cell surface of the infected cell, where virus assembly takes place. The cleavage allows hydrophobic sequences in the *N*-terminal domain to be involved in the postbinding fusion between virus and cell, which leads to cell entry and subsequent infection.

Retroviral syncytium induction (i.e., the formation of multinucleated giant cells by virus) was first shown in XC cells (a rat tumor cell line induced by Rous sarcoma virus *[6]*), following exposure to murine leukemia virus (MuLV), by Klement et al. *(7)*. These viruses are nonplaque forming and replicate in cultured mouse cells without producing any obvious cytopathology. Before the recognition of syncytium formation in XC cells and the subsequent development of plaque assays capitalizing on this observation *(8)*, bioassays and serological techniques were necessary to monitor MuLV infection.

This fusion phenomenon is not restricted to the murine C-type retroviruses, but extends to all classes of mammalian retrovirus in restricted cell types *(9)*. Other examples include the infection of human lymphoblastoid and lymphocytic cell lines with feline leukemia virus (FeLV) *(10)*, the endogenous feline retrovirus, RD114 *(11)*, the D-type simian retrovirus Mason Pfizer Monkey Virus (MPMV) *(12)*, HTLV-1 *(13,14)*, and HIV *(15,16)*.

The capacity of retroviral envelope glycoproteins to induce syncytia implies that membrane fusion occurs at neutral pH. Indeed, it has recently been shown that many retroviruses gain access to cells in a pH-independent manner *(17)*, following the binding of the virus to its receptor.

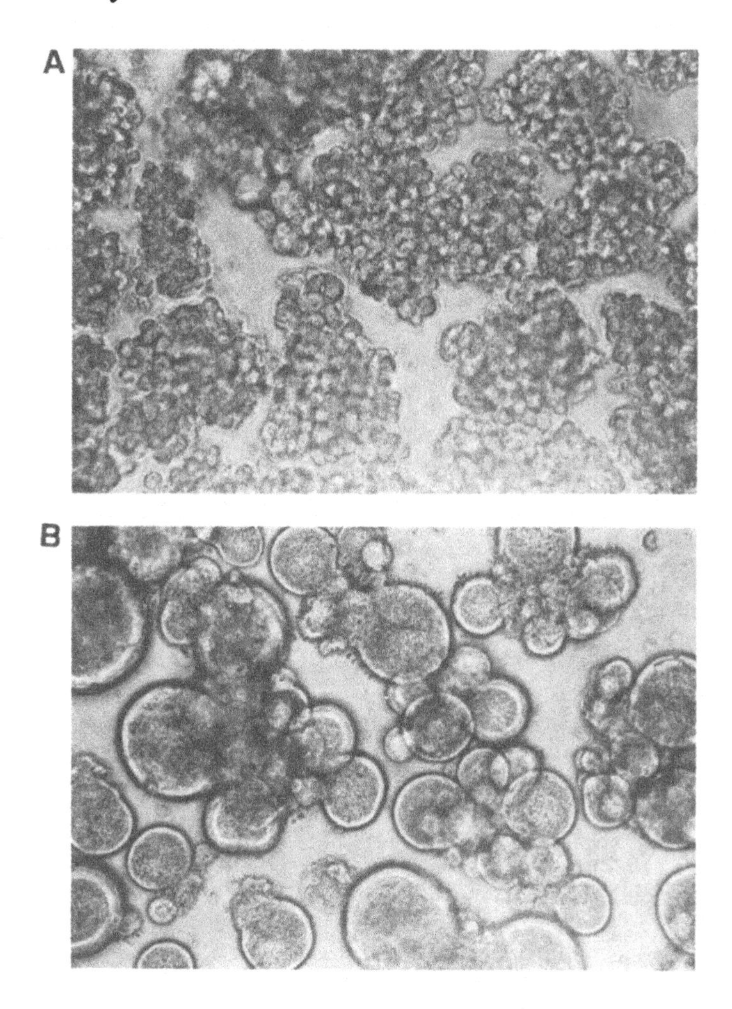

Fig. 1. Syncytial induction in the CD4+ human T-cell line, C8166 by HIV: **(A)** uninfected C8166 cells; **(B)** 2×10^5 cells infected with an equal vol (500 μL) of HIV-1 at a (moi = 3).

The restricted tropism of retrovirus in vitro and in vivo led in 1984 to the identification of the first retroviral receptor, namely the discovery by Dalgleish et al. *(15)* and Klatzmann et al. *(18)* that HIV binds with high affinity to the CD4 differentiation antigen on T-helper lymphocytes. Moreover, the marked cytopathic effect (cpe) that HIV induces in CD4-expressing T-cell lines (Fig. 1) can be exploited in rapid and facile tests of virus inhibition by neutralizing antiserum (*see* Chapter 8, this volume) or in antiviral drug assessment (Chapter 12).

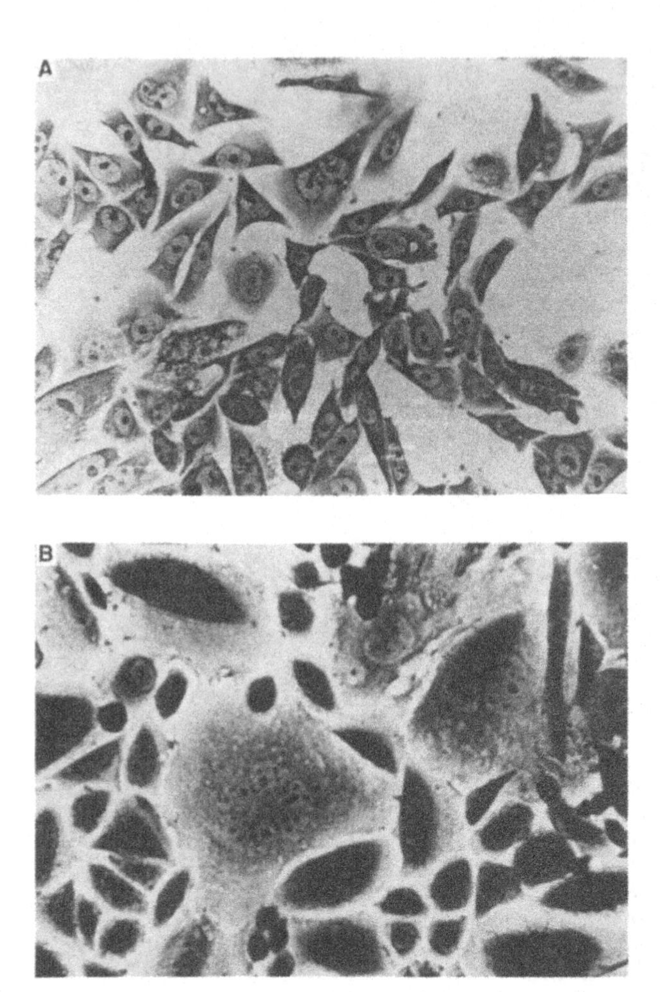

Fig. 2. Syncytial induction in cells of a human osteosarcoma cell line by HTLV-1: (A) uninfected HOS cells; (B) 10^5 HOS cells cocultivated for 24 h with an equal volume and number of HTLV-1-producing cells.

2. Materials and Methods

2.1. Syncytial Assays

Retrovirus-producing cells are added in equal volumes to uninfected, receptor-bearing cells and cocultivated overnight. In the example given in Fig. 2, the cultures were fixed and stained in 0.5% methylene blue with 0.16% basic fucsin in methanol. The proportion of nuclei in syncytia can be assessed after staining *(14,19)*.

2.2. Endpoint Titration Assays

Tissue Culture Infectious Dose (TCID) assays of retroviruses may be carried out by dilution beyond endpoint for the induction of a cpe that may include the appearance of large, multinucleated syncytia. For example, HIV plated in log-fold dilution on the CD4-bearing T-leukemic cell line C8166 may have a titer of approx 10^6 TCID within 7 d after infection (Fig. 1).

3. Notes

1. Fusion: In utilizing syncytial assays as a research tool, it may be important to distinguish between the fusion from within cells that results from proliferating virus and that which is independent of virus replication and occurs from without *(20)*. It has been demonstrated that retroviral replication within the indicator cell population is not essential for cell fusion. For example, the reverse transcriptase (RT) inhibitor azidothymidine (AZT) does not prevent HIV-induced cell fusion *(21)*. Inhibition of cellular macromolecule synthesis by cycloheximide, cytosine arabinoside, and tunicamycin failed to prevent syncytial formation in the F81 feline cell line infected with bovine leukemia virus (BLV) *(22)*. Furthermore, high-titer cell-free BLV is capable of inducing giant cell formation in these cells within 2 h of infection (in the absence of inhibitors). In neutralization assays, when the virus-blocking effects of antisera or drugs are being tested it may be necessary to separate this early polykaryocytosis (or "fusion from without"), which occurs rapidly when cells and virus are in close physical contact, from "fusion from within." The latter results much later—16 h–2 d, depending on the kinetics of virus infection. Since fusion from without is most likely caused by a fusion of the membrane of one virus particle with two cells, the parameters of the assay—cell density, virus concentration, and pretreatment of cells with DEAE-dextran or polybrene to enhance virus adsorption *(23, 24)*—can all be varied to minimize and maximize fusion from without.

   Thus, members of all three subfamilies of retrovirus, the foamy, onco-, and lentiviruses, can induce cell fusion in receptor-bearing cells in vitro. Whether syncytium formation is an important aspect of retroviral pathogenesis in vivo is not known. SIV-infected monkeys with simian AIDS show many syncytia on autopsy, especially in the brain *(25)*.
2. Enhancement of fusion: The frequency of cell fusion induced by retroviruses may be enhanced by employing several strategies.

   a. Low pH: Many enveloped viruses, among them influenza virus, Semliki Forest virus, and vesicular stomatitis virus, penetrate host cells in vivo by means of fusion between the virus and intracellular vesicular membranes, such as those of endosomes or lysosomes. The acidic environment within these vesicles is responsible for effecting in the viral envelope the conformational change that is a prerequisite to virus–host-cell-membrane fusion *(26)*. As for the influenza virus, which is the best characterized of the viruses that infect cells by a pH-dependent mechanism, a hydrophobic fusion domain in the envelope protein is exposed following the low-pH-induced conformational modification. It is thus possible to induce fusion in some model systems by mimicking the low-pH intracellular conditions. For example, Redmond et al. *(27)* have shown that mouse mammary tumor virus (the prototype B retrovirus) mediates membrane fusion of cultured cells when the pH is lowered.

b. Protease treatment: For one of the murine leukemia viruses (MLV), Andersen *(28,29)* has shown that the envelope glycoprotein, gp70, is cleaved during entry of the virus into mouse fibroblasts. Andersen and Skov *(30)* have further demonstrated that MLV-induced cell fusion of these cells is enhanced by treatment with the protease trypsin, suggesting that the cleavage of the virus envelope protein during entry and subsequent infection are dependent on the activity of proteolytic enzymes found in the low-pH environment of endosomal compartments within cells.

c. Amphotericin B: Pinter et al. *(31)* have demonstrated enhancement of MLV-induced fusion of mouse cells by the antibiotic amphotericin B, which is the active component of the antifungal agent, Fungizone.® At a final concentration of 2.5 µg/mL, the level of fusion obtained was about three orders of magnitude greater than that observed in control samples without the drug.

3. Receptor interference: Sommerfelt and Weiss *(19)* have recently classified retroviruses plating on human cells into different groups on the basis of the host-cell receptor used by the virus. This was done by means of cross-interference studies that exploited two well-established observations pertaining to virus-induced fusion:
   a. Cell fusion is blocked by competing viral glycoproteins *(32)*.
   b. Cells in culture that are actively producing virus will not produce syncytia unless uninfected cells are added to the culture *(12,33)*. Syncytial cross-interference assays devised to investigate another retroviral receptor, that for HTLV-1 *(34)*, were adapted

to highlight which virus-producing cells failed to form syncytia when cocultivated with a cell line producing a second retrovirus, i.e., to clarify whether both retroviruses recognized the same receptor *(19)*.

References

1. Scheid, A. and Choppin, P. W. (1974) Identification of biological activities of paramyzovirus glycoproteins. Activation of cell fusion, hemolysis, and infectivity by proteolytic cleavage of an inactive precursor protein of Sendai virus. *Virology* **57,** 475–490.
2. Scheid, A. and Choppin, P. W. (1976) Protease activation mutants of Sendai virus. Activation of biological properties by specific proteases. *Virology* **69,** 265–277.
3. Varsanyi, T. M., Jönvall and Norrby, E. (1985) Isolation and characterization of the measles virus F_1 polypeptide: Comparison with other paramyzovirus fusion proteins. *Virology* **147,** 110–117.
4. Temin, H. M. and Mitzutani, S. (1970) tRNA-dependent DNA polymerase in virions of Rous sarcoma virus. *Nature* **226,** 1211–1213.
5. Baltimore, D. (1970) tRNA-dependent DNA polymerase in virions of RNA tumor viruses. *Nature* **226,** 1209–1211.
6. Svoboda, J., Chyle, P., Simkovic, D., and Hilgert, I. (1963) Demonstration of the absence of infectious Rous virus in rat tumor XC, whose structurally intact cells produce Rous sarcoma when transferred to chicks. *Folia Biol. (Prague)* **9,** 77–81.
7. Klement, V., Rowe, W. P., Hartley, J. W., and Pugh, W. E. (1969) Mixed culture cytopathogenicity: A new test for growth of murine leukemia viruses in tissue culture. *Proc. Natl. Acad. Sci. USA* **63,** 753–758.
8. Rowe, W. P., Pugh, W. E., and Hartley, J. W. (1970) Plaque assay techniques for murine leukemia viruses. *Virology* **42,** 1136–1139.
9. Teich, N. M. (1985) Taxonomy of retroviruses, in *RNA Tumor Viruses,* vol. 1 (Weiss, R. A., Teich, N. M., Varmus, H. E., and Coffin, J., eds.), Cold Spring Harbor Laboratory, NY, pp. 25–207.
10. Rangan, S. R. S., Moyer, M. C., Cheong, M. P., and Jensen, E. M. (1972) Detection and assay of feline leukemia virus (FeLV) by a mixed-culture cytopathogenicity method. *Virology* **47,** 247–250.
11. Hampar, B., Rand, K. H., Lerner, A., del Villano, B. C., McAllister, R. M., Martos, L. M., Derge, J. G., Long, C. W., and Gilden, R. V. (1973) Formation of syncytia in human lymphoblastoid cells infected with type-C viruses. *Virology* **55,** 453–463.
12. Chatterjee, S. and Hunter, E. (1980) Fusion of normal primate cells: A common biological property of the D-type retroviruses. *Virology* **107,** 100–108.
13. Hoshino, H., Shimoyama, M., Miwa, N., and Sugimura, T. (1983) Detection of lymphocytes producing a human retrovirus associated with adult T-cell leukemia by syncytia induction assay. *Proc. Natl. Acad. Sci. USA* **80,** 7337–7341.
14. Nagy, K., Clapham, P. R., Cheingsong-Popov, R., and Weiss, R. A. (1983) Human T-cell leukemia virus type I: Induction of syncytia and inhibition by patients' sera. *Int. J. Cancer* **32,** 321–328.
15. Dalgleish, A. G., Beverley, P. C. L., Clapham, P. R., Crawford, D. H., Greaves, M. F., and Weiss, R. A. (1984) The CD4 (T4) antigen is an essential component of the receptor for the AIDS retrovirus. *Nature* **312,** 763–766.

16. Montagnier, L., Gruest, J., Chamaret, S., Dauguet, C., Axler, C., Guetard, D., Nugeyre, M. T., Barré-Sinoussi, F., Chermann, J. C., Brunet, J. B., et al. (1984) Adaptation of lymphadenopathy associated virus (LAV) to replication in EBV-transformed B lymphoblastoid cell lines. *Science* **225,** 63–66.
17. McClure, M. O., Sommerfelt, M. A., Marsh, M., and Weiss, R. A. (1990) On the pH-dependence of mammalian retrovirus infection. *J. Gen. Virol.* **71,** 767–773.
18. Klatzmann, D., Champagne, E., Chamaret, S., Gruest, J., Guetard, D., Hercend, T., Gluckman, J-C., and Montagnier, L. (1984) T-lymphocyte T4 molecule behaves as the receptor for human retrovirus LAV. *Nature* **312,** 767,768.
19. Sommerfelt, M. A. and Weiss, R. A. (1990) Receptor interference groups of twenty retroviruses plating on human cells. *Virology* **176,** 58–69.
20. Bratt, M. A. and Gallaher, W. R. (1969) Preliminary analysis of the requirements for fusion from within and fusion from without by Newcastle disease virus. *Proc. Natl. Acad. Sci. USA* **64,** 536–543.
21. Baba, M., Schols, D., Pauwels, R., Nakashima, H., and De Clercq, E. (1990) Sulfated polysaccharides as potent inhibitors of HIV-induced syncytium formation: A new strategy towards AIDS chemotherapy. *J. Acq. Immun. Defic. Synd.* **3,** 493–499.
22. Graves, D. C. and Jones, L. V. (1981) Early syncytium formation by bovine leukemia virus. *J. Virol.* **38,** 1055–1063.
23. Toyoshima, K. and Vogt, P. K. (1969) Enhancement and inhibition of avian sarcoma viruses by polycations and polyanions. *Virology* **38,** 414–426.
24. Zarling, D. A. and Keshet, I. (1979) Fusion activity of virions of murine leukemia virus. *Virology* **95,** 185–196.
25. Wiley, C. A., Schrier, R. D., Nelson, J. A., Lampert, P. W., and Oldstone, M. B. A. (1986) Cellular localization of human immunodeficiency virus infection within the brains of acquired immune deficiency syndrome patients. *Proc. Natl. Acad. Sci. USA* **83,** 7089–7093.
26. Marsh, M. and Helenius, A. (1989) Virus entry into animal cells. *Adv. Virus Res.* **36,** 107–151.
27. Redmond, S., Peters, G., and Dickson, C. (1984) Mouse mammary tumor virus can mediate cell fusion at reduced pH. *Virology* **133,** 393–402.
28. Andersen, K. B. (1985) The fate of the surface protein gp70 during entry of retrovirus into mouse fibroblasts. *Virology* **142,** 112–120.
29. Andersen, K. B. (1987) Cleavage fragments of the retrovirus protein gp70 during virus entry. *J. Gen. Virol.* **68,** 2193–2202.
30. Andersen, K. B. and Skov, H. (1989) Retrovirus-induced cell fusion is enhanced by protease treatment. *J. Gen. Virol.* **70,** 1921–1927.
31. Pinter, A., Chen, T. -E., Lowy, A., Cortez, N. G., and Silagi, S. (1986) Ecotropic murine leukemia virus-induced fusion of murine cells. *J. Virol.* **57,** 1048–1054.
32. Rand, K., Davis, J., Gilden, R. V., Oroszlan, S., and Long, C. (1975) Fusion inhibition: Bioassay of a type-C viral protein. *Virology* **64,** 63–74.
33. Ahmed, M., Korol, W., Larson, D. L., Harewood, K. R., and Mayyasi, S. A. (1975) Interactions between endogenous baboon type-C virus and oncogenic viruses. 1. Syncytium induction and development of infectivity assay. *Int. J. Cancer* **16,** 747–755.
34. Sommerfelt, M. A., Williams, B. P., Clapham, P. R., Solomon, E., Goodfellow, P. N., and Weiss, R. A. (1988) Human T cell leukemia viruses use a receptor determined by human chromosome 17. *Science* **242,** 1557–1559.

CHAPTER 8

Neutralization of Virus

Jane McKeating

1. Introduction

Viral-specific antibodies of any class that bind to particular epitopes on the surface protein of a virion are capable of neutralizing the infectivity of the virion. Classical neutralization results when antibody binds to the virion and thereby prevents infection of a susceptible cell. The mechanisms of virus neutralization are not fully understood. Neutralization is not simply a matter of coating the virion with antibody, nor indeed of blocking attachment to the host cell. In fact, neutralized virions generally bind to their receptor on susceptible cells *(1,2)*. The block to infection may occur at any point following adsorption and entry. Studies on the neutralization of influenza virus have shown that its neutralization by certain monoclonal antibodies is not mediated through inhibition of attachment, penetration, uncoating, or transport of the viral genome to the nucleus *(3)*, illustrating the complexity of virus neutralization.

1.1. Principles of Antibody Neutralization

The measurement of reduction of viral infectivity on a susceptible cell line is the most straightforward method for the detection and quantitation of neutralizing antibodies. Assays of infection inhibition can be broadly divided into "quantal" or quantitative assays. Quantal assays register whether infection has occurred. Rather than measure the number of infecting particles, quantal assays are all-or-none, and so are less precise than quantitative assays.

From: *Methods in Molecular Biology, Vol. 8: Practical Molecular Virology: Viral Vectors for Gene Expression*
Edited by: M. Collins © 1991 The Humana Press Inc., Clifton, NJ

Infection is measured by an end-point assay of viral production in target cells, and a range of diverse assays may be used. Infection can be measured by several parameters: the ability of the virus to cause degenerative or cytopathic effects (cpe) in the target cells, immunological staining for viral proteins *(4,5)*, and nonspecific chemical measurement of virus-induced cell death by the incorporation of [^{3}H]thymidine *(6)* or viral dyes, such as tetrazolium salts, leading to a color change *(7)*.

A quantitative assay is one in which each virus particle multiplies and give rise to a localized focus of infected cells, which, through a cpe, gives rise to a visible plaque after a number of days in culture. The classic example of the quantitative assay is the plaque assay, in which the susceptible cell line is made adherent, and, following binding of virus to cell, soft agar/agarose is layered over the cells, so that progeny viral particles are restricted to the immediate vicinity of the original infected cells. To render the plaques more readily visible, the cell monolayers are usually stained with a viral dye, such as neutral red; the living (uninfected) cells take up the stain and the plaques appear as clear areas against a red background. Some viruses, e.g., herpes and pox viruses, will produce plaques in cell monolayers maintained in liquid medium; because most of the virus is cell-associated, plaques form by direct cell-to-cell transmission of virus.

The methodology given here is for neutralization assays for the human immunodeficiency virus (HIV), the causative agent of AIDS. However, the techniques are applicable to other viral systems.

2. Materials, Cell Lines, and Media

2.1. Cell Lines

CD4$^+$ T-cell lines, such as H9 *(8)*, MT4 *(9)*, and C8166 *(10)*, should be maintained in RPMI 1640 medium supplemented with 10% fetal calf serum (FCS). Adherent HeLa cells transfected with the CD4 gene *(5)* should be maintained in DMEM supplemented with 10% fetal calf serum.

3. Methods

3.1. Quantal Assays of Neutralization

3.1.1. Preparation of Extracellular Virus Stock

1. Viral stocks for neutralization assays are prepared by mixing 1 part of HIV-infected T-cells with 4 parts of uninfected T-cells.
2. The cells are collected 48 h after cocultivation by centrifugation at 1500 rpm for 5 min and the extracellular virus harvested and filtered through a 0.45-µm filtration system.

3. The infectious titer of the extracellular virus is determined by performing tenfold serial dilutions of virus. Of each virus dilution, 50 µL is added to 100 µL of T-cells (2×10^5 cells/mL) in microtiter plates and kept at 37°C for 7 d. A minimum of eight replicates for each virus dilution is set up. Ultrafiltration can reduce virus particle aggregates.
4. The plates are scored for the presence or absence of syncytia, and the tissue-culture infectious dose ($TCID_{50}$) determined using the following Kärber formula:

$$\log TCID_{50} = 0.5 + \log \text{ of highest concentration of virus used } (\% \text{ of all wells showing CPE}/100/ \times n)$$

where n = number of dilutions/log interval, i.e., $n = 1$ for dilutions of a single-log interval.

3.1.2. Neutralization Reaction

1. Antibody containing serum must first be "inactivated" by heating at 56°C for 1 h to remove any nonspecific inhibitors.
2. The heat-inactivated serum is serially diluted and 10 µL mixed with a constant vol (40 µL) of a stock virus dilution of known $TCID_{50}$, usually 10^3, at 37°C for 1 h.
3. The virus/serum inoculum (50 µL) is then incubated with 100 µL of susceptible target cells at 2×10^5/mL in microtiter plates.
4. The cells are then incubated at 37°C in 5% CO_2 and the initial virus/serum inoculum is either left *in situ* or removed by repeated washing. The removal of inoculum is readily achieved only when the target cells are adherent.
5. After a period of incubation, the length of which is dependent on the replication cycle of the virus under test, the presence of viral infection is measured by some aspect of cpe. If the initial cell line chosen for the assay is susceptible to HIV-induced multinucleated syncytia (giant cells) (*see* Chapter 7), then direct viewing of the development of syncytia may be used as an indicator of infection. The titer of neutralization is taken, most conveniently to be the highest serum dilution to show absence of syncytia in the indicator cells, although the cutoff may be as arbitrary as with other assays (e.g., 80% syncytial reduction).

3.2. Quantitative Assays

3.2.1. Neutralization Reaction

The reaction is identical to Step 3.1.2. above.

3.2.2. Preparation of Cell Monolayers

1. Adherent cell lines: HeLa.CD4 *(5)* cells (5×10^4) are seeded in 35-mm wells and 1 d later are used for infection.
2. Suspension cell lines:
   a. Wells are treated with 50 µg/mL of poly-L-lysine (PLL) for 1 h at room temperature. Wells are washed three times with distilled water and left to dry.
   b. Cells, either MT4 *(9)* or C8166 *(10,11)*, are washed in serum-free medium three times and resuspended at a final cell concentration of $1–2 \times 10^6$/mL.
   c. Cells (1 mL) are allowed to adsorb to PLL-treated wells for 20 min at room temperature. Unbound cells are removed by aspiration.

3.2.3. Virus Infection and Agarose Overlay

1. Allow virus or virus–serum mixture in a vol of 500 µL to adsorb to cells for 1 h at 37°C.
2. Remove unbound virus and wash the cell monolayer with culture medium (RPMI). Carefully add 1 mL of agarose overlay, consisting of 0.9% sea-plaque agarose (ICN) in RPMI 1640 with 20% FCS.
3. Allow agarose to set, and incubate plates at 37°C for 3 d. Add a second agarose overlay as before, containing neutral red stain at a final concentration of 0.00032%.
4. Look at and count plaques. The titer of neutralization is taken to be the highest serum dilution leading to a 90% reduction in plaque formation.

In summary, antibody-mediated neutralization of viruses can be measured by a number of parameters, all of them serving as indicators of viral infection. Thus, the choice of assay will be dependent on the in vitro growth characteristics of the virus in question. The study of neutralizing antibodies is of considerable interest, since, for many viruses, there is a close relationship between possession of neutralizing antibody and immunity to infection or disease.

4. Notes

1. Time of appearance of HIV-induced syncytia in $CD4^+$ T-cell lines is dependent on the dose and strain of virus used and on the target cell. For example, an infectious dose of HTLV-IIIB of 10^3 $TCID_{50}$ can induce syncytia 24 h after infection of C8166 cells.
2. Since these assays are performed in liquid medium, the possibility that

the syncytia are clonal is unlikely. However, these assays on adherent cells may be made quantitative if the kinetics of viral replication ensure that the syncytia formed are clonal and generated as a result of "one hit" virus infection. Assays based on visible cpe depend on the cytopathogenicity of viral isolates. Some HIV isolates do not form syncytia, so one has to use other parameters as indicators of viral infection, as discussed earlier and reviewed in other chapters.

3. The use of pseudotype virus particles with the envelope of HIV and the core and genome of vesicular stomatitis virus (VSV) enables an indirect plaque assay to be used and is reviewed in Chapter 9.

References

1. Highlander, S. L., Cai, W. H., Person, S., Levine, M., and Glorioso, J. C. (1988) Monoclonal antibodies define a domain on herpes simplex virus glycoprotein B involved in virus penetration. *J. Virol.* **62,** 1881–1888.
2. Linsley, P. S., Ledbetter, J. A., Kinney-Thomas, E., and Hu, S. L. (1988) Effects of anti-gp120 monoclonal antibodies on CD4 receptor binding by the env protein of HIV-1. *J. Virol.* **62,** 3695–3702.
3. Taylor, H. P. and Dimmock, N. J. (1985) Mechanisms of neutralization of influenza virus by 1gM. *J. Gen. Virol.* **66,** 903–907.
4. Zagury, D., Bernard, J., Cheynier, R., Desportes, I., Leonard, R., Fouchard, M., Reveil, B., Ittele, D., Lurhuma, Z., Mbayo, K., Wane, J., Salaun, J., Goussard, B., Dechazal, L., Burny, A., Nara, P., and Gallo, R. C. (1988) A group-specific anamnestic immune reaction against HIV-1 induced by a candidate vaccine against AIDS. *Nature* **332,** 728–731.
5. Chesebro, B. and Wehrly, K. (1988) Development of a sensitive quantitative focal assay for human immunodeficiency virus infectivity. *J. Virol.* **62,** 3778,3779.
6. Harada, S., Purtilo, D., Koyanagi, K., Sonnabend, J., and Yamamoto, N. (1986) Sensitive assay for neutralizing antibodies against AIDS-related viruses. *J. Immunol. Meth.* **92,** 177–181.
7. Montefiori, D. C., Robinson, W. E., Schuffman, S. S., and Mitchell, W. M. (1988) Evaluation of anti-viral drugs and neutralising antibodies to human immunodeficiency virus by a rapid and sensitive microtiter infection assay. *J. Clin. Microbiol.* **26,** 231–235.
8. Popovic, M., Sarngadharan, M. G., Read, E., and Gallo, R. C. (1984) A method for detection, isolation, and continuous production of cytopathic human T-lymphotropic retroviruses of the HTLV family from patients with AIDS and pre-AIDS. *Science* **224,** 493–500.
9. Harada, S., Koyanagi, Y., and Yamamoto, N. (1985) Infection of HTLV-III/LAV in HTLV-I carrying MT2 and MT4 and application in a plaque assay. *Science* **229,** 563–566.
10. Sodroski, J. G., Rosen, C. A., and Haseltine, W. A. (1984) *Trans*-acting transcriptional activation of the long terminal repeat of human T-lymphotropic viruses in infected cells. *Science* **225,** 381–385.
11. McKeating, J. A., McKnight, A., McIntosh, K., Clapham, P. R., Mulder, C., and Weiss, R. A. (1989) Evaluation of human and simian immunodeficiency virus plaque and neutralization assays. *J. Gen. Virol.* **70,** 3327–3333.

CHAPTER 9

Vesicular Stomatitis Virus Pseudotypes of Retroviruses

Paul R. Clapham

1. Introduction

Pseudotype viruses are phenotypically mixed virions containing the genome or nucleocapsid of one enveloped virus and the surface or envelope (*env*) glycoproteins of another. This chapter will concentrate on vesicular stomatitis virus (VSV) pseudotypes retaining the VSV nucleocapsid but with alternative *env* glycoproteins derived from retroviruses.

Many retroviruses infect cells with minimal cytopathology, and quantitative infectivity assays based on cell death cannot be used. VSV is a highly cytopathic virus, and pseudotypes can provide quantitative plaque-forming assays specific for retrovirus infection. VSV pseudotype attachment and entry into cells is dependent on the adopted retroviral *env* glycoproteins. After entry, the VSV genomic RNA contained in the nucleocapsid is transcribed, and a VSV cytocidal cycle of replication ensues.

The requirement of pseudotypes for cell receptors specific for the sequestered retrovirus glycoproteins has been exploited to define the receptor based cell tropisms of several retroviruses. Examples include bovine leukemia virus (BLV) *(1)* and the human retroviruses HTLV-I and HTLV–II *(2)*, as well as HIV *(3)*. VSV pseudotypes have also been used to study neutralizing antibodies directed to retroviral glycoproteins.

Pseudotype infectivity is inhibited by antibodies targeted to the natural neutralization epitopes on the adopted *env* glycoproteins and this system has

From: *Methods in Molecular Biology, Vol. 8: Practical Molecular Virology: Viral Vectors for Gene Expression*
Edited by: M. Collins © 1991 The Humana Press Inc., Clifton, NJ

been used to identify linear and conformational neutralization epitopes on BLV *(4,5)*.

VSV pseudotypes were first reported by Choppin and Compans *(6)*, who demonstrated phenotypically mixed VSV particles with the surface glycoproteins of the paramyxovirus SV5 and, subsequently, with the retrovirus coats of murine leukemia and avian myeloblastosis viruses *(7)*. VSV pseudotypes of most retrovirus groups including B, C, and D-type viruses *(8)* as well as the human retroviruses (HTLV and HIV) *(3,9)*, have been described.

The molecular events involved in the acquisition of membrane glycoproteins onto budding enveloped viruses are not understood; however, it is known for Semliki forest virus (a togavirus) that antiidiotype antibodies directed to antibodies made against the cytoplasmic domain of the *env* glycoprotein react with the underlying membrane protein *(10)*. This result implies a tight association and recognition sequences between these two proteins. Despite the fact that phenotypic mixing of surface *env* glycoproteins between very diverse virus families can occur, almost all nonviral cell-surface proteins are excluded from budding virions *(11)*.

VSV particles do not incorporate different retrovirus glycoproteins to the same efficiency. In general, the *env* glycoproteins of avian retroviruses are most efficiently assembled to an extent at which up to 1% of the VSV progeny can be demonstrated to contain foreign envelope proteins *(12)*. The formation of VSV pseudotypes containing the envelopes of mammalian retroviruses occurs less readily with a maximum efficiency of 0.001–0.01%.

To measure the infectivity of a pseudotype, the vast excess of VSV wt particles must first be eliminated by neutralizing with a hyperimmune anti-VSV serum. Serum derived from experimentally infected animals are capable of neutralizing over 10 logs of VSV infectivity. Other strategies involve the use of VSV_{IND} mutants containing lesions in the G glycoprotein. Treatment of thermolabile mutants at 45°C removes nearly six logs of VSV infectivity. Alternatively, temperature sensitive G glycoprotein maturation mutants have been used to produce or assay pseudotypes at the nonpermissive temperature (39.5°C) at which the G glycoprotein fails to reach the cell surface.

2. Materials

2.1. Viruses and Cell Lines

Details of viruses and cell lines described are included in the text.

2.2. Culture of Cell Lines

Adherent nonhematopoietic cell lines are cultured in minimal essential medium (MEM) containing 4–10% heat-inactivated fetal calf serum (HI-FCS) (Gibco) and hematopoietic cells in RPMI 1640 plus 10% HI-FCS.

2.3. Agar or Agarose Medium

Agar medium is made from 120 mL of 2.4% Difco agar, maintained in the molten state at 44°C, mixed with an equal vol of 2× MEM containing 2% HI-FCS and 10% tryptose phosphate broth (prewarmed to 44°C). Agarose medium is made by mixing 40 mL of molten 2% seaplaque agarose (ICN) at 40°C with 40 mL of 2× MEM and 20 mL HI-FCS (both prewarmed to 40°C).

3. Methods

3.1. Production of VSV Stocks

Generally, the Indiana strain of VSV has been used; however, in countries where the use of VSV may be restricted, other rhabdoviruses can be used. For instance, Chandipura virus was shown to be as efficient as VSV_{IND} at assembling BLV pseudotypes *(1)*; however, different rhabdoviruses have different cell tropisms, and Chandipura does not replicate in human hematopoietic cell lines as well as VSV_{IND} (Neil Nathenson, personal communication). VSV_{IND} wt stocks can be produced in a variety of cell lines. However, mink lung (CCL64) cells generally give high yields of progeny.

1. Infect monolayers of mink cells that are just subconfluent at a low multiplicity of infection or MOI (0.1–0.01 plaque-forming units, or PFU, per cell), for 1 h at 37°C.
2. Wash once with growth medium.
3. Incubate at 37°C until the cytopathic effect (cpe) is clearly apparent (usually overnight).
4. Harvest the supernatant, clarify by spinning at low-speed (1500 rpm for 5 min in a bench centrifuge), divide into 1-mL aliquots, and store at –70°C.

3.2. Titration of VSV Stocks

VSV stocks are quantified again using mink lung cells:

1. Seed 10^6 cells in 2 mL of growth medium per 30-mm-diameter well and grow overnight.
2. Thaw a VSV aliquot, and make serial tenfold dilutions in complete phosphate-buffered saline containing 1% HI-FCS (PBS/1% FCS).
3. Remove the medium from each well and add 0.2 mL of each virus dilution.
4. After 1 h incubation at 37°C, remove nonadsorbed virus particles by aspiration and wash the cell layer once with 2 mL PBS/1% FCS.
5. Add 2 mL of agar medium maintained at 44°C.

6. Allow the agar to set on the bench and incubate the dishes at 37°C. Plaques are visible without staining after 1–2 d and are counted visually. Exactly the same procedures are used for production and titration of the conditional lethal mutants described above except they are carried out at the permissive 33°C temperature.

3.3. Production of VSV Pseudotype Stocks

VSV pseudotypes of retroviruses are most conveniently produced from chronically infected virus-producer cell lines. Budding VSV particles incorporate avian retrovirus coat glycoproteins efficiently enough to produce pseudotypes with titers high enough to be exploited for most purposes. VSV pseudotype titers of mammalian retroviruses including the human HIV and HTLV retroviruses, are generally low; it is thus important to use highly productive retrovirus-infected cells.

1. Infect retrovirus producer cells in exponential growth with VSV_{IND} at an MOI of 10 PFU/cell in a cell concentration of 5×10^6 cells mL.
2. After 1 h of adsorption at 37°C, wash the infected cells once in growth medium, resuspend at a cell density of $1–2 \times 10^6$ cells/mL, and incubate at 37°C.
3. The time of harvest for putative pseudotype progeny will depend on the susceptibility of the producer cells to VSV replication. For HTLV-I producing human osteosarcoma (HOS) cells, harvest the progeny virus 12–18 h after infection. The optimal time of harvest does not necessarily correspond with any particular stage of cytopathology (cpe).
4. Clarify supernatants containing progeny virus by a low-speed centrifugation.
5. If the supernatant is very alkaline (pink), adjust the pH by carefully adding sterile 1% acetic acid dropwise until the medium is orange (but not yellow).
6. VSV pseudotypes of most retroviruses can be frozen and stored at –70°C.

3.4. Titration of VSV Pseudotype Stocks

Pseudotype titers are quantified by similar plaque-forming assays described for VSV wt titration. Obviously, assay cells must have appropriate receptors; however, mink lung (CCL64) cells bear receptors for many retroviruses, including HTLV-I but not HIV *(9,13)*.

1. Before adsorption to assay cells, treat the putative pseudotype virus with an anti-VSV hyperimmune neutralizing serum for 1 h at 37°C.
2. Seed 10^6 assay cells in growth medium in 30-mm wells the night before they are needed and incubate at 37°C.

3. Add anti-VSV serum (0.1 mL) to 0.1 mL of pseudotype and 0.8 mL of PBS/1% FCS (0.1 mL pseudotype and 0.9 mL PBS/1% FCS are the control) and incubate at 37°C for 1 h.
4. These mixtures are already at a 10^{-1} dilution of virus; then make further serial dilutions: 10^{-2} and 10^{-3} for anti-VSV treated virus and 10^{-2} to 10^{-8} for untreated.
5. Remove growth medium from assay cells and add 0.2 mL of appropriate pseudotype dilutions immediately (10^{-1}, 10^{-2}, and 10^{-3} for α-VSV-treated and 10^{-6}, 10^{-7}, and 10^{-8} for untreated). At this stage a polycation may be included to enhance plating efficiency (*see* Note 2). The remaining procedure is the same as used for assaying VSV_{IND} wt. Pseudotype titers of about 10^4 PFU/mL should be achievable.

3.5. *HTLV-I and HTLV-II Pseudotypes* (see *Note 1*)

HTLV-I peudotypes are best produced from the few infected nonlymphoid cell lines available: HOS/PL or HT1080/PL *(14)*. Pseudotypes produced from these cells will plate efficiently on a range of different adherent cell lines *(2)*, making detection and quantitation easy. Pseudotypes made through HTLV-I-transformed T-lymphocytes plate well on fewer cell lines (Heb7a are good indicators, mink lung less so), implying that during infection of nonlymphoid cell lines, there is selection of variant HTLV-I viruses that infect by nonlymphoid cell receptors more efficiently.

Only transformed T-lymphocyte lines are available as HTLV-II producers, and the indicator cells must be chosen carefully—again Heb7a are good indicators.

3.6. *HIV Pseudotypes*

HIV pseudotypes can be produced from many of the productive systems available; however, H9-infected T-lymphocytes have been the most consistent in our hands. The receptor for HIV, the CD4 antigen, is present on lymphocyte and monocyte cell lines and has been expressed on some nonlymphoid human cell lines, such as HeLa *(15,16)*. Assay of HIV pseudotypes on $CD4^+$ lymphocyte cell lines is tricky, since most of these cell lines grow in suspension. Adherent sublines of CEM and GHI are available; however, these are sensitive to agar medium. To overcome this problem, either the 2.4% Difco agar can be replaced by 2% seaplaque agarose (ICN), or a plaque-indicator cell layer, such as mink lung, can be added. In the latter situation, the adherent CEM or GHI cells are seeded thinly into wells ($3–5 \times 10^5$ cells/30-mm well) and the assay carried out in the normal way until removing the virus inocula and washing. After this stage, 10^6 mink lung cells in 2 mL of growth medium are added and allowed to adhere (usually for about 1.5 h) before adding agar-containing medium. Although the CEM or GHI cells may not survive under this overlay, they will survive long enough for at least one round of VSV replication and production of progeny VSV, which will infect surrounding mink lung cells and form a plaque. The advantage of using agar overlay with mink lung cells as a plaque indicator is that plaques are easy to see. This procedure can also be used to assay on suspension cells. The suspension cells are first stuck down using poly-L-lysine. Wells are treated at room temperature with poly-L-lysine (50 µg/mL) in water for 1 h and washed three times in water. Assay cells are washed twice in serum-free medium (plus HEPES) and 5×10^5 cells/2 mL added per well. After 20 min on a flat surface, the cells should be stuck down and the assay carried out as described above, using a plaque indicator mink lung cell layer. Alternatively, suspension cells can be stuck down with poly-L-lysine at a high density to form a monolayer; in this situation an agarose medium overlay must be used, and not agar.

3.7. Testing Specificity of Pseudotypes

There are a number of ways to check that pseudotypes are really pseudotypes and not unneutralized background or clumped VSV wt. Neutralization by specific sera to the retrovirus providing the coat protein, e.g., HTLV-I$^+$ human patient's serum *(9)*, or inhibition of infection by mAbs to the cell receptor if available, e.g., anti-CD4 for HIV, will specifically eliminate and identify the pseudotype fraction.

Alternatively, the phenomenon of viral interference can be utilized *(8)*. Titration of pseudotypes on cells infected and uninfected with the appropriate retrovirus should give efficient plating on uninfected cells and only background plaques on infected cells, where the receptors are either blocked or downregulated.

Knowledge of the host range of the cell can also be exploited; e.g., VSV (HTLV-I) pseudotypes plate on mink lung, rat XC, and human HOS cells, but not on bovine MDBK or bat lung cells *(2)*.

3.8. Detection of Neutralizing and Nonneutralizing Antibodies to Retroviral env Glycoproteins

VSV pseudotype assays are easily adapted to measure neutralizing antibodies to retroviruses *(9)* by mixing antibody dilutions with a set number of pseudotype PFUs and looking for reduction in plaque numbers. Plaque counts are compared to controls in which pseudotype was mixed with appropriate control serum or ascites. Pseudotype assays can also be arranged to measure nonneutralizing antibodies to *env* glycoproteins. Antibody-bound pseudotypes can be precipitated with *Staphylococcus aureaus* bearing protein and eliminated by a low-speed centrifugation. The remaining pseudotype in the supernatant is then mixed with α–VSV_{IND} and assayed in the normal way.

4. Discussion

VSV pseudotypes have been exploited to study retroviral interaction with and entry into cells. The interpretation of results depends on how accurately pseudotypes parallel attachment and penetration in natural retrovirus infection. Only one example of a discrepancy between pseudotype and nonpseudotypes has been reported: mAbs directed to the gp41 transmembrane glycoprotein of HIV neutralize nonpseudotype particles, but not VSV(HIV) pseudotypes *(17)*. This result suggests that some neutralization epitopes may be expressed differently on pseudotyped glyproteins. The antibodies used in this case were all IgMs. IgMs are large pentameric structures, perhaps too large to interact with an epitope on gp41, close to the pseudotype membrane, where α-VSV-coated VSV G glycoproteins would be predominant.

Recently, VSV pseudotypes of mammalian retroviruses have been used to study the mode of uptake in cells *(18)*. Enveloped viruses can enter cells either by direct fusion at the cell surface, e.g., Sendai, or by receptor-mediated endocytosis, e.g., influenza, in which the low pH of the endosome is required for a conformational change in the hemagglutinin molecule to occur before membranes of the virus and endosome can fuse. Even though VSV is pH-dependent, pseudotype particles take on the entry properties of the adopted coat glycoproteins, so that VSV (HIV) pseudotypes, for instance, are pH-independent. These observations suggest that pseudotypes are good models of natural retroviral infection.

5. Notes

1. VSV pseudotypes of most retroviruses can be frozen and stored at –70°C; however, HTLV pseudotypes are fragile and need to be snap-frozen and stored in liquid nitrogen to preserve their infectivity.
2. The adsorption and infectivity of many retroviruses are enhanced by the presence of such polycations as polybrene or DEAE-dextran; either of these may be included in the assay. Generally, 0.2 mL of either polybrene (20 µg/mL) or DEAE-dextran (25 µg/mL) are added to the cell layers in 30 mm wells directly before the addition of 0.2 mL of pseudotype dilution.

Acknowledgments

I wish to thank Robin Weiss for introducing me to pseudotypes, and Jan Zavada for generously supplying us with anti-VSV sheep serum.

References

1. Zavada, J., Cerny, L., Zavadova, Z., Bozonova, J., and Altstein, A. D. (1979) A rapid neutralization test for antibodies to bovine leukemia virus with the use of rhabdovirus pseudotypes. *J. Natl. Cancer Inst.* **62,** 95–101.
2. Weiss, R. A., Clapham, P., Nagy, K., and Hoshino, H. (1985) Envelope properties of human T-cell leukemia viruses. *Curr. Top. Micobiol. Immunol.* **115,** 235–246.
3. Dalgleish, A. G., Beverly, P. C. L., Clapham, P. R., Crawford, D. H., Greaves, M. F., and Weiss, R. A. (1984) The CD4 (T4) antigen is an essential component of the receptor for the AIDS retrovirus. *Nature* **3,** 763–767.
4. Bruck, C., Portetelle, D., Burny, A., and Zavada, J. (1982) Topographical analysis by monoclonal antibodies of BLV-gp51 epitopes involved in viral functions. *Virology* **122,** 353–362.
5. Portetelle, D., Dandoy, C., Burny, A., Zavada, J., Siakkou, H., Gras-Masse, H., Drobecq, H., and Tartar, A. (1989) Synthetic peptides approach to identification of epitopes on bovine leukemia virus envelope glycoprotein gp5l. *Virology* **169,** 34–41.
6. Choppin, P. W. and Compans, R. W. (1970) Phenotypic mixing of envelope proteins of the parainfluezna virus and vesicular stomatitisvirus. *J. Virol.* **5,** 609–616.
7. Zavada, J. (1972) VSV pseudotype particles with the coat of avian myeloblastosis virus. *Nature New Biol.* **240,**122.
8. Weiss, R. (1984) Experimental biology and assay of RNA tumor viruses, in *RNA Tumor Viruses* (Weiss, R., Teich, N.,Varmus, H., and Coffin, J., eds.), CSH, New York, pp. 209–261.
9. Clapham, P., Nagy, K., and Weiss, R. A. (1984) Pseudotypes of human T-cell leukemia virus type 1 and neutralization by patients' sera. *Proc. Natl. Acad. Sci. USA* **81,** 2886–2889.
10. Vaux, D. J. T., Helenius, A., and Mellman, I. (1988) Spike–nucleocapsid interaction in Semliki Forest virus reconstructed using network antibodies. *Nature* **336,** 36–42.
11. Zavada, J. (1982) The pseudotypic paradox. *J. Gen. Virol.* **63,** 15–54.
12. Weiss, R. A. and Bennett, P. L. P. (1980) Assembly of membrane glycoproteins studied by phenotypic mixing between mutants of vesicular stomatitis virus and retroviruses. *Virology* **100,** 252–274.
13. Sommerfelt, M. A. and Weiss, R. (1990) Receptor interference groups among twenty retroviruses plating on human cells. *Virology* **176,** 58–69.
14. Clapham, P., Nagy, K., Cheingsong-Popov, R., Exley, M., and Weiss, R. A. (1983) Productive infection and cell free transmission of human T-cell leukemia virus in a nonlymphoid cell line. *Science* **222,** 1125–1127.
15. Maddon, P. J., Dalgleish, A. G., McDougal, J. S., Clapham, P. R., Weiss, R. A., and Axel, R. (1986) The T4 gene encodes the AIDS virus receptor and is expressed in the immune system and the brain. *Cell* **47,** 333–348.
16. Chesebro, B. and Wehrly, K. (1988) Development of a sensitive quantitative focal assay for human immunodeficiency virus infectivity. *J. Virol.* **6,** 3779–3788.
17. Dalgleish, A. G., Chanh, T. C., Kennedy, R. C., Kanda, P., Clapham P. R., and Weiss, R. A. (1988) Neutralisation of diverse HIV-1 strains by monoclonal antibodies raised against a gp41 synthetic peptide. *Virology* **165,** 209–215.
18. McClure, M. O., Sommerfelt, M. A., and Weiss, R. A. (1990) The pH independence of mammalian retrovirus infection. *J. Gen. Virol.* **71,** 767–773.

CHAPTER 10

Cell Lineage Studies Using Retroviruses

Jack Price

1. Introduction

In order to study cell lineage in any given experimental situation, a method by which a label can be introduced into a precursor cell is required. By means of this label, the subsequent fate of the cell and its progeny can be followed. A retrovirus, as an agent of gene transfer, can be used as a genetic label, providing an indelible, heritable marker of lineage. Unlike most artificial means of gene transfer, the use of retroviruses in transferring genes into cells is both efficient and accurate. This has made retroviruses invaluable in studying cell lineage in vivo, particularly in mammalian embryos, which are too small and inaccessible to be amenable to study with conventional lineage labels.

Retroviruses have been used to study cell lineage in two different ways. One method has been to make use of the fact that every retrovirus integrates randomly into the host-cell chromosomal DNA. Thus, every retroviral integration generates a unique site and can be recognized by the particular endonuclease restriction fragment to which it gives rise on Southern blots *(1,2)*. Consequently, precursor cells can be infected either in vivo or in vitro, and then allowed to develop normally. Subsequent analysis of different tissues or cell populations will reveal whether the same infected cells have contributed to more than one such group of cells—that is, whether the same retroviral insertion sites are found in more than one population of cells.

This approach has two major virtues and two obvious limitations. The virtues are that, first, the insertion site provides an unequivocal marker of

From: *Methods in Molecular Biology, Vol. 8:*
Practical Molecular Virology: Viral Vectors for Gene Expression
Edited by: M. Collins © 1991 The Humana Press Inc., Clifton, NJ

clonality and, second, expression of viral genes is not required. The limitations are that small clones cannot be detected, so many rounds of cell division are required between infection and detection, and that the method addresses only lineage relationships between populations of cells; individual cells within populations cannot be distinguished. Unfortunately, these limitations exclude some of the more interesting cell lineage questions. Most of the methods involved in cell lineage studies of this type are common to other uses of retroviruses and are covered in other chapters in this volume. Consequently, this approach will not be considered further in this chapter.

The approach to cell lineage that will be considered here employs a retroviral vector that encodes a marker gene, the expression of which can be detected histochemically in infected cells *(3,4)*. Infected cells and the clones to which they give rise will express the marker gene and, consequently, can be distinguished from uninfected cells by a histochemical stain. In practice, virus is introduced into an embryo so that a small number of cells become infected. Then, at some later developmental stage, the infected tissue is sectioned and stained histochemically, allowing the investigator to discover what has become of the infected cells. This method has the advantage of permitting single cells to be detected so that a precise identification of the infected cells is possible. Also, clones of any size can be detected, which means that lineage can be studied very late in development, when precursor cells have only one or two more cycles of division to complete. There are, however, two potential disadvantages that must be born in mind. First, constitutive expression of the marker gene is required. If under any circumstances expression were regulated, the results would be distorted. This means that care needs to be exercised in the choice of the viral construct to be used for any given experiment. Genes driven off the retroviral enhancer, for example, will not be expressed by cells of the preimplantation mouse embryo *(5)*.

The second problem is that this method has no intrinsic marker of clonality. Each clone of cells will stain similarly, and there is no certain way in which one can be distinguished from another. In practice, this may or may not be a major problem. Generally, in these experiments, relatively low titers of virus are used so that infections are rare. In this way, marked clones are scattered throughout a tissue, and distinguishing between them becomes less of a problem. Difficulties may arise, however, in situations in which cell mixing causes cells within a clone to become well-separated.

The most successful lineage studies so far have used viruses in which the marker gene is β-galactosidase *(6–10)*. This has been the gene of choice for a variety of reasons:

1. It is a readily available cDNA of reasonable size (ca. 3 kb) *(11)*.
2. It is a much studied gene for which good enzyme assays *(12)* and histo-

chemical methods already exist *(13)*.

3. There is some reason to believe that the β-galactosidase enzyme can exist in cells without disturbing their normal metabolism or distorting their subsequent development. (There are various reports demonstrating this point. *See*, for example, refs *14–16*.)

Although vectors containing other genes have been developed and will certainly be important in future experiments of this type, this chapter will concentrate on the use and detection of vectors containing β-galactosidase. The experimental procedures will be considered in three parts: administration of the virus, preparation of the tissue, and β-galactosidase histochemistry.

2. Materials

2.1. Administration of Virus

1. Viral supernatant, produced according to the protocols elsewhere in this volume and concentrated, if necessary, according to the protocols in Chapter 3.
2. Polybrene, stock of 1 mg/mL. Dilute 1:100 into the virus before use (*see* Note 1).
3. Anesthetic. For *in utero* injections into rat or mouse embryos, I recommend a combination of hypnorm™ (Janssen Pharmaceuticals Ltd., Oxford, UK) and hypnovel™ (Roche Welwyn, Garden City, UK), at a dose of 2.7mL/kg of a mixture of 1:1:2 hypnorm:hypnovel:water injected interperitoneally.
4. Suture thread: 5/0 prolene monofilament (Ethicon Ltd., Edinburgh UK).

2.2. Tissue Preparation

1. Phosphate-buffered saline (PBS), pH 7.4: 0.17*M* NaCl, 3m*M* KCl, 10 m*M* Na_2HPO_4, 1.8 m*M* KH_2PO_4.
2. Fixative: 2 g of paraformaldehyde in 100 mL of 0.1*M* PIPES buffer (pH 6.9) + 2 m*M* magnesium chloride + 1.25 m*M* EGTA.
3. Sucrose solution: 15 g/100 mL of fixative solution.
4. OCT embedding compound (Tissue Tek, Miles Laboratories, Elkhart, IN).
5. Glass slides dipped into a solution of 1% gelatin plus a pinch of chrome alum.

2.3. Histochemistry

1. Detergent Solution: 0.01% sodium desoxycholate + 0.02% NP 40 + 2 m*M* magnesium chloride in PBS.
2. X-gal buffer: detergent solution containing 5 m*M* potassium ferricyanide and 5 m*M* potassium ferrocyanide.

3. X-gal (5-bromo-4-chloro-3-indolyl β-galactopyranoside): Make an X-gal stock solution of 40 mg/mL in dimethylformamide and keep it at 4°C in the dark. Use only glass, not plasticware, in the preparation of this solution; something that affects the solubility of X-gal seems to be absorbed from plastic. Add this stock to X-gal buffer to give an X-gal solution with a final concentration of 1 mg/mL.
4. 50% alcohol, 70% alcohol, and absolute alcohol.
5. Histoclear™ (National Diagnostics, Somerville, NJ).
6. DPX mountant (BDH, Poole, UK).

3. Methods

3.1. Administration of Virus

Clearly, the precise protocol for this part of the procedure will vary greatly among experiments, depending on exactly what tissue and what stage of embryonic development are under study. What follows are a number of general guidelines.

1. Concentrated or unconcentrated virus (containing polybrene at 10 µg/mL) can be injected into embryonic structures without causing more than superficial damage and without any evidence of adverse effects on cells. For example, we have injected virus into brain ventricles of embryonic rats from embryonic day 14 (E14) onward with good rates of success. We do this with the animal under hypnovel/hypnorm anesthesia. Following laparotomy, a fiberoptic light probe is inserted behind the uterus. With this transillumination, embryos are clearly visible from E14 onward, so virus can be injected using a hand-held Hamilton syringe and a 30-G needle through the uterine wall into any of a variety of structures. We have found that we can easily inject 1 or 2 µL of virus into brain ventricles. Following suture of the abdominal wall and skin, the pregnant animal recovers quickly, and in most cases gives birth normally.

   We have also successfully injected virus into neonatal rodents under cryoanesthesia. If very young animals are cooled in an ice box for several minutes, their body temperature falls, their hearts stop, and they become completely anesthetized. This state can be maintained long enough for viral injections into the animal to be made. The pups recover very quickly upon subsequent warming.
2. Some loss of titer will be experienced from the value given by 3T3 cells in culture to that found with in vivo injections. This is not surprising, given that some virus will be lost from the site of injection and that cells in vivo are not necessarily going to be as available to the virus as they would be in a culture dish. Experience suggests that the in vivo titer will

be between 5 and 25% of the theoretical figure, but it could well be considerably lower in more demanding experiments.
3. Retroviruses seem to have poor penetration in tissue; they will not, for instance, cross basement membranes. Consequently, virus needs to be put very close to the site of the dividing cells that are to be marked.

3.2. Tissue Preparation

Tissue can be stained for β-galactosidase as single cells in culture, as tissue sections, or as whole mounts. The following material concentrates on the preparation of tissue sections, since this is the most involved, but note is made of how to deal with the other preparations.

1. Animals should be perfused with fixative (*see* Note 2). The tissue of interest should then be removed and immersed in the fixative at 4°C. If perfusion is not possible or is undesirable, the tissue can be removed fresh and fixed just by immersion. In either case, the length of fixation will depend on the size of the piece of tissue. We would typically fix brains overnight after perfusion (*see* Note 3).
2. The tissue is immersed in fixative containing 15% sucrose until it sinks. The sucrose acts as a cryoprotectant.
3. Mount and orient the tissue in OCT and freeze either on dry ice or by plunging the tissue block into isopentane cooled in liquid nitrogen. For larger blocks of tissue, the dry-ice method is probably superior. It is slower, but the tissue is less likely to crack.
4. Cut sections of the desired thickness on a cryostat. We have worked successfully with section thicknesses of 10–75 µm. Collect the sections on gelatin/chrome alum coated slides, allow them to dry in air, and keep the slides stored at –20°C.

3.3 β-Galactosidase Histochemistry

1. Fix the tissue sections on the slides by immersing them for 10 min in ice-cold 0.5% glutaraldehyde (or any other paraformaldehyde or glutaraldehyde combination). This is not absolutely necessary, but will ensure even fixation, especially if the tissue was originally fixed by immersion only (*see* Note 4).
2. Wash twice for 10 min on ice in PBS + 2 m*M* magnesium chloride.
3. Incubate the slides for 10 min in detergent solution on ice.
4. Incubate for 2–3 h in X-gal solution at 37°C (*see* Note 5).
5. Wash several times in PBS, and then in water, and counterstain if appropriate (we use 2% Orange G in 1% phospho-tungstic acid for 30 s). Next, dehydrate through 50 and 70% alcohol into absolute alcohol, then clear

in histoclear, and finally mount in DPX mountant. The sections can then be viewed under an interference contrast microscope. The retrovirally marked cells should have turned a deep shade of indigo blue (*see* Note 6).

4. Notes

1. Polybrene is a positively charged molecule that aids the binding of the virus to the cell membrane. It is essential for good rates of infection.
2. The β-galactosidase enzyme is insensitive to either paraformaldehyde or glutaraldehyde over a wide range of concentrations. It is, however, completely inactivated by every alcohol-, acetone-, or chloroform-based fixative that we have tried. This includes commercial formalin preparations that contain a low concentration of methanol. EGTA and Mg^{2+} ions are included in the fixative because they enhance the staining of the lacZ gene product and reduce the staining of the endogenous enzyme.
3. If it is required that the staining be done on tissue whole mounts rather than on sections, fix the tissue by perfusion and immersion as described above, wash extensively with PBS + 2 m*M* magnesium chloride, incubate overnight in X-gal buffer minus X-gal at 4°C, and then incubate overnight in X-gal solution at 37°C. The tissue can then be washed and either prepared for cryostat sectioning as described above or prepared for conventional wax embedding.
4. If cultured cells are to be stained, a much simpler protocol can be followed. Fix in 0.5% glutaraldehyde in PBS for 15 min at room temperature, wash several times with PBS, and incubate in X-gal solution for 30 min (longer if required). If cultures are stained on tissue-culture plastic, they can be viewed with an inverted microscope without mounting; if the cultures are grown on cover slips, they can be mounted without dehydration in glycerol.
5. For X-gal staining, we have found 5 m*M* to be the optimum concentration for both the ferricyanide and ferrocyanide, and we invariably use this concentration when staining cultured cells. When staining sections, however, it appears that increasing the concentration decreases the diffusion of the reaction product that otherwise occurs. We have found that concentrations of 20–30 m*M* ferricyanide and ferrocyanide give better results, although the reaction is slightly slower at these concentrations. It should also be noted that both ferricyanide and ferrocyanide can be replaced by 0.5 mg/mL of nitro blue tertazolium to give a faster reaction generating a product that is more intensely blue, but more diffuse *(17)*. The 1 mg/mL concentration of Xgal is used routinely in our laboratory, but lower concentrations have been used by others. The optimum concentration is best arrived at empirically for different situa-

tions. The X-gal solution, once made, is stable and can be kept either at room temperature or at 4°C, and can be reused for several weeks. A substitute for X-gal ("bluo-gal") has recently been marketed by Bethesda Research Laboratories (Bethesda, MD) and is claimed to diffuse less than X-gal.

6. Unfortunately, the blue reaction product fades with time. Also, this fading is uneven; some staining fades quickly, whereas other structures in the same section retain the staining apparently unchanged. We have been unable to prevent this. It is also worth noting that X-gal staining is not compatible with immunofluorescence. The blue reaction product seems to absorb UV light.

Acknowledgments

None of the methods described in this chapter originate with me. All the methods are based on published sources. More important, many of the small improvements and useful hints that are part of any reliable technique have come originally from colleagues and collaborators, past and present. I gratefully acknowledge this assistance.

References

1. Soriano, P. and Jaenisch, R. (1986) Retroviruses as probes for mammalian development: Allocation of cells to the somatic and germ cell lineages. *Cell* **46**, 19–29.
2. Lemischka, I. R., Raulet, D. H., and Mulligan, R. C. (1986) Developmental potential and dynamic behaviour of hematopoietic stem celle. *Cell* **45**, 917–927.
3. Sanes, J. R., Rubenstein, J. L. R., and Nicolas, J.-F. (1986) Use of a recombinant retrovirus to study postimplantation cell lineage in mouse embryos. *EMBO J.* **5**, 3133–3142.
4. Price, J., Turner, D., and Cepko, C. (1987) Lineage analysis in the vertebrate nervous system by retrovirusmediated gene transfer. *Proc. Natl. Acad. Sci. USA* **84**, 156–160.
5. Jaenisch, R., Fan, H., and Croker, B. (1975) Infection of preimplantation mouse embryos and of newborn mice with leukemia virus: tissue distribution of viral DNA. *Proc. Natl. Acad. Sci. USA* **72**, 4008–4012.
6. Turner, D. and Cepko. C. (1987) Cell lineage in the rat retina: A common progenitor for neurons and glia persists late in development. *Nature* **328**, 131–136.
7. Price, J. and Thurlow, L. (1988) Cell lineage in the rat cerebral cortex: a study using retroviral-mediated gene transfer. *Development* **104**, 473–482.
8. Gray, G. E., Glover, J. C., Majors, J., and Sanes, J.R. (1988) Radial arrangement of clonally related cells in the cicken optic tectum: Lineage analysis with a recombinant retrovirus. *Proc. Natl. Acad. Sci. USA* **85**, 7356–7360.
9. Luskin, M. B., Perlman, A. L., and Sanes, J. R. (1988) Cell lineage in the cerebral cortex of the mouse studied in vivo and in vitro with a recombinant retrovirus, *Neuron* **1**, 635–647.
10. Walsh, C. and Cepko, C. L. (1988) Clonally related cortical cells show several migration patterns. *Science* **241**, 1342–1345.

11. Kalnins, A., Otto, K., Ruther, U., and Muller-Hill, B. (1983) Sequence of the lacZ gene of *Escherichia coli*. *EMBO J.* **2**, 593–597.
12. Norton, P. A. and Coffin, J. M. (1985) Bacterial β-galactosidase as a marker of Rous sarcoma virus gene expression and replication. *Mol. Cell Biol.* **5**, 281–290.
13. Dannenberg, A. M. and Suga, M. (1981) Histochemical stains for macrophages in cell smears and tissue sections: β-galactosidase, acid phosphatase, non-specific esterase, and cytochrome oxidase, in *Methods for Studying Mononuclear Phagocytes* (Adams, D. O., Edelson, P. J., and Koren, M. S., eds.) Academic, New York, pp. 375–396.
14. Lis, J. T., Simon, J. A., and Sutton, C. A. (1983) New heat shock puffs and β-galactosidase activity resulting form transformation of *Drosophila* with an hsp70-lacZ hybrid gene. *Cell* **35**, 403–410.
15. Hall, C. V., Jacob, E. P., Ringold, G. M., and Lee, F. (1983) Expression and regulation of *Escherichia coli* lacZ gene fusions in mammalian cells. *J. Mol Applied Genetics* **2**, 101–109.
16. Alan, N. D., Cran, D. G., Barton, S. C., Hettle, S., Reik, W., and Surani, M. A. (1988) Transgenes as probes for active chromosomal domains in mouse development. *Nature* **333**, 852–855.
17. McGadey, J. (1970) A tetrazolium method for nonspecific alkaline phosphatase. *Histochemie* **23**, 180–184.

CHAPTER 11

Retroviral Vectors as Insertional Mutagens

Joop Gäken and Farzin Farzaneh

1. Introduction

A critical step in the life-cycle of retroviruses is the integration of the double-stranded DNA copy of their RNA genome into the genome of the host cell *(1)*. Although the provirus DNA sequences flanking the site of integration are precisely determined and characteristic of each virus *(2)*, the vast majority of the integrations are the product of nonhomologous recomination events, resulting in the pseudorandom integration of the provirus into the host-cell genome *(3,4)*. Retroviruses can therefore act as agents of insertional mutagenesis. The insertion of the provirus into the genome could, in principle, result in either gene activation or gene inactivation. The insertional inactivation would be the result of provirus integration within the coding or regulatory sequences of the gene of interest, thus disrupting the expression of a functional gene product *(5–13)*. Insertional activation could be the product of the integration of viral-promoter enhancer elements, contained within the long terminal repeat (LTR) sequences in the vicinity of a silent gene, resulting in the increased transcription of that gene *(14–26)*. In addition to these direct *cis*-acting effects, there can be indirect *trans*-regulatory effects resulting from the presence of the viral genome within the cell, but irrespective of its position of integration. This could be the product of genes or other regulatory elements encoded by the virus. The position-indepen-

From: *Methods in Molecular Biology, Vol. 8: Practical Molecular Virology: Viral Vectors for Gene Expression*
Edited by: M. Collins

dent effect(s) of retrovirus integration would be easy to identify; they would be present in all cells or in a vastly larger number of cells than would be compatible with low-frequency integrations into specific genomic domains. The study of these possible *trans*-regulatory effects, which are, in general, believed to be absent in the nononcogenic retroviruses, are not described in this section.

1.1. Mutation Frequency

In both insertional activation and inactivation, the frequency of mutations are directly proportional to the number of integration events and the target size. The target size is determined by the total size of the genome and the average size of the domain within which provirus integration would result in the selected phenotype. Mammalian cells contain $2–3 \times 10^6$ kbp of DNA/haploid genome, and most genes span between 2 and 50 kbp. Assuming that the integration of the provirus anywhere within this domain would result in the insertional inactivation, a single copy gene would represent a target approx $10^{-5}–10^{-6}$ of the total genome. Therefore, the random integration of each provirus would increase the inactivating mutation frequency in the gene of interest by $10^{-5}–10^{-6}$. Since the background frequency of inactivating mutations in most genes is of the same order of magnitude, many more than a single integration per cell is necessary in order to increase the frequency of insertional mutants significantly above background. However, in practice, the actual frequency of such mutations may differ from this predicted rate. For example, in a population of superinfected F9 embryonal carcinoma cells with an average of 25 integrations/cell, the inactivating mutation frequency in the hypoxanthine-guanine phosphoribosyl transferase (*hgprt*) gene is increased from approx 10^{-7} (spontaneous) to 10^{-6} (in the infected cells); an increase about 50-fold lower than expected *(12)*. This may be because the cellular sites of provirus integration are not truly random; the *hgprt* gene may therefore, represent a "cold spot" for retroviral integration. On the other hand, because the majority of the provirus integrations are into actively transcribed genomic sequences *(27)*, the frequency for other genomic sites, particularly the actively transcribed domains, may in practice be higher than the predicted rate.

As in other forms of mutagenesis, it is difficult to isolate diploid cells with inactivating mutations in codominant genes, unless there is some form of actual or functional haploidy in that gene. This is because, with an average mutation frequency of approx 10^{-6}/haploid copy, the frequency of mutations in both copies of an otherwise functional gene would be about 10^{-12}. However, if (a) only one functional copy of the gene is available in the cell (a situation sometimes referred to as functional haploidy; for example, that in the *hgprt* gene in female mammalian cells, caused by X-chromosome inactivation), (b) there is a high enough gene-dosage effect (in the absence of dose compensation), or (c) there is partial genotypic haploidy (as in a number of established cell lines, e.g., HL-60 "promyelocytes"), it would be possible to isolate such insertionally inactivated mutants. There are, however, examples of codominant genes in diploid cells in which the spontaneous frequency of mutation is very much higher than the mathematically predicted rate. An example of this is the autosomal gene adenosine phosphoribosyl transferase (*aprt*). Interestingly, the frequency of both spontaneous and retroviral insertional inactivation of *aprt* in mouse embryonal carcinoma cells is only 100-fold lower than the frequency in the *hgprt* locus *(12)*. This suggests either that the *aprt* locus is a "hot spot" for mutagenesis or that there is some form of functional haploidy in this autosomal gene.

The frequency of insertional activations resulting in the expression of a previously silent gene would also be affected by similar factors; however, for most genes, the target size for an insertional activation (as opposed to inactivation) may in fact be small. This is because integrations within only a limited domain in the vicinity of the coding sequences is likely to result in the *cis*-activation of the target gene. However, because insertional activations are dominant, the cointegration of the provirus in both copies of a diploid gene is not necessary. Therefore, the selection of retroviral insertional activation of a silent gene is made comparatively easier.

1.2. Multiplicity of Infection

In principle, there are two types of retroviral vectors that could be used for insertional mutagenesis: replication-defective and replication-competent. With replication-competent vectors, only a few integrations per cell can be obtained *(28)*. This is because the expression of the viral envelope gene in the infected cell (and presumably the saturation of the cellular receptors for the envelope protein) would eventually prevent the reinfection of the cell with additional copies of the virus. An exception to this general rule is presented by embryonal carcinoma cell lines, which in general are refractory to the expression of viral genes *(29)*, thus permitting superinfection to very high copy numbers *(30)*. In contrast to the replication-competent retroviruses, the replication-defective vectors lack the envelope gene; they can therefore be used for superinfection of target cells, thus resulting in multiple provirus integrations in a single cell. The multiplicity of infection (moi) can be increased by concentrating the vector that is shed in the culture media of the packaging cells (e.g., centrifugation at 12,000*g* for 15 h); or alternatively by the coculture of the packaging and target cells. The highest moi is obtained with the coculture procedure. However, it is important to have prevented the replication of the packaging cells in order to avoid the subsequent contamination of the target cells by the packaging cells. This can be achieved by the lethal UV or γ-irradiation of the packaging cells or by their treatment with the DNA synthesis inhibitor, mitomycin C. In our experience, the highest moi is obtained in coculture with packaging cells treated with mitomycin C. However, when using replication-defective vectors it is important to ensure the absence of helper viruses (*see* Chapter 2, this vol). This is important both as a health and safety consideration (particularly when working with amphotropic packaging cells) and for the subsequent stability of the generated genotype and the ability to identify the actual gene whose mutagenesis has resulted in the appearance of the selected phenotype.

The superinfection of the target cells increases the mutation frequency and thus facilitates the isolation of mutants. However, this poses another problem, namely the subsequent difficulty in identifying, among all the integrated copies present, the single integration that has resulted in the activation, or inactivation, of the gene(s) responsible for the generation of the selected phenotype. This problem will be discussed further in Section 1.5., which deals with the identification of the mutated genes.

1.3. Mutant Selection Procedures

As in other forms of mutagenesis, the availability of a reliable and efficient means of mutant selection is very important. This usually means either the selective killing of the cells not treated with mutagen or the growth advantage of the mutated cells. Examples include the isolation of *hgprt*-cells by their ability to grow in the presence of the nucleotide analogs, e.g., 8-azaguanine *(31)*, or the selection of growth-factor-independent cells from a population of cells dependent on the exogenous supply of these factors (e.g., the isolation of factor-independent cell lines from the myeloid precursor cell line D35 *[26]*). The selection of insertional mutants in genes the products of which are expressed on the cell surface, can also be achieved by immune lysis of the wild-type cells, as for the isolation of mutants in the beta 2-microglobulin locus using allele-specific antibodies *(11)*. Although in theory the isolation of mutants by screening procedures may also be possible, in practice this will usually be very difficult. One notable exception to this would be the isolation of mutants that are products of the insertional activation of cell-surface genes. Such mutants could be isolated by fluorescence-activated cell sorting, enabling the population to be enriched for cells with increased levels of expression of the corresponding gene product.

1.4. Cloning of the Site of Provirus Integration

There are two strategies available for the cloning of the cellular sequences that have been disrupted by the insertion of the provirus sequences: (a) construction of genomic libraries; (b) PCR-mediated amplification.

1.4.1. Construction of Genomic Libraries

Provided that the vector used for mutagenesis carried within its genome a functional tRNA suppressor gene (e.g., the *supF* tyrosine suppressor tRNA gene), it would be possible to make subgenomic libraries very highly enriched for the cellular sequences flanking the sites of provirus integration *(32,33)*. For this purpose, the total genomic DNA is isolated from the clonal populations of the infected cells and digested with restriction enzymes in such a way as to generate intact DNA fragments containing the *supF* gene and flanking cellular sequences. This is achieved by the digestion of the genomic DNA with either a vector "noncutter," or single cutters that cut between the *supF* gene and the end of the 5' or 3' LTR. The mutant-cell DNA is then cloned in phage vectors that carry amber mutations in genes that are essential for the phage proliferation (e.g., λ gtwes or EMBL 3A). In such a vector, in a suppressor-free host, only those clones that contain a functional *supF* gene in the insert will proliferate. Therefore, each of the generated phage colonies will contain cellular sequences flanking one of the sites of provirus integration.

1.4.2. PCR-Mediated Amplification

The polymerase chain reaction (PCR) is a powerful method for the primer-directed amplification of specific DNA sequences in vitro *(34)*. In each cycle of the PCR, lasting only a few minutes, the specific DNA sequence contained within the chosen primers is duplicated (*see* Fig. 1). This exponential amplification results in an increase of up to 1,000,000-fold, in such a sequence within 20–30 cycles, lasting only a few hours. In a procedure first described by Silver and Keerikatte *(35)*, it is possible to capture one of the two cellular flanking sequences into the adjacent LTR (*see* Fig. 2); PCR can then be used to amplify specifically the cellular sequences flanking the site of provirus integration. For this purpose, mutant-cell DNA is first digested with a frequently cutting restriction endonuclease, such as *Rsa* I, with a 4-base recognition sequence. The digested DNA is ligated and then redigested with another endonuclease (e.g., *Eco* RV) with a 6-base recognition site between the site recognized by the frequent cutter and the cellular site of integration. The cellular DNA originally bordering this LTR will now be contained within the two LTR fragments generated by the infrequent cutter; this will not be true of the cellular sequences flanking the other LTR. The use of primers directed against provirus sequences 5' and 3' of the second restriction enzyme then allow the amplification of the cellular sequence originally flanking one of the two LTR sequences, depending on the choice of enzymes used. Therefore, for each integrated copy of the provirus, one fragment, representing one of the two cellular flanking sequences, will be amplified. The presence within the PCR synthetic oligonucleotide primers of appropriate recognition sequences for restriction enzymes would allow the simple cloning of the ampli-

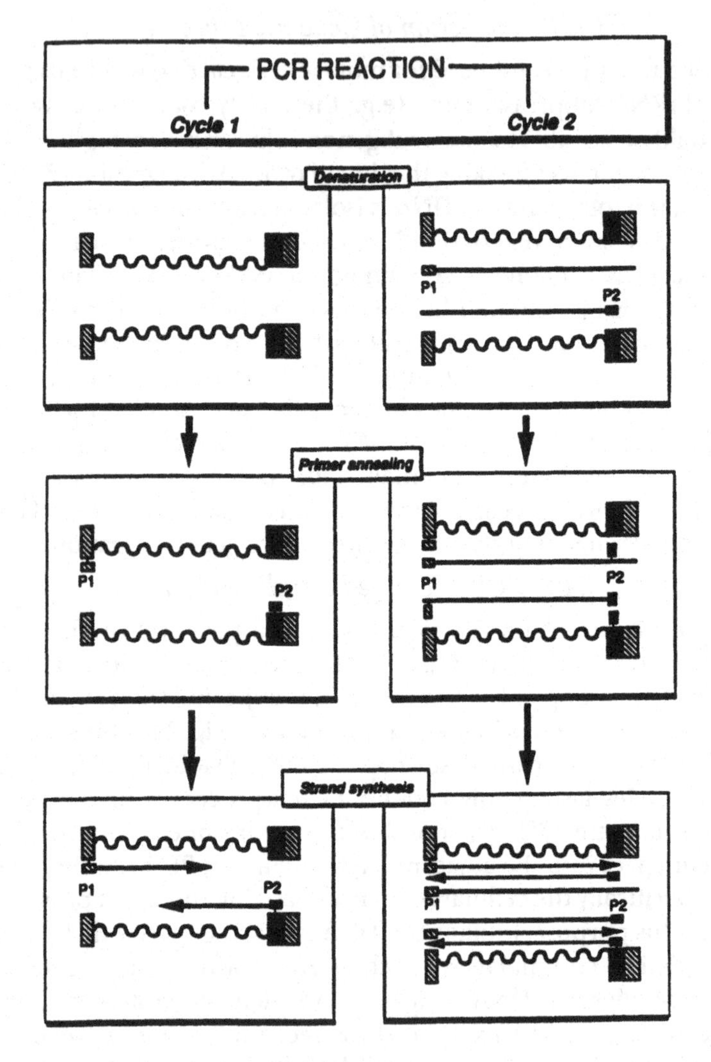

Fig. 1. Polymerase chain reaction.

fied sequences in a suitable vector. Alternatively, the amplified fragments could be resolved by size fractionation and then directly sequenced *(36)*.

1.5. Identification of the Gene of Interest

As already discussed, retroviral insertional mutagenesis usually requires the superinfection of the cells in order to achieve mutation frequencies significantly higher than the background rate. The identification of the single

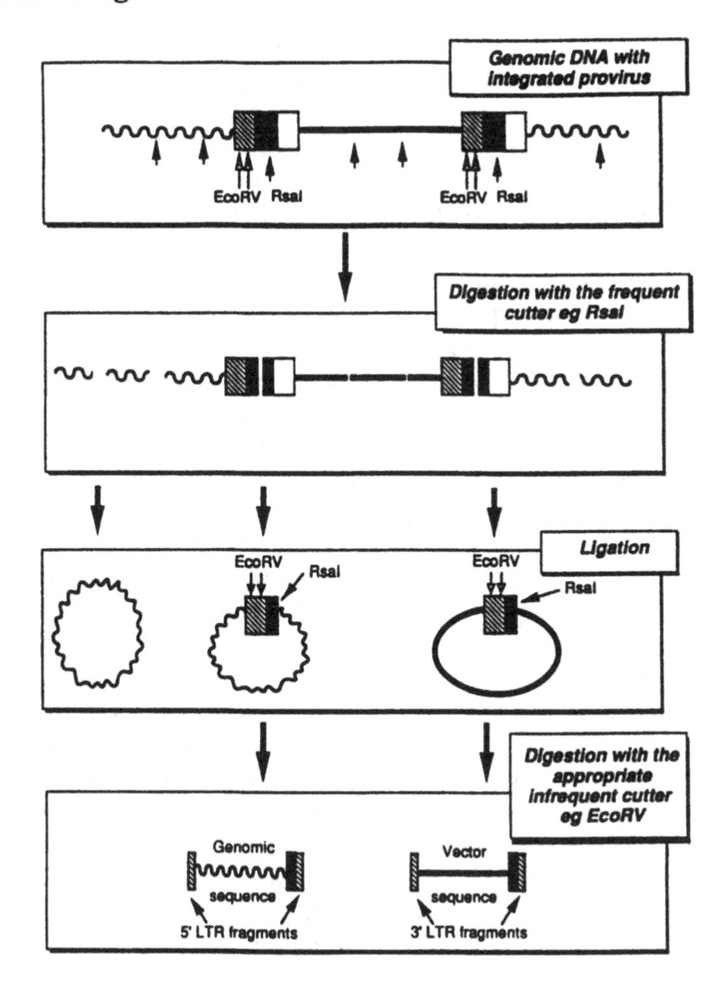

Fig. 2. Capture of flanking DNA sequences.

integration responsible for the appearance of the selected phenotype, among the many integrations present in a superinfected cell, is by far the most difficult aspect of the practical use of retroviral vectors as agents of insertional mutagenesis. This is a particularly severe problem when the objective is the isolation of insertionally inactivated genes, and it is the reason for the relatively modest success thus far in the experimental use of retroviral insertional mutagenesis for the identification of new genes. Of course, the natural transduction and insertional activation of protooncogenes by retroviruses has been responsible for the identification of the vast majority of the oncogenes (*see* ref. *1*).

The precise nature of the strategy for identifying genes mutated by retroviral insertion would depend, to a very large extent, on the predicted nature of the gene and the characteristics of the function that is being studied. However, thus far, this has largely depended on the identification of common sites of provirus integration in independent clones of cells sharing a selected phenotype, and/or the previous knowledge of the gene being mutated, as was the case in the isolation of mutants both in the *hgprt (12,13)* and the beta 2-microglobulin loci *(11)*. Here we describe another approach, the use of nuclear run-on probes for library screening, a procedure applicable to the isolation of genes involved in inducible cellular functions, such as differentiation, damage-inducible DNA repair, and the like.

1.5.1. Identification of Common Sites of Provirus Integration

The identification of the integration event that has resulted in the appearance of the selected phenotype could be aided by the detection of a common site of provirus integration in independently arising clones of cells with the selected phenotype *(12,21)*. The identification of such a site may be possible by the screening of Southern blots prepared with DNA from independently isolated clones of mutant-cells sharing the selected phenotype. Should a common site of integration be identified, it could be isolated by the appropriate digestion and size fractionation of the genomic DNA and its subsequent cloning either in a phage vector, as described above, or by direct amplification, using the PCR. In practice, however, it may be difficult to identify such common sites of provirus integration, because both insertional activation and inactivation could result from provirus integration at many different sites within the coding and regulatory sequences of a given gene. This problem is compounded by the usual presence of preferred sites of provirus integration, that are irrelevant to the development of the selected phenotype.

1.5.2. Library Screening by Nuclear Run-On Probes

In appropriately isolated nuclei, it is possible to elongate and radioactively label, to a high specific activity, RNA transcripts that have already been initiated at the time of nuclei isolation; no new transcripts are initiated. This procedure enables the synthesis of probes to all actively transcribed genes and offers a powerful means for the screening and identification of genes that are differentially expressed during inducible cellular processes (e.g., the induction of differentiation). A combination of two such probes made from the wild-type cells before and after treatment with the inducing agent could be used for the screening of the corresponding subgenomic library containing the flanking sequences. The differential pattern of hybridization to the two probes would then allow the rapid identification of the integration events that are likely to have resulted in the appearance of the selected phenotype.

The detected pattern could reveal valuable information about the nature of the cellular sequence disrupted by the integration event. For instance, those clones in the subgenomic library that do not hybridize to either of the two probes are likely to represent provirus integrations into nontranscribed, silent, genomic sequences. Similarly, a clone that hybridizes to both probes would represent a constitutively active gene unlikely to be involved in the inducible cellular function being studied. However, clones hybridizing to only one of the two probes would represent a likely candidate for a differentially regulated gene disrupted by the provirus integration.

This approach offers a powerful tool for the identification of the gene(s) of interest among the many sites of provirus integration in mutant-cell lines generated by superinfection. A similar approach has recently been used by Scherdin et al. *(27)* to demonstrate the preferential integration of provirus into actively transcribed genomic domains. Section 3. below describes the application of this procedure to the identification of genes involved in the inducible differentiation of a "promyelocytic" cell line (HL-60).

2. Materials

All solutions are autoclaved before use, unless otherwise stated.

1. Mitomycin C: Stock solution 200 µg/mL in serum-free medium (filter-sterilized).
2. Sodium pyruvate: 100 m*M* sterile stock solution.
3. Noble agar: 5% (w/v) in double-distilled water.
4. PCR buffer: 10 m*M* Tris HCl (pH 8.3), 50 m*M* KCl, 2 m*M* $MgCl_2$, and 0.01% (w/v) gelatin.
5. NZ medium: 10 *g* NZ amine, 5 *g* yeast extract, 5 *g* NaCl, and 2 *g* $MgSO_4$ per liter adjusted to pH 7.5 with NaOH.
6. NZ medium: 10 *g* NZ amine, 5 *g* NaCl, and 2 *g* $MgS0_4$ per liter adjusted to pH 7.5 with NaOH.
7. Maltose: 10% (w/v) stock solution in distilled water, filter sterilized.
8. NZ top agar: 10 *g* NZ amine, 5 *g* NaCl, 2 *g* $MgS0_4$, and 6.5 *g* Difco agar per liter; adjust to pH 7.5 with NaOH. For NZ top agarose, use 6.50 *g* of agarose.
9. NZ agar: 10 *g* NZ amine, 5 *g* NaCl, 2 *g* $MgS0_4$, and 11 *g* Difco agar per liter; adjust to pH 7.5 with NaOH. For NZ agarose, use 11 *g* of agarose.
10. SM buffer: 50 m*M* Tris HCl (pH 7.5), 100 m*M* NaCl, 10 m*M* $MgCl_2$, and 0.01% (w/v) gelatin.
11. Polybrene stock solution: 30 mg/mL polybrene in H_2O, filter sterilized.
12. DMSO stock solution: 25% (v/v) DMSO in serum-free medium, filter sterilized.

13. Denaturing solution: 0.5*M* NaOH, 1.5*M* NaCl.
14. Neutralizing solution: 0.5*M* Tris HCl (pH 7.4), 1.5*M* NaCl.
15. 2 × SSC solution: 0.3*M* NaCl, 0.03*M* sodium citrate (pH 7.0).
16. Phosphate-buffered saline (PBS): 137m*M* NaCl, 2.7m*M* KCl, and 10 m*M* phosphate buffer (pH 7.4), obtainable in tablet form (Sigma).
17. Run-on lysis buffer: 10 m*M* Tris HCl (pH 7.9), 140 m*M* KCl, 5 m*M* $MgCl_2$, and 1 m*M* dithiothreitol.
18. Run-on reaction buffer: 20 m*M* Tris HCl (pH 7.9), 20% glycerol, 140 m*M* KCl, 10 m*M* $MgCl_2$, and 1 m*M* dithiothreitol.
19. Run-on extraction buffer: 10 m*M* Tris HCl (pH 7.4), 1% SDS, 20 m*M* EDTA.
20. HSB buffer: 10 m*M* Tris HCl (pH 7.4), 0.5*M* NaCl, 50 m*M* $MgCl_2$, and 2 m*M* $CaCl_2$).
21. NETS buffer: 10 m*M* Tris HCl (pH 7.4), 100 m*M* NaCl, 10 m*M* EDTA.
22. Prehybridization and hybridization buffer: 10 m*M* TES [*N*-Tris (hydroxymethyl)methyl-2-aminoethanesulfonic acid] (pH 7.4), 0.2% SDS, 10 m*M* EDTA, 0.3*M* NaCl, 5 × Denhardt's solution, 100 µg/mL tRNA, and 100 µg/mL poly(A).
23. 100 × Denhardt's solution: 2% BSA, 2% polyvinylpyrrolidone, and 2% Ficoll.

3. Methods

In the study described here, the amphotropic packaging cell line M3P, producing the retroviral vector M3Pneo-sup (developed and kindly provided by Carol Stocking, University of Hamburg), is used (*see* Note 1). The packaging cells were generated by the infection of PA317 cells *(37)* with the supernatant of the ecotropic packaging cell line GP + E-86 *(38)*, containing the M3Pneo-sup vector. For the insertional mutagenesis of the human promyelocytic leukemic cell line HL-60, these cells are cocultured with the amphotropic packaging cell line. The object is the isolation of cells resistant to the induction of differentiation by a number of agents, including phorbol esters (inducers of monocytic differentiation) and retinoic acid (inducer of myeloid differentiation). The differentiation-resistant cells can readily be isolated from the differentiation-competent cells by their ability to continue to grow and form discrete colonies of cells in soft-agar cultures; the competent cells will terminally differentiate, become postmitotic, and fail to develop colonies.

3.1. Infection and Mutagenesis Procedure

1. Seed M3P cells at a density of approx 5 × 10^5 cells/10-cm plate, in Dulbecco's Modified Eagle's Medium (DMEM) supplemented with 10% fetal bovine serum (FBS).

2. Take subconfluent plates of M3P cells approx 2 d after plating.
3. Add mitomycin C to a final concentration of 10 µg/mL and incubate M3P cells for another 4 h at 37°C, in order to prevent the further proliferation of the M3P cells. The cells treated with mitomycin C will continue to release virus for several days at an approx titer of 10^5/mL (as assayed on NIH3T3 fibroblasts) (*see* Note 3).
4. Take off the medium containing mitomycin C and wash cells three times with sterile PBS.
5. Add 10 mL of RPMI medium (10% FBS) containing 10^6 HL-60 cells in total.
6. Add polybrene and DMSO, to a final concentration of 3 µg/mL and 0.3%, respectively, and cocultivate the HL-60 and M3P cells for 4 d.
7. Collect the HL-60 cells (centrifuge at 500 *g*, for 5 min at room temperature). Resuspend in fresh medium at 2–3 × 10^5/mL and add to fresh plates of packaging cells treated with mitomycin C for a second 4-d period of coculture.
8. Collect the infected HL-60 cells and determine the average number of infected cells and the average number of integrations per cell. The number of infected cells can be estimated by the standard soft-agar cloning of HL-60 cells in the presence of 1mg/mL G418, for 2–3 wk, with media changes every 4–5 d (*see* Section 3.2. for soft agar cloning). The average number of provirus integrations per cell can be determined by standard dot- and/or Southern-blot hybridization using a viral probe (other than the LTR, to avoid possible cross-hybridization with endogenous retroviruses).
9. The infection cycle (two 4-d periods) described above routinely results in an average of 1 provirus/cell; the infection cycle may be repeated for greater numbers of integrated provirus (*see* Note 4). Because of the high efficiency of infection, preselection of the infected cells is not necessary, and it is possible to proceed directly to the mutant selection.

3.2. Selection of Differentiation Resistant HL-60 Mutants

1. Adjust the density of infected HL-60 cell to approx 2 × 10^5 cells/mL in fresh RPMI supplemented with 20% FBS and 1m*M* sodium pyruvate.
2. Autoclave a solution of 5% noble agar (in distilled water) and maintain at 60°C.
3. Add the required vol of the agar solution to the suspension of HL-60 cells (maintained at 37°C) to the final concentration of 0.3%. Mix well and rapidly plate in tissue-culture dishes (5–10 mL/6-cm dish or 10–20

mL/10-cm dish). Allow the agar to set at room temperature, and then incubate at 37°C.

4. The following day, add an equal vol of medium containing twice the required final concentration of the differentiation-inducing agent, combination of agents, or the selectable marker (G418).
5. Feed cultures by changing the liquid medium every 4–5 d, leaving the agar part of the culture intact.
6. Differentiation-resistant colonies should appear 2–3 wk later (*see* Note 5).
7. Using a glass Pasteur pipet, remove the soft agar colony and expand, initially in microwell plates.

3.3. Cloning of the Site of Provirus Integration

Two alternative procedures can be used: (a) PCR-mediated amplification of the site of provirus integration; (b) selective cloning of the cellular flanking sequences by using the *supF* gene contained within the vector.

3.3.1. PCR-Mediated Amplification

The oligonucleotide primers used are 5'-(CTAGA GCTC)C CTACA GGTGG GGTC TTTCA-3' (P1), and 5'-(AGCTC TAGA)G CTCAG GGCCA AGAAC AGAT-3' (P2). The amplified fragments contain 71 bp of the 5' LTR sequences (bases 2270–2289 and 2509–2745) *(39)*, and the cellular sequences between this LTR and the next genomic *Rsa* I site. In addition, a 311-bp viral sequence will be amplified from the 3' LTR and the flanking viral sequences (*see* Note 6). The P1 primer contains a *Sac* I, and the P2 an *Xba* I, restriction site for the subsequent ease of directional cloning of the amplified sequences.

1. Isolate total genomic DNA from individual clones of the differentiation-resistant cells (*see* elsewhere in this vol, for details of genomic DNA isolation). All subsequent restriction digests, ligations, and PCR amplifications are carried out in the same PCR buffer.
2. Dissolve 500 ng of total genomic DNA in 50 µL of the PCR buffer, and digest to completion with the four-cutter *Rsa* I (5 U for 1 h at 37°C).
3. Heat-inactivate the restriction enzyme by incubating at 65°C for 10 min.
4. Ligate the digested DNA (approx 500 ng of DNA in 50 µL of PCR buffer and 5 U of T4 DNA ligase overnight at 16°C). Heat-inactivate ligase as in Step 3.
5. Digest with *Eco* RV to linearize the circles (*see* vol 2, this series, on the use of restriction enzymes). The 5' LTR sequence now flanks the cellular sequences 5' of the site of provirus integration.

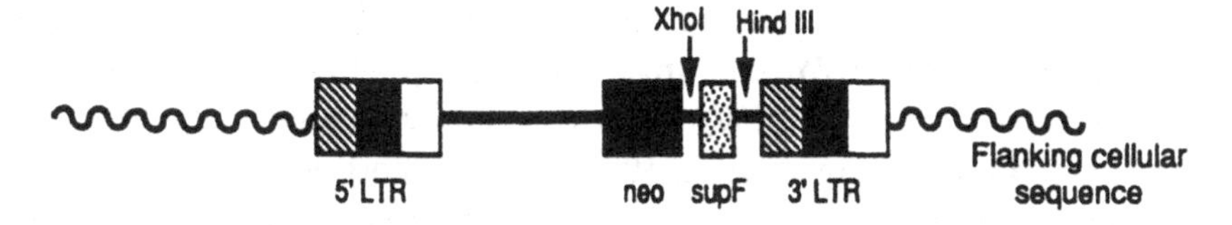

Fig. 3. Integrated M3Pneo-sup provirus.

6. Add 20 n*M* of each of the dNTPs, 100 pmol of each of the two primers, and 2 U of Taq-polymerase (Cetus). Adjust the vol to 100 µL with the PCR buffer and overlay with 50 µL of mineral oil.
7. Place the tubes in an automated heating block and amplify the DNA over 60 cycles, adding another 2 U of Taq-polymerase after 30 cycles. The PCR conditions are dependent on the actual oligonucleotide primers used and need to be optimized empirically. The best condition for the primers used here are annealing, 63°C for 1.0 min; extension, 72°C for 3.0 min; and denaturation, 92°C for 1.5 min.
8. After PCR amplification, a small vol (10 µL) should be run on a gel to confirm the presence of PCR amplified DNA bands (*see* Note 7).
9. The amplified DNA sequences can now either be cloned into a suitable vector using the *Sac* I and *Xba* I restriction sites or be directly sequenced after gel separation of each of the amplified sequences. The latter would be appropriate only if clear size separation of the PCR amplified flanking cellular sequences has been possible. The screening of the cloned fragments will be done as described for the *supF* mediated cloning.

3.3.2. supF-*Mediated Cloning*

1. Digest to completion the total genomic DNA isolated from independent differentiation-resistant clones of HL-60 cells. The restriction enzyme used should be either a vector noncutter or a single cutter that cuts between the *supF* gene and the 5' or 3' LTR. This is to ensure the release of a DNA fragment containing an intact *supF* gene and the largest possible length of flanking cellular sequences. In the case of the vector used here, *Xho* I and *Hind*III are most suitable (*see* Fig. 3).
2. Ligate the digested DNA into λ gtwes or EMBL 3A (or any other suitable phage vector) and package the phage particles (a phage-packaging kit could be used). *See* Chapter 20, this volume, for details of the ligation procedure.
3. Infect the suppressing host (e.g., *E. coli* LE392) to enable the amplification of the library. For this purpose, prepare plating bacteria by centri-

fuging (3000*g*, for 10 min) 50 mL of an overnight culture (LE392) in NZY medium. The NZY medium should be supplemented with maltose (final concentration of 0.2%), in order to enhance phage attachment to the bacteria. Wash the pellet in 10 m*M* $MgSO_4$ and resuspend in 1/2 vol (25 mL) 10 m*M* $MgSO_4$. These bacteria can be stored at 4°C for up to 2 wk. Infect the plating bacteria by mixing 1–2 mL of the suppressing host with a 100 µL aliquot of the packaged DNA mixture. Leave at room temperature for 15 min.

4. Add 30 mL of NZ top-agar at 45°C, mix well, and pour onto 22- × 22-cm NZ agar plate. Incubate overnight at 37°C.
5. Scrape off the top agar, add 30 mL of SM buffer and 0.3 mL of chloroform. Stir gently at room temperature for 20 min.
6. Centrifuge the slurry (5000*g* for 10 min). Collect the supernatant, add a drop of chloroform, and store at 4°C. The library can be stored for a few yr in this form. Determine the phage titer, using serial dilutions, essentially as described in 3 above. The described amplification of the phage library in the suppressing host should give a titer of approx 10^{10}/mL.
7. At least 10^8 phages, representing a minimum of 10^6 independent recombinants, are plated on a suppressor-free host (e.g., *E. coli* MC1061). The preparation of plating bacteria and infection procedure are as in 3 above. This should generate a library of 10^3–10^4 recombinants, each containing the *supF* gene and the cellular sequences flanking the site of provirus integration.
8. The library should be isolated and the actual titer determined as described in Steps 5 and 6 above.

3.4. Library Screening

1. The cloned libraries should be plated on suitable hosts, such that there are approx 400 plaques/10-cm plate. This is a convenient number for the subsequent isolation of discrete plaques and will adequately represent each of the individual sites of integration, even in superinfected cells with multiple copies of integrated provirus. For the library screening, use NZ agarose plates and NZ top agarose instead of agar in order to reduce the background in subsequent hybridizations.
2. Make replica filters by gently layering successive filters onto the plates containing the plaques. Leave each filter for approx 5 min. In order to avoid the peeling off of the top agarose, these plates should have been kept at 4°C for 1 h. The orientation of the filters should be marked asymmetrically with a needle dipped in radioactive India ink (the needle must pierce the agarose to ensure marking of subsequent filters in the

same position). Two such replica filters should be made for each of the agar plates to be screened. The original plates, from which the replicas were made, are sealed with Parafilm® and stored at 4°C.

3. Remove the replica filters, and incubate in denaturing solution for 2 min, and in neutralizing solution for 5 min.
4. Wash the filters in 2 × SSC for 2 min, air-dry, and then fix (bake for 2 h at 80°C for nitrocellulose or GeneScreen™).
5. The prehybridization and hybridization to the run-on probes are described below.

3.5. Preparation of the Nuclear Run-On Probes

Two run-on probes should be made: (a) Wild-type cells in the absence of the inducer, and (b) Wild-type cells in the presence of the inducer.

All procedures should be carried out on ice or at 4°C, unless otherwise specified (*see* Note 8).

1. Centrifuge approx 3–5 × 10^7 cells at 500*g* for 10 min, wash twice with ice-cold PBS, and resuspend in 1 mL of the run-on lysis buffer.
2. Add NP-40, to a final concentration of 0.5%, mix gently, and leave on ice for 5 min.
3. Recover the nuclei by centrifugation at 2000*g* for 2 min and wash once in the run-on reaction buffer.
4. Resuspend pellet in 100 µL of 2 × run-on reaction buffer. The nuclei can now be snap-frozen in liquid nitrogen and stored at –70°C, or used directly for the run-on labeling.
5. Add together: 100 µL of the suspension of nuclei (approx 10^7 nuclei), 250 µCi 5'-[α-^{32}P]UTP; 1 m*M* each of ATP, CTP, and GTP; 20 µg of creatine kinase; 50 U RNasin (Boehringer); and water, to a final vol of 200 µL.
6. Incubate at 37°C for 20 min with occasional mixing.
7. Add 100 U of RNase-free DNase 1 and incubate for 5 min at 37°C.
8. Add 2 mL run-on extraction buffer and 20 µg/mL proteinase K, and incubate for 45 min at 37°C.
9. Add 2.2 mL of the high-salt buffer (HSB).
10. Isolate the labeled RNA by hot phenol extraction. Add an equal vol of phenol saturated with NETS buffer and an equal vol of chloroform/isoamylalcohol (24:1). Heat to 65°C for 5–10 min, shaking occasionally. Chill and centrifuge for 2 min at 2000*g*.
11. Remove the organic phase (bottom phase) and leave the interphase and the aqueous phase behind, repeat the phenol-chloroform extraction.
12. Add an equal vol of chloroform isoamylalcohol, mix thoroughly, and centrifuge for 2 min at 2000*g*.

13. Remove the upper aqueous phase, add 50 μg tRNA, and then add sodium acetate (pH 5.2) to a final concentration of 0.3*M*. Precipitate the RNA with 2.5 vol of cold 100% ethanol. Centrifuge (15 min at 10,000*g*), wash with cold 70% ethanol, and briefly dry the RNA pellet under vacuum.
14. Dissolve the RNA in 250 μL double distilled water.
15. Take a 1-μL aliquot and count the Cerenkov radiation in the tritium channel of a liquid scintillation counter to measure the total radioactivity in the labeled RNA.
16. Denature the RNA for 20 min on ice in 0.2*M* NaOH. The RNA is now ready to be used in run-on probing.

3.6. Hybridization of Labeled RNA

1. Incubate the nitrocellulose filters in a shaking water bath for 4 h at 65°C in the prehybridization buffer.
2. Add the denatured RNA (approx 5×10^6 cpm/mL) and incubate in a shaking water bath at 65°C for 36–40 h.
3. Wash the filters twice in 2 × SSC at room temperature for 5 min.
4. Wash the filters twice in 2 × SSC at 65°C for 30 min.
5. Incubate the filters with RNase A (0.4 μg/mL) and RNase T1 (10 U/mL) in 2 × SSC at 37°C for 30 min.
6. Wash the filters twice with 2 × SSC at room temperature for 5 min.
7. Incubate the filters with proteinase K (50 μg/mL) in 2 × SSC, 0.5% SDS at 37°C for 30 min.
8. Wash the filters twice with 2 × SSC at room temperature for 5 min.
9. Air dry the filters and expose to X-ray films for autoradiography (*see* Note 9).

4. Notes

1. In the generation and choice of the amphotropic cell lines used for insertional mutagenesis in human cell lines, it is important to ensure as far as possible the absence of helper virus production. The ecotropic and amphotropic cell lines generated by Markowitz et al. *(38,40)* or Danos and Mulligan *(41)* are particularly suitable. This is because the separation of *gag*, *pol*, and *env* genes, required for packaging, has substantially reduced the chance of recombinations that result in the generation of helper or helper-independent virus.
2. The efficiency of infection of many human cell lines by the amphotropic vectors is very low. The isolation of the murine ecotropic retrovirus

receptor *(42)* allows the possibility of establishing human cell lines stably expressing the receptor and therefore, susceptible to very efficient infection with ecotropic vectors *(42)*. Naturally, this would also reduce the possible health and safety hazards that may be associated with the use of amphotropic vectors.

3. Treatment with mitomycin C can be replaced by γ-irradiation (2000 rads) of the virus producer cell line. Both procedures result in a similar titer of the vector in the culture supernatant.
4. The coculture of HL-60 and producer cells (M3P), as described, can be repeated many times (we have performed such cocultures for up to 10 cycles). The optimum concentration of polybrene and DMSO has to be established empirically for each cell line. The recommended 3 μg/mL polybrene and 0.3% DMSO increases the infection frequency of HL-60 four- to fivefold in each cycle of 4 d.
5. In HL-60 cells infected as described here, with an average of one integration per cell, the frequency of appearance of stable *hgprt*-cells increases 50-fold from a background frequency of approx 0.1×10^{-6} to one of about 5×10^{-6}.
6. In the PCR-amplification procedure, the detection of a 311-bp band generated by the amplification of the viral sequences flanking the 3' LTR serves as an internal control. The presence of this band in the agarose gels of the PCR products indicates that the amplification has been successful. Other amplified sequences in the PCR products would represent the different cellular sequences flanking the 5' LTRs at each of the provirus integration sites.
7. There is complete homology between the chosen PCR primers and a number of murine endogenous proviruses (e.g., Moloney mink cell focus-forming virus). Because of this property, the described PCR procedure will capture not only cellular sequences flanking the M3Pneo-sup vector, but also the flanking sequences to other murine endogenous viruses, that may be released by the murine packaging cells and cause insertional mutations in the target cell *(43)*.
8. The use of siliconized plastic- and glassware is recommended for the preparation of the run-on probe, in order to prevent loss of RNA by adhesion to the vessels.
9. If the library screening with the run-on probe produces a faint signal, or when integration into genes with transcripts of low abundance are suspected, a more sensitive screening procedure should be adopted. For this purpose, the phage plaques containing the cellular sequences flanking the site of provirus integration should be isolated and expanded.

The inserted DNA should then be isolated from either plate or suspension phage lysates. The isolated DNA samples can then be screened on slot blots with the run-on probes as already described. For details of phage isolation, expansion, and lysate preparation, *see* Vol. 2, this series.

Acknowledgments

The HL-60 mutagenesis studies in this laboratory are supported by grants from the Cancer Research Campaign, Leukemia Research Fund, and the Medical Research Council. We are grateful to Barbara Skene and Carol Stocking for invaluable advice in the retroviral insertional mutagenesis studies.

References

1. Varmus, H. (1988) Retroviruses. *Science* **240,** 1427–1435.
2. Varmus, H. and Swanstrom, R. (1982) Replication of retroviruses, in *RNA Tumour Viruses* (Weiss, R., Teich, N., Varmus, H., and Coffin, J., eds.), Cold Spring Harbor Laboratory Press, Cold Spring Harbor, NY, pp. 369–512.
3. Hughes, S. H., Shank, P. R., Spector, D. H., Kung, H. G., Bishop, J. M., Varmus, H. E., Vogt, P. K., and Breitman, M. L. (1978) Proviruses of avian sarcoma virus are terminally redundant, coextensive with unintegrated linear DNA and integrated at many sites. *Cell* **15,** 1397–1410.
4. Shimotohno, K. and Temin, H. M. (1980) No apparent nucleotide sequence specificity in cellular DNA juxtaposed to retrovirus proviruses. *Proc. Natl. Acad. Sci. USA* **77,** 7357–7361.
5. Jenkins, N. A., Copeland, N. G., Taylor, B. A., and Lee, B. K. (1981) Dilute (d) coat colour mutation of DBA/2J mice is associated with the site of integration of an ecotropic MuLV genome. *Nature* **293,** 370–374.
6. Copeland, N. G., Jenkins, N. A., and Lee, B. K. (1983) Association of the lethal yellow (*Ay*) coat color mutation with the ecotropic murine leukemia virus genome. *Proc. Natl. Acad. Sci. USA* **80,** 247–249.
7. Copeland, N. G., Hutchison, K. W., and Jenkins, N. A. (1983) Excision of the DBA ecotropic provirus in dilute coat/color revertants of mice occurs by homologous recombination involving the viral LTRs. *Cell* **33,** 379–387.
8. Schnieke, A., Harbers, K., and Jaenisch, R. (1983) Embryonic lethal mutation in mice induced by retrovirus insertion into the alpha 1 (I) collagen gene. *Nature* **304,** 315–320.
9. Jaenisch, R., Harbers, K., Schnieke, A., Lohler, J., Chumakov, I., Jahner, D., Grotkopp, D., and Hoffmann, E. (1983) Germ line integration of Moloney murine leukemia virus at the Movl3 locus leads to recessive lethal mutation and early embryonic death. *Cell* **32,** 209–216.
10. Wolf, D. and Rotter, V. (1984) Inactivation of p53 gene expression by an insertion of Moloney murine leukemia virus-like DNA sequence. *Mol. Cell Biol.* **4,** 1402-1410.
11. Frankel, W., Potter, T. A., Rosenberg, N., Lenz, J., and Rajan, T. V. (1985) Retroviral insertional mutagenesis of a target allele in a heterozygous murine cell line. *Proc. Natl. Acad. Sci. USA* **82,** 6600–6604.

12. King, W., Patel, M. D., Lobel, L. I., Goff, S. P., and Chi Nguyen-Huu, M. (1985) Insertion mutagenesis of embryonal carcinoma cells by retroviruses. *Science* **228,** 554–558.
13. Kuehn, M. R., Bradley, A., Robertson, E. J., and Evans, M. J. (1987) A potential animal model for Lesch-Nyhan syndrome through introduction of HPRT mutations into mice. *Nature* **326,** 295–298.
14. Hayward, W. S., Neel, B. G., and Astrin, S. M. (1981) Activation of a cellular *onc* gene by promoter insertion in ALV-induced lymphoid leukosis. *Nature* **290,** 475–480.
15. Varmus, H. E.,Quintrell, N., and Oritz, S. (1981) Retroviruses as mutagens: Insertion and excision of a nontransforming provirus alters expression of a resident transforming provirus. *Cell* **25,** 23–36.
16. Nusse, R. and Varmus, H. E. (1982) Many tumors induced by the mouse mammary tumor virus contain a provirus integrated in the same region of the host cell genome. *Cell* **31,** 99–109.
17. Fung, Y. K. T., Lewis, W. G., Crittenden, L. B., and Kung, H. J. (1983) Activation of the cellular oncogene c-*erbB* by LTR insertion: Molecular basis for induction of erythroblastosis by avian leukosis virus. *Cell* **33,** 357–368.
18. Peters, G., Brookes, S., Smith, R., and Dickson, C. (1983) Tumorigenesis by mouse mammary tumour virus: Evidence for a common region for provirus integration in mammary tumours. *Cell* **33,** 369–377.
19. Tsichlis, N., Strauss, P. G., and Hu, L. F. (1983) A common region for proviral DNA integration in MoMuLV-induced rat thymic lymphomas. *Nature* **302,** 445–449.
20. Wagner, E. F., Covarrubias, L., Stewart, T. A., and Mintz, B. (1983) Lethalities in mice homozygous for human growth hormone gene sequences integrated in the germ line. *Cell* **35,** 647–655.
21. Dickson, C., Smith, R., Brookes, S., and Peters, G. (1984) Tumorigenesis by mouse mammary tumor virus: Proviral activation of a cellular gene in the common integration region *int–2*. *Cell* **37,** 529–536.
22. Steffen, D. (1984) Proviruses are adjacent to c-*myc* in some murine leukemia virus-induced lymphomas. *Proc. Natl. Acad. Sci. USA* **81,** 2097–2101.
23. Shen-Ong, G. L. C., Potter, M., Mushinski, J. F., Lavu, S., and Reddy, E. P. (1984) Activation of c-*myb* locus by viral insertional mutagenesis in plasmacytoid lymphosarcomas. *Science* **226,** 1077–1080.
24. Lemay, G. and Jolicoeur, P. (1984) Rearrangement of a DNA sequence homologous to a cell-virus junction fragment in several Moloney murine leukemia virus-induced rat thymomas. *Proc. Natl. Acad. Sci. USA* **81,** 38–42.
25. Cuypers, H. T., Selten, G., Quint, W., Zijlstra, M., Maandag, R. E., Boelens, W., van Wezenbeek, P., Melief, C., and Berns, A. (1984) Murine leukemia virus induced T-cell lymphomagenesis: Integration of proviruses in a distinct chromosomal region. *Cell* **37,** 141–150.
26. Stocking, C., Löliger, C., Kawai, M.,Suciu, S., Gough, N., and Ostertag, W. (1988) Identification of genes involved in growth autonomy of haematopoietic cells by analysis of factor-independent mutants. *Cell* **53,** 869–879.
27. Scherdin, U., Rhodes, K., and Breindl, M. (1990) Transcriptionally active genome regions are preferred targets for retrovirus integration. *J. Virol.* **64,** 907–912.
28. Kozak, C. A. (1985) Susceptibility of wild mouse cells to exogenous infection with xenotropic leukemia viruses: Control by a single dominant locus on chromosome 1. *J. Virol.* **55,** 690–695.

29. Sarma, P. S., Cheong, M. P., and Hartley, J. W. (1967) A viral influence test for mouse leukemia viruses. *Virology* **33**, 180–184.
30. Stewart, C. L., Stuhlmann, H., Jahner, D., and Jaenisch, R. (1982) *De novo* methylation, expression, and infectivity of retroviral genomes introduced into embryonal carcinoma cells. *Proc. Natl. Acad. Sci. USA* **79**, 4098–4102.
31. Hooper, M. L. (1985) *Mammalian Cell Genetics.* Wiley, New York.
32. Reik, W., Weiher, H., and Jaenisch, R. (1984) Replication-competent Moloney murine leukemia virus carrying a bacterial suppressor tRNA gene: Selective cloning of proviral and flanking host sequences. *Proc. Natl. Acad. Sci. USA* **82**, 1141–1145.
33. Lobel, L. I., Patel, M., King, W., Nguyen-Huu, M. C., and Goff, S. P. (1985) Construction and recovery of viable retroviral genomes carrying a bacterial suppressor transfer RNA gene. *Science* **228**, 329–332.
34. Saiki, R. K., Scharf, S., Faloona, F., Mullis, K. B., Horn, G. T., Erlich, H. A., and Arnheim, N. (1985) Enzymatic amplification of beta-globin genomic sequences and restriction site analysis for diagnosis of Sickle cell anaemia. *Science* **230**, 1350–1354.
35. Silver, J. and Keerikatte, V. (1989) Novel use of polymerase chain reaction to amplify cellular DNA adjacent to an integrated provirus. *J. Virol.* **63**, 1925–1928.
36. Wong, C., Dowling, C. E., Saiki, K., Higuchi, R. G., Erlich, H. A., and Kazazian, H. H. (1987) Characterization of β-thalassaemia mutations using direct genomic sequencing of amplified single copy DNA. *Nature* **330**, 384-386.
37. Miller, A. D. and Buttimore, C. (1986) Redesign of retrovirus packaging cell lines to avoid recombination leading to helper virus production. *Mol. Cell Biol.* **6**, 2895–2902.
38. Markowitz, D., Goff, S., and Bank, A. (1988) A safe packaging line for gene transfer: separating viral genes on two different plasmids. *J. Virol.* **62**, 1120–1124.
39. Stacey, A., Arbuthnott, C., Kollek, R., Coggins, L., and Ostertag, W. (1984) Comparison of myeloproliferative sarcoma virus with Moloney murine sarcoma virus variants by nucleotide sequencing and heteroduplex analysis. *J. Virol.* **50**, 725–732.
40. Markowitz, D., Goff, S., and Bank, A. (1988) Construction and use of a safe and efficient amphotropic packaging cell line. *Virology* **167**, 400–406.
41. Danos, O. and Mulligan, R. C. (1988) Safe and efficient generation of recombinant retrovirus with amphotropic and ecotropic host ranges. *Proc. Natl. Acad. Sci. USA* **85**, 6460–6464.
42. Albritton, L. M., Tseng, L., Scadden, D., and Cunningham, J. M. (1989) A putative murine ecotropic retrovirus receptor gene encodes a multiple membrane-spanning protein and confers susceptibility to virus infection. *Cell* **57**, 659–666.
43. Scadden, D. T., Fuller, B., and Cunningham, J. M. (1990) Human cells infected with retrovirus vectors acquire an endogenous murine provirus. *J. Virol.* **64**, 424–427.

CHAPTER 12

Assessment of Compounds for Anti-HIV Activity

Clive Patience, John Moore, and Mark Boyd

1. Introduction

Since the identification of HIV (human immunodeficiency virus) as the retrovirus responsible for AIDS (acquired immunodeficiency syndrome), huge efforts have been made to identify compounds possessing antiviral activity. Many assay systems are available to measure various stages of the viral life-cycle. These are very convenient for assessing the antiviral activity of compounds. The assay systems often use T-cell lines expressing CD4 on their surface, since this molecule is an essential component of the receptor for the virus *(1)*. The basic principle when testing for anti-HIV compounds involves the interaction of virus with test material under suitable culture conditions and the subsequent measurement of a parameter reflecting the amount of virus present and, thus, the degree of viral replication.

When deciding on, or developing, an assay system, consideration must be given to the test material in question. If the material is a pure substance, targeted at a specific site of action, e.g., a virus-specific enzyme, such as the RNA-dependent DNA polymerase (reverse transcriptase, or RT), then purified-enzyme kinetic assays may be appropriate. For such assays, the rate of conversion of a suitable substrate for the RT in the presence of the putative anti-HIV agent is measured. In most cases, however, such an assay may not be

From: *Methods in Molecular Biology, Vol. 8:*
Practical Molecular Virology: Viral Vectors for Gene Expression
Edited by: M. Collins © 1991 The Humana Press Inc., Clifton, NJ

appropriate. Even pure molecules may require metabolic processing before becoming active. Also, in many cases the test material may consist of unknown structures, of which the site, or sites, of action are unknown. For such cases, more appropriate assays will involve the assessment of viral replication in the aforementioned CD4-positive T-cell lines. Since such assays are also appropriate for purified test materials, this chapter will concentrate on such methods.

In all assays, it is crucial to include an appropriate control substance, the activity of which can be compared to that of the test material. Control compounds may be related to the test materials by having similar structures or sites of action. If, however, no appropriate control is immediately obvious, the inclusion of several control substances will never detract from the data.

Once the principle of the assay system has been decided on, it becomes necessary to assess how the infection of the cells will take place and with which virus. For HIV, there are well-characterized viruses readily available (such as HTLV-III$_{RF}$ and HTLV-III$_{b}$), many of which are suitable for antiviral assays. Infection of the cells can take on many forms. For example, cell lines chronically infected with the virus (i.e., those carrying the viral genome in a ds-DNA form integrated into the cellular genome) can be treated with the test material, and one can then look for a reduction in the amount of virus produced. More commonly, uninfected cells are exposed to an inoculum of cell-free virus in the presence of test material and an assay for antiviral activity is performed.

As an end point for antiviral activity, one or more parameters may be examined. These include infectious virus, levels of virus-specific enzymes (e.g., RT), levels of viral components (e.g., p24), or the measure of viral cytopathic effect (cpe). Infectious virus levels can easily be determined from culture supernatant by simple titration, but, since this is very labor-intensive, it is not practical when a significant number of compounds are to be tested. The following three assay methods all lend themselves to automation, thus reducing the labor required, and all have the added advantage of being performed in a 96-well microtiter format, thus reducing costs. The assays described will be those methods measuring the following parameters as a measurement of antiviral activity: inhibition of virus-induced CPE, levels of viral p24, and levels of viral RT. In all antiviral assays, it should be noted that, if a reduction in viral levels is seen, the effect may not be the result of a specific antiviral effect, but may be the result of material being toxic to the assay cells. Thus, in all cases, assays measuring the cytotoxicity of the test materials, e.g., by tritiated thymidine uptake or viable cell count, should be run in parallel.

1.1. Colorimetric Assay for Inhibition of Virus-Dependent CPE

When HIV is seeded onto uninfected C8166 cells, viral replication takes place, resulting in the formation of giant multinucleate syncytial cells, which ultimately die. MTT (3-[4, 5-Dimethythiazol-2-yl]-2, 5-diphenyltetrazolium bromide) is a yellow substance that, when acted upon by mitochondrial dehydryogenases present in viable cells, turns into a blue formazan product *(2)*. If this blue product is released from the cells, then the number of viable cells in a well can be judged. Thus, in wells in which virus replication has left no viable cells, blue color will not be produced; in wells in which viral replication has been inhibited, the viable cells will produce the blue product. Cells no longer viable as a result of cytotoxicity of the test material will not produce any blue product. Thus it is possible within a single titration to determine the minimum cytotoxic concentration of a material and also the minimum antiviral concentration, giving an indication of the material's therapeutic ratio. A significant therapeutic ratio is necessary if a substance is to be of use as a potential treatment for patients.

1.2. Elisa for Determining Levels of Viral p24 Antigen

Determination of viral proteins in culture supernatants provides another parameter that reflects the degree of viral replication. Compounds restricting virus replication will thus be reflected by reduced levels of viral proteins in culture supernatants. Specific determination of levels of viral p24, a protein present in HIV, will thus be a method of identifying compounds with antiviral activity. Commercial kits are widely available, but described in the following pages is a method that does not use such kits, providing a more economic, but still sensitive, assay for p24 antigen of HIV-1 or HIV-2.

The assay is a twin-site "sandwich" ELISA. Briefly, p24 antigen is captured from a detergent lysate of virions onto a polyclonal antibody absorbed on a solid phase. Bound p24 is detected with an alkaline phosphatase-conjugated anti-p24 monoclonal antibody using the AMPAK™ ELISA amplification system.

1.3. Reverse Transcriptase Assay

Reverse transcriptase (E.C.2.7.7.7, RNA-directed DNA polymerase) activity can be detected in the supernatant of HIV-infected cells in culture. This greatly simplifies the assay conditions *(3)*, since none of the problems that can be caused by contaminating cellular DNA polymerases present in cell lysates will be encountered *(4)*.

The first step in the assay is to release the enzyme from the virion core, which is accomplished by using a mild detergent that does not inactivate the enzyme. Once released, the enzyme activity is measured by the level of production of polymerized DNA from nucleotides. For this, reverse transcriptase requires an RNA template to copy and a primer that provides the free 3' OH group essential for all DNA polymerases. The control for this assay is the replacement of the RNA template by a DNA template, since reverse transcriptase cannot copy this template under these assay conditions. The amount of RNA-directed DNA polymerization is determined by harvesting the reaction mixture onto a positively charged membrane to which the nucleic acid will bind. This harvesting is carried out in the presence of salt, in a salt concentration at which nucleotides will not bind or will do so only very inefficiently.

2. Materials

2.1. Colorimetric Assay

1. M8166 cells (a clone of C8166 cells obtained from R. C. Gallo) in exponential growth phase in culture medium at 4×10^5 cells/mL.
2. Culture medium: RPMI 1640 + 10% (v/v) fetal calf serum.
3. Virus: HTLV-III_{RF} containing 100 $TCID_{50}$/100 µL (as determined by this assay method).
4. Microtiter trays, 96-well (Nunc).
5. MTT (Sigma) dissolved at 7.5 mg/mL in phosphate-buffered saline (PBS).
6. Acidified isopropanol (i.e., 2 mL conc HCl/500 mL isopropanol) + 10% (v/v) Triton X-100.
7. Microtiter plate shaker.
8. Microtiter plate photometer capable of measuring absorbance at 690 and 540 nm.

2.2. Elisa

1. Immulon II microELISA plates (Dynatech).
2. D7320, D7330 antibodies (Aalto Bio Reagents, Dublin).
3. 100 m*M* $NaHCO_3$, pH 8.5.
4. TBS: (Tris-buffered saline) 144 m*M* NaCl, 25 m*M* Tris HCl), pH 7.5.
5. Skimmed milk, e.g., Marvel (Cadbury Ltd.).
6. Empigen (Calbiochem).
7. EH12 E1-AP.
8. Sheep serum (Advanced Protein Products).
9. Tween 20.

10. AMPAK™ amplification system (Novo Nordisk, Cambridge, UK).
11. Recombinant p24 antigen (American Biotechnology Ltd.).

2.3. Reverse Transcriptase Assay

The following items and reagents are required for the reverse transcriptase assay.

1. 1*M* Tris-HCl, pH 8.0.
2. 0.1*M* Dithiothreitol (DTT): A stock solution of 1.0*M* DTT can be made in 10 m*M* sodium acetate, pH 5.2, and stored in aliquots at –20°C.
3. 0.1*M* $MgCl_2$.
4. 1*M* KCl.
5. 1% (v/v) Triton X-100.
6. 6 m*M* glutathione: Store in aliquots at –20°C.
7. 10 m*M* ethyleneglycol-*bis*(β-aminoethylether) *N,N,N′,N′*,-tetra acetic acid (EGTA).
8. Template primer (1 mg/mL): One each of poly(rA)oligo$(dT)_{12}$ and poly (dA) oligo$(dT)_{12}$ (Pharmacia). Store in aliquots at –20°C.
9. 2 × SSC. A stock solution of 20 × SSC can be made by dissolving 175.3g sodium chloride and 88.2 g sodium nitrate in dH_2O and adjusting the pH to 7.0. Make the volume up to 1 L.
10. DE81 paper (Whatman).
11. 95% (v/v) ethanol.
12. Cell harvester (Skatron).
13. Microtiter plate shaking incubator (Dynatech).
14. Microtiter plates (96-well).
15. [Me-^{3}H]-Thymidine 5′-triphosphate, ammonium salt (Amersham, TRK 424). This is supplied as a 1/1 mixture of ethanol and water, which is dried by evaporation overnight and then resuspended in dH_2O to give a specific activity of 0.1 mCi/mL.

3. Methods

3.1. Colorimetric Assay

1. Sample preparation is performed in culture medium to produce 50 µL/well of test material at four times the test concentrations required by the operator. If serial dilutions are required, a convenient method is to add 75 µL of the maximum test concentration to the top well of a microtiter plate and 50 µL of culture medium to all other wells. Serially transfer 25 µL from well to well down the plate, thus producing 1:3 dilutions.

2. Add 50 µL of cell suspension to every well.
3. Add 100 µL of virus dilution (or culture medium for the cell control) to every well.
4. Incubate the plates humidified for 96 h at 37°C (with 5% CO_2 in air).
5. Add 20 µL of MTT to every well and incubate for a further hour.
6. Remove 150 µL of culture supernatant from every well and replace with 100 µL of the Triton–isopropanol mixture.
7. Shake until all the blue formazan crystals have dissolved.
8. Read the absorbency of every well at both 690 and 540 nm, subtracting the former from the latter.
9. The percentage of virus inhibition can be calculated by:

   % protection = [(ODt)HIV – (ODc)HIV]/[(ODc)MOCK – (ODc)HIV]

   where (ODt)HIV = OD of treatment well using infected cells, (ODc)HIV = OD of untreated well using infected cells, and (ODc)MOCK = OD of untreated well using uninfected cells.

3.2. Elisa

3.2.1. Sample Preparation

To inactivate viable HIV, empigen is added to bring the final concentration to 1% (v/v). This is kept at 56°C for 30 min. Samples can then be frozen, if required. The ELISA works most sensitively at an empigen concentration of 0.05–0.1% (v/v). Thus the inactivated samples must initially be diluted at least 1/20 in TBS. If higher dilutions of sample are required, then dilutions should be performed in TBS + 0.05% (v/v) empigen to maintain the assay sensitivity.

3.2.2. Method

1. D7320 (for HIV-1) or D7330 (for HIV-2) is diluted to 5–10 µg/mL in 100 m*M* $NaHCO_3$ and 100 µL is added to the inner 60 wells of the immulon II microtiter plates. The plates are sealed and incubated overnight at room temperature.
2. Wash plates twice with 200 µL of TBS.
3. Add 200 µL of 2% w/v skimmed milk in TBS, and keep for 30 min at room temperature to block nonspecific binding sites.
4. Wash twice with 200 µL of TBS.
5. Add a 100 µL sample, and keep for 3–4 h at room temperature (always including a p24 calibration titration or known positive culture).
6. Wash twice with 200 µL of TBS.
7. Dilute EH12E1-AP to approx 0.5 µg/mL in TBS, 20% (v/v) sheep se-

rum, 0.5% Tween 20, 2% skimmed milk. Add 100 µL/well and incubate for 1 h at room temperature.

8. Develop following the manufacturer's directions with the AMPAK ELISA amplification kit.
9. Measure the absorbance at 490 nm.

3.3. Reverse Transcriptase Assay

Typical results from 20 µL of culture supernatant from cells infected with HTLV-3_{RF} are 5000–10,000 cpm with poly(cA)oligo$(dT)_{12}$ and 20-80 cpm with poly(dA)oligo(dT)12 as the template primers. The most common problem encountered is insufficient washing of the reaction mixture through the DE81 mat. This can easily be diagnosed if the background counts are high (>100 cpm). If, however, even upon extensive washing the background is still high, then it may be that the SSC concentration is higher than it should be, since this will stabilize the binding of shorter polymers and nucleotides to the DE81 filter.

3.3.1. Preparation of the Reaction Cocktail

Mix *on ice* 5 µL of 1*M* Tris-HCl, pH 8.0; 5 µL of 0.1*M* DTT; 5 µL of 0.1*M* $MgCl_2$; 15 µL of 1M KCl; 5 µL of 1% (v/v) Triton X-100; 5 µL of 6 m*M* glutathione; 5 µL of 10 m*M* EGTA; 5 µL of the appropriate template primer, and 50 µL of dH_2O. Note that 80 µL of reaction cocktail is required for each well on the 96-well plate. Typically each sample will be run in triplicate (at least) on each of the template primers.

3.3.2. Assaying HIV Reverse Transcriptase

1. In a 96-well plate, mix, *on ice*, 80 µL of reaction cocktail and 20 µL of fresh or frozen cell supernatant. Note that samples can be stored frozen for many weeks at –70°C with little loss of activity. Leave on ice for 10 min and then add 20 µL of 0.1 mCi/mL [Me-^{3}H] thymidine 5' triphosphate.
2. Incubate at 37°C for 2–4 h.
3. Prewet a DE81 filter with 2× SSC in the cell harvester. Harvest the reaction mixture through the mat with 2× SSC for approx 30 s. Wash the mat for 10–30 s with 95% (v/v) ethanol and air-dry.
4. Cut up the mat, place into nonaqueous scintillation fluid, and count for ^{3}H.

4. Notes

4.1. Colorimetric Assay

1. Good cell growth is essential for this assay. If problems with growth are encountered, they may be overcome by adding 200 µL of sterile deionized water to all the outer wells of the microtiter plate and using the

inner 60 wells only for antiviral tests.

2. Care must be taken not to remove cells when taking the 150 µL of supernatant from each well. This is made easier by the fact that M8166 cells grow in a very clumped form and that they are easily visible, since they will be stained blue by the formazan product.

4.2. p24 Elisa

1. Antibodies (All antibodies should be reconstituted and stored frozen.): D7320 from Aalto Bio Reagents (Dublin) is a mixture of three polyclonal sheep antibodies raised against peptides from the HIV-1 (LAV-1) sequence and then affinity-purified against the immunogenic peptides. The amino acid sequences used are
   - SALSEGATPQDLNTML aa 173–188
   - GQMREPRGSDIA aa 236–237
   - LDIRQGPKEPFRDYV aa 283–297

   These sequences are substantially conserved between HIV-1 isolates.

   D7330 from Aalto Bioreagents is similar to the above, but is raised to a single, conserved peptide AEWDVQHPIPGPLPAGQLREPR, from the HIV-2 (ROD) sequence.
2. The assay can routinely detect 1–3 pg/well of HIV-1 p24; the HIV-2 assay being somewhat less sensitive.
3. EH12E1 is an mAb raised against HIV-1 that was isolated by Bridget Ferns, Richard Tedder, and colleagues *(5)* and mapped to a complex epitope incorporating two distinct peptide sequences *(6)*. These are GHQAAMQMLKETINEEAAEWDRVHPVHAGPIAPGQ (aa 193–227) and NPPIPVGEIYKRWII (aa 253–267). These regions of p24 are conserved between HIV-1 strains and, substantially, between HIV-1 and HIV-2. The alkaline phosphatase conjugate of EH12E1 (EH12E1-AP) was prepared by Novo Nordisk and is available from the MRC reagent program (NIBSC).

   If EH12E1-AP is unavailable, EH12E1 can be used at 1/1000 dilution of ascites fluid in TMT/SS, followed by 2 × TBS washes and detection of mAb with rabbit antimouse Ig (Dakopatts Ltd.) at 1/1000 in TMT/SS.

References

1. Dalgleish, A. G., Beverley, P. C. L., Clapham, P. R., Crawford, D. R., Greaves, M. F., and Weiss, R. A. (1984) The CD4 (TL) antigen is an essential component of the receptor for the AIDS retrovirus. *Nature* **312,** 763–767.

2. Pauweh, R., Balzarini, J., Baba, M., Snoeck, R., Schols, D., Herdewijn, P., Resmyter, J., and De Clercq, E. (1988) Rapid and automated tetrarolium-based colorimetric assay for detection of anti-HIV compounds. *J. Virol. Methods* **20,** 309–321.
3. Goff, S., Traktman, P., and Baltimore, D. (1981) Isolation properties of Moloney murine leukemia virus mutants: Use of a rapid assay for release of virion reverse transcriptase. *J. Virol.* **38,** 239–248.
4. Boyd, M. T., Maclean, N., and Oscier, D. G. (1989) Detection of retrovirus in patients with myeloproliferative disease. *Lancet* **i,** 814–817.
5. Spence, R. P., Jarvill, W. M., Ferns, R. C., Tedder, R. S., and Parker, D. (1989) The cloning and expression in *E. Coli* of sequences coding for p24, the core protein of human immunodeficiency virus and the use of the recombinant protein in characterizing a panel of monoclonal antibodies against the viral p24 protein. *J. Gen. Virol.* **70,** 2843–2851.
6. Ferns, R. B., Partridge, J. C., Spence, R. P., Hunt, N., and Tedder, R. S. (1989) Epitope location of 13 anti-*gag* HIV-1 monoclonal antibodies and their cross reactivity with HIV-2. *AIDS* **3,** 829–834.

CHAPTER 13

Analysis of Protein Factors that Interact with Adenovirus Early Promoters

Helen C. Hurst

1. Introduction

On account of their relatively small, defined genomes, viruses have long been a convenient model system in which to study eucaryotic gene expression. Usually, viral infection is followed by early gene expression, which allows the subsequent processes of viral DNA replication and late gene expression to proceed. All these processes must be precisely regulated and involve both virally encoded and host proteins.

The human adenoviruses have been extensively studied, in particular, in regard to the expression of the early genes (E1A, E1B, E2A, E3, and E4). This group of RNA polymerase II transcribed promoters are interesting from a transcription regulation standpoint, since all can be stimulated by one protein product of the first expressed viral gene, namely, E1A. The host heat-shock Hsp70 gene is also transcriptionally activated by E1A. This transactivation clearly results from an increase in the level of transcription initiation at these promoters, but the precise mechanism remains to be elucidated (reviewed, *1*). The E1A protein itself has no intrinsic ability to bind specifically to DNA *(2)*, so the approach has been to identify transcription factors that do recognize the early promoters and that may also interact with E1A. This approach has proved useful on another series of E1A-inducible genes (the

From: *Methods in Molecular Biology, Vol. 8: Practical Molecular Virology: Viral Vectors for Gene Expression*
Edited by: M. Collins

viral VA genes) that are transcribed by RNA polymerase III. In general, Pol III transcription is better understood than Pol II transcription, which requires many more factors. One factor required for Pol III transcription, TF IIIC, appears to have two forms, one of which has much greater transcriptional activity than the other. In adenovirus-infected cells, the abundance of TF IIIC remains constant; however, a greater proportion is present in the more active form *(3)*. It is therefore proposed that ElA can affect this accumulation of active TF III C, possibly through a phosphorylation event *(4)*.

In an attempt to find a parallel mechanism for Pol II transcribed genes, several workers have examined the ability of nuclear proteins present in normal and adenovirus-infected cells to bind specifically to the early promoters. Several factors have been described. Some show an abundance or an activity increase in infected cells (e.g., E2F *[5]* and E4F *[6]*). Others show constitutive binding but, intriguingly, bind to several E1A-inducible promoters (e.g., ATF), and their transcriptional activity in infected cells may be altered *(7)*. There is also a school of thought that finds the precise nature of the TATA box to be the major difference between E1A-inducible and noninducible genes *(8)*.

Obviously, all these binding studies must be complemented by appropriate functional assays. These include transfection assays with promoter deletions and chimeric promoters. For example, a portion of the E4 promoter containing two ATF sites can convert an E1A-insensitive promoter to an E1A-inducible one *(9)*. Also, in vitro transcription assays using the nuclear binding extracts allow the classical biochemical approach of reconstructing these interactions in the Eppendorf tube. However, the determination of where nuclear factors bind to a promoter has recently revealed much useful information about eucaryotic transcriptional regulation. The rest of this chapter describes the basic techniques required to embark on this type of study.

2. Materials

2.1. Preparation of Nuclear Extracts

1. Buffer A: 10 m*M* HEPES–KOH, pH 8.0, 1.5 m*M* $MgCl_2$, 10 m*M* KCl, 0.5 m*M* DTT.
2. Buffer B: 20 m*M* HEPES–KOH, pH 8.0, 450 m*M* KCl, 1.5 m*M* $MgCl_2$, 0.2 m*M* EDTA, 25% glycerol, 0.5 m*M* DTT, 0.5 m*M* PMSF.
3. Dialysis buffer (D): 20 m*M* HEPES–KOH, pH 8.0, 20% Glycerol, 100 m*M* KCl, 1.5 m*M* $MgCl_2$, 0.2 m*M* EDTA, 0.1% LDAO (lauryl diamine oxide, Calbiochem), 1 m*M* sodium molybdate, 0.5 m*M* PMSF, 0.5 m*M* DTT. Add DTT and PMSF from stock solutions stored at –20°C (DTT, 1 *M* in water, PMSF, 100 m*M* in methanol) just before use.

2.2. DNase Footprinting

1. 6% native acrylamide gel: For a 40-mL gel (16 cm long), use 6 mL of stock acrylamide (38:2, acrylamide:bis acrylamide) and 4 mL of 10× TBE (108 g/L trizma base, 55 g/L boric acid, 9.5 g/L EDTA). Polymerize with 40 µL of TEMED and 250 µL of 10% ammonium persulfate.
2. M & G buffer: 2.5*M* ammonium acetate, 1 m*M* EDTA, 0.1% SDS, 10 m*M* $MgCl_2$.
3. Stop buffer: 0.3*M* sodium acetate, 10 m*M* EDTA, 10 µg/1 mL tRNA carrier.
4. Formamide sample buffer: 98% deionized formamide, 20 m*M* EDTA, 0.1% bromophenol blue, 0.1% xylene cyanol.
5. DNase buffer: 20 m*M* HEPES–KOH, pH 8.0, 50 m*M* KCl, 20% glycerol, 2 m*M* DTT, 2 m*M* $MgCl_2$.
6. 2× PK stop: 100 m*M* Tris-HCl, pH 7.5, 2% SDS, 20 m*M* EDTA, 400 µL/mL proteinase K (Sigma).

2.3. Methylation Protection

Dimethyl sulfate (DMS) from BDH. DMS is not soluble in water, but a "dilution" can be made if the mix is vortexed hard before withdrawal of each aliquot.

2.4. Gel Retention Assays

1. 10× T_4 kinase buffer: 500 m*M* Tris-HCl, pH 7.5, 100 m*M* $MgCl_2$, 50 m*M* DTT, 10 m*M* spermidine, 10 m*M* EDTA.
2. Gel retention buffer (GRB): 25 m*M* HEPES–KOH, pH 8.0, 1 m*M* EDTA; 5 m*M* DTT, 10% glycerol, 50–300 m*M* KCl. The optimal salt concentration for observing complex formation can be determined by doing the assay at different KCl concentrations.
3. Gel retention gel: Stock acrylamide is 44:0.8 acrylamide:bis acrylamide. Pour 0.5× TBE gel (concentration, 6–8%; length, ~16 cm): a comb with narrow slots (0.5 cm) gives the neatest results. The running buffer is 0.5× TBE.

3. Methods

3.1. Preparation of Nuclear Extracts

The cells most commonly used for studying adenoviruses are human HeLa cells, which can be grown in suspension and as monolayers. Suspension cultures are ideal for preparing nuclear extracts from either normal or adenovirus-infected cells.

1. In either case, grow cells in RPMI with 5% fetal calf serum to a density of 5–8 × 10^5 cells/mL.
2. Concentrate cells for infection 10-fold and then incubated with titred adenovirus at 20 PFU/cell for 1 h at 37°C.
3. After infection, dilute the cells back to 5 × 10^5 cells/mL and culture for a further 6 h (early infection) or 18 h (late infection) before harvesting.

Crude nuclear extracts can be prepared in a number of ways; the method described here, which is based on that of Dignam et al. *(10)*, has the added advantage that the extract can also be used for in vitro transcription assays.

1. Harvest suspension or monolayers of cells, wash in PBSA, and note the packed cell volume (PCV). All further procedures are carried out at 4°C.
2. Resuspend the cells in Buffer A using 5 × PCV.
3. After 10 min on ice, repellet the cells at 2000 rpm for 10 min and resuspend in 2× PCV of Buffer A. (NOTE: The cells will have swollen, so ignore the new packed volume and use the original PCV.)
4. Now lyse the cells with 10–15 strokes of pestle B of an all-glass Dounce homogenizer. The extent of lysis should be monitored under the microscope.
5. Pellet the nuclei at 25,000*g* (e.g., 15k in a Sorvall SS34 rotor) for 20 min.
6. Carefully remove the cytoplasmic supernatant. This should be kept to assay later, since some factors leach appreciably from the nuclei, even under the low-salt conditions used so far.
7. Resuspend the nuclear pellet using 10 strokes of pestle B in the Dounce homogenizer. Use 3 mL of buffer B/10^9 starting cells.
8. Stir or rotate the suspension gently for 30–60 min, and then clear it by centrifugation at 25,000*g* for 30 min.

The supernatant represents the crude nuclear extract and should contain about 10 mg/mL protein (Biorad Protein Assay). For most purposes the extract can be used at this concentration, but it can be concentrated by precipitation with ammonium sulfate (AS).

1. Add solid AS at 0.33 g/mL of extract and slowly stir to dissolve at 4°C.
2. Spin the cloudy suspension for 15 min at 25,000*g* and resuspend the pellet in one-tenth of the original volume.
3. Whether precipitated or not, the crude extract should be dialyzed vs two changes (100 vol each) of Buffer D, frozen in aliquots, and stored at –80°C. Multiple rounds of freezing and thawing the extract do not appear to affect the binding activity of most nuclear proteins, but this should be checked. The addition of molybdate or fluoride helps to pre-

vent the action of phosphatases, and ethylene glycol added (to 5–20%) can help the activity of more unstable proteins. However, if extracts are to be used for in vitro transcription, then thaw them only once, since RNA polymerase II is very unstable.

For most purposes, the dialyzed crude extract can be used directly for binding assays (*see below*). However, some factors bind reproducibly only when partially purified. This can be done on a Heparin-agarose column. To interact with DNA, nuclear factors must have a net positive charge so they will bind to weakly cationic resins, such as Heparin-agarose (Biorad). The resin is easy to use:

1. Pour a column and equilibrate with 0.1*M* KCl dialysis buffer.
2. The crude extract can then be applied; use about 1 mL of resin for 1–2 mg of protein, and keep the flow-through in case some of the protein does not bind.
3. Then wash the column with about 10 column volumes of dialysis buffer.
4. Binding proteins are usually eluted in one step using 0.4*M* KCl in buffer D. Again use about 10 column volumes of elution buffer, but this time collect fractions equivalent to one-tenth to one-half of the column volume, so that the peak can be collected in as small a volume as possible. For a large column, it is a good idea to have a UV monitor connected to the column outflow. Otherwise, directly measure the OD_{280} of each fraction vs the elution buffer as blank.
5. Pool the peak fractions as the 0.4*M* fraction. Washing again at 1.0*M* KCl using the same procedure is a good idea. Both pooled, eluted fractions must be dialyzed back to 0.1*M* KCl for binding assays; also check the amount of binding activity in the flow through.

3.2. DNase Footprinting

Two techniques are commonly used to look at DNA–protein interactions: DNase footprinting and gel retention assays (GRA). The latter are very simple and useful when studying one binding activity in isolation (*see below*). However, when starting to look at a new promoter, apart from determining if nuclear proteins can bind to the chosen probe (e.g., after partial purification over Heparin-agarose), GRA initially will not yield much information. Hence, it is best to start with footprinting. However, in order to ascertain which regions of a gene may be interesting to footprint, a classical deletion analysis of the gene should be done first, so that the likely functional promoter and enhancer sequences can be identified. These can then be subcloned for convenient labeling and excision as footprinting probes.

3.2.1. Making the Probe

The ideal starting material is a subcloned promoter fragment of 200–300 bp with differing unique restriction sites at each end.

1. Digest 25 µg of plasmid with the 3' restriction enzyme chosen to give a 5' overhang.
2. Phenol-extract the completed digest, ethanol-precipitate the DNA, and resuspend at 1 µg/µL. This linearized DNA can be used as a stock to make probes.
3. To label the "top" strand of the probe, incubate 5 µg of linear DNA with 50 µCi of high-specific-activity ^{32}P α dNTP and M-MLV reverse transcriptase (BRL) with the buffer provided in a 15-µL reaction volume for 60 min at 37°C.
4. Separate the DNA from unincorporated label on a G50 spun column and then recut the DNA at its 5' unique restriction site.
5. Load the sample onto a small native 6% acrylamide gel and run for 90 min in 1× TBE at 200 V.
6. Cover the gel in Saran™ wrap and mark with radioactive or fluorescent ink.
7. Autoradiograph the gel briefly; the probe should be so hot that a 1- to 2-min exposure is adequate.
8. The film should reveal two labeled bands: the vector and the promoter fragment. Cuthe latter from the gel using the film as a template and the ink markers to orient the film with respect to the gel.
9. Carefully put the acrylamide slice into an Eppendorf tube and mash thoroughly with a pipet tip.
10. Add 1.0 mL of M & G buffer, mix, and allow the probe to elute at 37°C for several hours.
11. The buffer must then be filtered from the spent acrylamide. This can be done conveniently by putting a circle of GF-C filter cut with a No. 5 cork borer at the base of a 2-mL syringe. Add the probe slurry and let the buffer drip, under gravity, into a 5-mL polyallomer ultracentifuge tube.
12. Wash through with a further 0.5 mL of M & G buffer.
13. Add 3.5 mL of ethanol to the tube, mix, and allow the DNA to precipitate on dry ice; do not add any carrier.
14. To recover the DNA, spin at 30,000 rpm for 30 min in an SW55 rotor or equivalent. All the counts should be pelleted; the ethanol can be poured off, and the probe air-dried. Assume that about 25% of the probe is lost during the purification, and resuspend it at ~5 fmol/µL in water.

The probe is now ready to use in DNAase footprinting assays. However, to locate the position of any footprints relative to the probe, it is a good idea to use some of it in a Maxam and Gilbert A & G sequencing reaction *(11)*.

1. To 5 µL of the probe, add 5 µL of water and 3 µL of 10% formic acid, and incubate for 12 min at 37°C.
2. Add 300 µL of stop buffer and mix and precipitate the DNA with 750 µL of ethanol on dry ice.
3. Pellet, wash, and dry the DNA and resuspend in a fresh 1:10 dilution of piperdine, made using ice-cold water.
4. Piperdine-cleave the DNA for 30 min at 95°C, and then add 60 µL of 3*M* NaOAC, 440 µL of water, 5 µL of carrier tRNA, and mix *thoroughly* to dilute the piperdine.
5. Precipitate the DNA on dry ice by adding 650 µL of isopropanol.
6. Wash the final pellet and resuspend in formamide sample buffer at ~1000 Cherenkov counts/µL and use 5 µL as a marker lane on the final DNase gel (*see below*).

3.2.2. DNase 1 Incubations

DNase footprinting does not lend itself to precise recipes; conditions for each probe have to be determined empirically. The following protocol provides a guide.

1. Combine 5–15 fmol of labeled probe with 150–200 µL of crude or partially purified nuclear extract and 2–5 µL of nonspecific DNA competitor (*see* Note 1).
2. Make the volume up to 45 µL with DNase buffer. For control incubations with probe alone (free DNA), omit the nuclear extract.
3. Incubate for 20 min at room temperature.
4. Meanwhile, make dilutions of DNase 1 (Amersham, UK) in DNase buffer. For the free DNA assays, dilute to 0.1 µL/mL; for extract incubations, use a 5 µL/mL dilution.
5. Add 3–5 µL of the appropriate DNase 1 dilution to each tube *in turn* for 30–60 s.
6. Stop the reaction by adding 50 µL of 2× PK stop and incubate for 10 min at 37°C.
7. Extract samples with 100 µL of phenol/chloroform and ethanol precipitate with 3 µg of tRNA as carrier.
8. Cherenkov count the dried pellets, resuspend them in formamide sample buffer at ~1000 cpm/µL, and load 5000 cpm in all tracks on a 50% urea, 10% acrylamide sequencing gel. Any footprints will appear as gaps in

the random DNase 1 cutting pattern when incubations with extract are compared to the free DNA pattern. The position of these "gaps" within the promoter can be determined relative to the A & G marker lane for the same probe.

3.3. Methylation Protection Assays

As a result of steric considerations, the gaps in DNase 1 footprinting patterns are larger than the actual factor binding site, which is usually a 6- to 10-bp sequence in the center, though flanking sequences may modulate binding affinity. To help identify this binding motif, methylation protection assays can be performed. In outline, the end-labeled DNase footprinting probes are again incubated with protein and subsequently treated with dimethylsulfate (DMS), which methylates guanine residues at position N7 *(11)*. Those G residues in close proximity to bound protein are protected from methylation. Equal counts of DMS-treated DNA from incubations with (bound) and without (free) protein are cleaved using piperidine and resolved on sequencing gels. Bands representing the protected G residues will be either diminished in intensity or absent from the "bound" lanes when compared with free-DNA lanes.

As with DNase footprinting, the precise conditions will have to be determined empirically.

1. Start with the same 45-µL DNase footprinting incubations and, after 20 min, treat each one with 2 µL of DMS (diluted fivefold in water) for 45–60 s.
2. Stop the reactions using 1 µL of β-mercaptoethanol, and then add 50 µL of 2× PK stop.
3. Continue to process the samples and run them on a gel with an A & G marker lane, exactly as described for DNase footprinting.

3.4. Gel Retention Assays

Once a binding site has been identified, it is convenient to have oligonucleotides made to these sequences. Most people use oligos that are 20–30 bp long and incorporate restriction enzyme site overhangs in order to facilitate cloning and end-labeling for gel retention assay probes.

3.4.1. Annealing and Labeling Oligos

1. Complementary single-stranded deprotected oligos are readily annealed in 1× T_4 kinase buffer *(12)*. Combine equal amounts (up to 500 µg of each strand) of each oligo with the buffer in a total vol of 150 µL.

2. Incubate the tube for 2 min at 90°C, 10 min at 37°C, and 5 min at room temperature.
3. Ethanol-precipitate, wash, and resuspend the ds oligo at 1 µg/µL in 10 m*M* Tris-HCl, pH 8.0.
4. End-label 100 ng of ds oligo with 20 µCi of ^{32}P α dNTP and reverse transcriptase, as described for the DNase footprinting probe. The labeling reaction can be done in 10 µL and then diluted to 100 µL with 10 m*M* Tris-HCl, pH 8.0, before loading onto a G50 spun column to remove unincorporated label. The resulting probe can be assumed to be at 1 ng/µL.

3.4.2. Gel Retention Assays

It is convenient to set up GRA tubes in a rack on a bed of ice. Each component can be added as a separate drop on the side of the tube and the incubation started by a brief spin.

1. A typical assay contains 0.1 ng of labeled oligo, 5–10 µg of crude protein, 1–2 µg of nonspecific competitor, with the volume made up to 25 µL using GRB. The first assays should be optimized for protein, competitor, and salt concentration.
2. Incubate the tubes for 20 min at room temperature.
3. Load, without adding dye, onto a 0.5× TBE native gel; and run for 90 min at 200 V. Then fix and dry the gel for autoradiography. The ideal result is one in which the labeled ds oligo probe is in excess and running about two-thirds down the gel. Nearer the top of the gel should be a discrete band corresponding to the retarded protein/probe complex. The specificity of this complex can be tested in further assays by adding an excess (20- to 100-fold) of cold ds oligo as a specific competitor in the incubation step. The presence of cold oligo should abolish the complex band on the final autoradiograph. Some retarded bands may remain; these can be assumed to be nonspecific interactions. Further competition experiments with GRA may include challenging with oligos to similar binding sites found in other promoters to determine relative binding affinities. In addition, a series of ds oligos can be made incorporating point mutations within the binding site; the relative ability of these mutant oligos to compete for complex formation will define the binding site sequence more clearly.

4. Notes

1. Nonspecific competitor: This is required in binding incubations using crude nuclear extract to help titrate out nonspecific protein–DNA inter-

actions. The nature and amount of nonspecific competitor can be crucial, but there are no hard and fast rules as to what will give the most consistant results. The most commonly used are natural DNA (such as sonicated calf-thymus or salmon-sperm DNA, or a restriction digest of plasmid DNA) or synthetic polymers (such as poly dA/dT or poly dI/dC, from Pharmacia). In general, the synthetic polymers give the cleanest results; however, although some factors "reveal themselves" if homopolymers are used, others are best studied using heteropolymers, i.e., d(A-T)/d(A-T) or d(I-C)/d(I-C), so both kinds should be tried. In all cases, a solution of the competitor in water at 1 μg/μL is the most convenient to use.

2. DNase footprinting and methylation protection: These assays really have to be worked out empirically for each probe. Always use more protein than you think you will need; all the probe must be bound to obtain good patterns. Most factors will bind from crude extract, but some (e.g., AP2) only reveal themselves when partially purified (e.g., over Heparin-agarose). Other factors, such as Sp1, will footprint well but do not work well in GRA. As with GRA, the KCl concentration in the buffer can be varied to optimize the binding of factors. If great difficulty is experienced, a variation to try is to stop the DNase or methylation reactions (by adding either EDTA [to 20m*M*] or 1 μL of β-mercaptoethanol, respectively) and then loading the incubation onto a gel retention gel. Once the gel is run, it can be autoradiographed while wet, as for DNase probe preparation. In this way, free and protein-complexed probe are separated on the gel and can be separately excised and recovered from the acrylamide by incubation in PK stop followed by filtration through a GF-C filter, as described previously. Phenol-extract and precipitate the DNA with the addition of carrier tRNA and load equal Cherenkov counts for bound and free probe in adjacent tracks of a sequencing gel.
3. Competition and cooperative binding: Competition studies with GRA have been described above, but these can also be coupled with DNase footprinting, again using an excess of cold ds oligo. The advantage of doing it this way is that the cooperative binding of adjacent factor-binding sites may be investigated. For example, if binding at one footprint site is competed by adding cold oligo representing only that binding sequence, binding at another nonrelated site on the same probe may be either unaffected or also diminished. The latter indicates some level of cooperative binding between the two sites.
4. Gel retention assays: In some cases, GRA with crude nuclear extract may reveal more than one complex, indicating that several factors have the capacity to bind in vitro to the oligonucleotide binding site probe. This

phenomenon is proving surprisingly common and would not be revealed by DNase footprinting studies alone. In this case, quite careful competition studies need to be done to decide which factor is really binding to the promoter site of interest. In addition, raising the salt concentration in the GRB will test which factor is binding more strongly to that site. The fact remains, though, that to be sure that any protein is of functional significance, it will have to be purified and tested in vitro in transcription assays *(10)*. GRA also allows a simple way of following a binding factor through its purification, which also uses the binding-site oligo coupled to CNBr-activated Sepharose to form an affinity column *(12)*.

References

1. Berk, A. J. (1986) Adenovirus promoters and E1A activation. *Annu. Rev. Genet.* **20,** 45–79.
2. Ferguson, D., Kippl, D., Andrisani, O., Jones, N., Westphal, H., and Rosenberg, M. (1985) E1A 13S and 12S mRNA products made in *E. Coli* both function as nucleus-localized transcription activators but do not directly bind DNA. *Mol. Cell. Biol.* **5,** 2653–2661.
3. Yoshinaga, S., Boulanger, P. A., and Berk, A. J. (1987) Resolution of human transcription factor TF IIIC into two functional components. *Proc. Natl. Acad. Sci. USA* **84,** 3585–3589.
4. Hoeffler, W. K., Kovelman, R., and Roeder, R. G. (1988) Activation of transcription factor IIIC by the adenovirus E1A protein. *Cell* **53,** 907–920.
5. Kovesdi, I., Reichel, R., and Nevins, J. R. (1986) Identification of a cellular transcription factor involved in E1A transactivation. *Cell* **45,** 219–228.
6. Raychandhuri, P., Rooney, R., and Nevins, J. R., (1987) Identification of an E1A-inducible cellular factor that interacts with regulatory sequences within the adenovirus E4 promoter. *EMBO J.* **6,** 4073–4081.
7. Lee, K. A. W., Hai, T-Y., SivaRamon, L., Thimmappaya, D., Hurst, H. C., Jones, N. C., and Green, M. R. (1989) A cellular transcription factor ATF activates transcription of multiple E1A-inducible adenovirus early promoters. *Proc. Natl. Acad. Sci. USA* **84,** 8355–8359.
8. Simon, M. C., Fison, T. M., Benecke, B. J., Nevins, J. R., and Heintz, N. (1988) Definition of multiple, functionally distinct TATA elements, one of which is a target in the HSP 70 promoter for E1A regulation. *Cell* **52,** 723–729.
9. Lee, K. A. W. and Green, M. R., (1987) A cellular transcription factor E4F1 interacts with an E1A-inducible enhancer and mediates constitutive enhancer function in vitro. *EMBO J.* **6,** 1345–1353.
10. Dignam, J. D., Lebowitz, R. M., and Roeder, R. G. (1983) Accurate transcription initiation by RNA polymerase II in a soluble extract from isolated mammalian nuclei. *Nucleic Acids Res.* **11,** 1475–1489.
11. Maxam, A. M. and Gilbert, W. (1980) Sequencing end-labelled DNA with base specific chemical cleavages. *Methods Enzymol.* **65,** 499–560.
12. Kadonaga, J. T. and Tjian, R. (1986) Affinity purification of sequence-specific DNA binding proteins. *Proc. Natl. Acad. Sci. USA* **83,** 5889–5893.

CHAPTER 14

COS Cell Expression

John F. Hancock

1. Introduction

COS-1 cells were created by transforming an established line of monkey epithelial cells, CV-1, with a defective mutant of SV40 *(1)*. The SV40 mutant used carried a small deletion within the origin of replication and, although this construct transformed CV-1 cells, which are permissive for lytic growth of SV40, no infectious virus was produced after prolonged culture. One transformed cell line, COS-1, was fully characterized and found to contain the complete early region of the SV40 genome. COS-1 cells express nuclear large T and all proteins necessary for replication of appropriate circular genomes. This was first demonstrated by showing that COS-1 cells could support the replication of early region mutants of SV40. More important, however, the introduction of any plasmid containing an SV40 origin of replication into COS-1 cells results in rapid replication of the plasmid to high copy number. Coincidently, of course, the transfected cells will express any gene on the plasmid that is driven by a suitable eukaryotic promoter. The combined effect of these phenomena is transient high-level expression of the encoded protein.

Using a COS-cell expression system has several advantages over generating stable cell lines. First, the whole process is rapid, taking just a few days from transfection to assay. Second, no selection for transfectants is required, since a high proportion of the cells take up the plasmid DNA and express the transfected gene. Third, uniformly high expression of wild-type and mutant

From: *Methods in Molecular Biology, Vol. 8:*
Practical Molecular Virology: Viral Vectors for Gene Expression
Edited by: M. Collins © 1991 The Humana Press Inc., Clifton, NJ

forms of proteins can be expected as long as the expressed protein is stable and nontoxic. The advantage over prokaryotic expression systems is that the expressed protein should undergo normal posttranslational processing and hence be localized normally within the cell or be secreted into the culture medium.

These characteristics of COS-cell transient expression have been exploited in a wide variety of ways, for example, expression cloning of growth factor receptors and lymphocyte antigens *(2,3)*, functional confirmation of cDNA predicted protein sequences *(4)*, and analysis of the posttranslational processing of various cellular proteins and oncogene products *(5,6)*.

1.1. Principle

The introduction of DNA into eukaryotic cells may be achieved by electroporation, calcium phosphate precipitation, or DEAE-dextran transfection. Each method has advantages and disadvantages. Calcium phosphate precipitation is best employed for the generation of stably transfected cell lines, whereas DEAE-dextran techniques are most applicable to transient expression sytems. A DEAE-dextran method will be described here that is simple to use and has proved to give highly reproducible results.

The basis of the method is to mix together plasmid DNA in the absence of any carrier DNA with a solution of DEAE-dextran. It is presumed that the the DNA forms high-mol-wt complexes with the dextran that stick to the COS cells when the mixture is applied to a cell monolayer. These complexes are internalized, probably by endocytosis, although this has never been formally demonstrated. Once inside the cell, the plasmid DNA is assembled into nucleosome-containing minichromosomes *(7)* and is rapidly replicated. The plasmid DNA does not become integrated into the host-cell genome with any significant frequency. The efficiency of DNA uptake by the cells is much improved by including a DMSO shock step that stimulates endocytosis *(8)*. Some investigators advocate the inclusion of a chloroquine treatment of cells to further increase the efficiency of DEAE-dextran transfections. Our experience is that this increases the toxicity of the procedure without significantly improving the proportion of successfully transfected COS cells. For example, without including such a chloroquine incubation, up to 20% of COS cells consistently take up DNA and express novel protein at high levels when assayed by immunofluorescence *(9)*. Protein expression is detectable 24 h after the DNA is added to the cells and reaches a peak at around 60–72 h after transfection.

The only constraint on plasmid design is the presence of an SV40 origin of replication. Otherwise, the SV40 enhancer/early promoter or the CMV promoter have both proven to be highly efficient for cDNA expression in

COS cells *(10,11)*. The DNA is applied in high concentration to the cells and the preparation should therefore be as pure as possible, preferably by banding on cesium chloride gradients *(12)*.

2. Materials

1. Dulbecco's Modified Eagle's Medium (DMEM).
2. Donor calf serum.
3. Phosphate-buffered saline, magnesium- and calcium-free (PBS).
4. Trypsin (1:250) in Puck's saline.
5. DEAE-dextran stock: Thoroughly dissolve DEAE-dextran (mol wt = 500,000 Sigma D9885) to a final concentration of 50mg/mL in sterile 0.1*M* Tris-HCl, pH 7.0. Filter-sterilize the solution and store as 100–200 µL aliquots at –20°C. Thaw the requisite number of aliquots immediately before use.
6. HEPES-buffered saline (HBS): For 2× HBS, make up a solution of 280 m*M* NaCl, 50 m*M* HEPES, 1.5 m*M* Na_2HPO_4, and adjust to pH 7.1–7.2. Sterilize the stock solution by filtration and store at –20°C in 20-mL aliquots.
7. Shock solution: Make up a solution of 10% DMSO in 1× HBS and filter-sterilize. Shock solution is stable at room temperature.
8. TE Buffer: 10 m*M* Tris-HCl, pH 7.5; 1 m*M* EDTA.

3. Method

1. Prior to transfection grow the COS cells in DMEM containing 10% v/v donor calf serum (DC10) so they are just confluent. A 175-mm tissue-culture flask grown to this density contains approx 15×10^6 COS cells. If the cells are allowed to grow to too high a density, trypsinization becomes difficult and the cells detach in clumps.
2. On the day of transfection (day 1) trypsinize the cells from the culture flask. Remove the culture medium and wash the cell monolayer twice with 25 mL of warm PBS. Then incubate the cell monolayer in 3 mL trypsin solution at 37°C until the cells are rounded and starting to float free from the flask (approx 1–2min). Add 7 mL of DC10 and titrate well to obtain a good single-cell suspension.
3. Count the cell suspension and plate the COS cells at 6×10^5 per 60-mm tissue-culture dish in a total volume of 5 mL of DC10. Incubate the plates at 37°C for 2 h.
4. Toward the end of the incubation, make up the DNA solutions for transfection. Aliquot 1.5 mL of DMEM (containing no serum) into sterile

bijoux. To this, add 5 μg of plasmid DNA in TE buffer, mix, and then add 7.5 μL of stock DEAE-dextran and mix again. No precipitate should be visible. If concentrated (>3 μg/mL) plasmid solutions are added to DMEM containing DEAE-dextran, the DNA can form large clumps. This is avoided by following the addition sequence indicated.

5. Check that the cells have reattached and are evenly spread over the plate. Remove the DC10 and wash the cells twice with 5 mL of prewarmed DMEM (containing no serum). The cells are only weakly adherent at this stage, so wash carefully.
6. Aspirate the final wash and add the DNA mixture. Swirl each plate gently to ensure that the mixture is equally distributed over the plate surface. Incubate at 37°C for 4 h. Ensure that the plate is completely flat in the incubator, so that the small volume of medium covers all the cells.
7. Check that the cells are still attached. They may have started to round up during this incubation, but only a tiny minority should be free-floating, since COS cells are well able to tolerate serum-free conditions.
8. Aspirate the DNA mixture and carefully add 1 mL of warm shock solution. DMSO is toxic at a 10% concentration and must be added very carefully to avoid washing the cells completely off the dish. Trickle the shock solution slowly down the side of the dish and rotate it once to distribute the solution over all the cells; then quickly return the dish to the incubator. The cells must be exposed to the shock solution for no longer than 2 min, therefore, start timing as soon as the first drop of shock solution is added and have the cells ready for the next step after exactly 2 min.
9. Aspirate the shock solution and gently wash the cells once with 5 mL of warm PBS. Replace the PBS with 5 mL of DC10 and incubate at 37°C. The cells always look very sick at this stage, with many being partially rounded up rather than flat and firmly adherent.
10. On day 2 check that the cells have recovered from the transfection procedure. The majority should now be firmly attached and have relatively normal morphology, although vacuolation can be pronounced. Expect a small proportion of free-floating dead cells; however, there should be no area of the dish that is bare of cells if the shock step was carried out correctly. There is no need to medium change the cells (*see* Note 4).
11. On day 4, 64–72 h after the DMSO shock, assay the cells for protein expression. The cells should be just confluent at this stage.

4. Notes

1. These transfections can readily be scaled up. For example, seed 4×10^6 cells in 30 mL of DC10 onto a 140-mm plate and, after 2 h, replace with 12 mL of DMEM containing 40 µg plasmid and 60 µL of DEAE-dextran stock. Keep all incubation times unchanged, but increase the volumes of the washes to 20 mL, and use 5 mL of shock solution in step 8 of Section 3.
2. If required, the COS cells may, on day 2, be reseeded to smaller tissue-culture dishes or onto cover slips for immunofloresence studies. Aspirate the DC10, wash twice with PBS, trypsinize, and take the cells up in an appropriate volume of DC10, e.g., 8 mL for four 30-mm dishes. Aliquot the cells and return to the incubator. It is much easier to reseed cells in this way than to attempt to transfect COS cells directly onto very small dishes or directly onto cover slips. This procedure also ensures that nearly identical aliquots of cells are available for later assays, since any variation in transfection efficiency between plates is abolished.
3. Cells can be pulse-labeled on day 4, before lysis, and assayed or labeled overnight by switching the cells to labeling medium on day 3, 48 h after the DMSO shock. In the case of secreted proteins, replace the culture medium with a small volume of fresh medium (e.g., 1.5 mL for a 60-mm dish) on day 3, 48 h after the DMSO shock, and, after an overnight incubation, assay for activity in the medium. Repeat this procedure on day 4 by replacing the overnight medium with a small volume of fresh medium and incubating for a further 8–12 h to collect a second sample for assay.
4. This protocol has been found to work well for a wide variety of constructs in different vectors. Transfection efficiencies assayed by immunofluoresence are at least 10%, and more typically 20%. If these levels are not achieved, an attempt may be made to further optimize the procedure for an individual clone of cells and expression vector. The transfection efficiency is dependent on the DNA and DEAE-dextran concentrations. Therefore, set up a series of control transfections varying both the plasmid concentration (between 1–15 µg/mL) and the final DEAE-dextran concentration (between 100–500 µg/mL) and assay each set of combinations. Certain investigators also add chloroquine diphosphate to the COS cells, along with the DNA-dextran mixture, in an attempt to improve transfection efficiencies. If this step is to be included, add chloroquine to the DNA-dextran mixture (to a final concentration of 100 µ*M*) before applying the mixture to the washed COS-

cell monolayer. It is then important to check the cells every hour of the subsequent 4-h incubation to ensure that excessive death is not occurring.

5. References

1. Gluzman, Y. (1981) SV40 transformed simian cells support the replication of early SV40 mutants. *Cell* **23,** 175–182.
2. D'Andrea, A. D., Lodish, H. F., and Wong, G. G. (1989) Expression cloning of the murine erythropoietin receptor. *Cell* **57,** 277–285.
3. Aruffo, A. and Seed, B. (1987) Molecular cloning of a CD28 cDNA by a high effficiency COS cell expression system. *Proc. Natl. Acad. Sci. USA* **84,** 8573–8577.
4. Yamasaki, X., Taga, T., Hirata, Y., Yawata, H., Kawanishi, Y., Seed, B., Taniguchi, T., Hirano, T., and Kishimoto, T. (1988) Cloning and expression of the Human Interleukin-6 (BSF-2/IFN-2) Receptor. *Science* **241,** 825–828.
5. Jing, S. Q. and Trowbridge, I. S. (1987) Identification of the intermolecular disulphide bonds of the human transferrin receptor and its lipid attachment site. *EMBO J.* **6,** 327–331.
6. Hancock, J. F., Magee, A. I., Childs, J., and Marshall, C. J. (1989) All *ras* proteins are polyisoprenylated but only some are palmitoylated. *Cell* **57,** 1167–1177.
7. Reeves, R., Gorman, C., and Howard, B. (1985) Minichromosome aasembly of non-integrated plasmid DNA transfected into mammalian cells. *Nucleic Acids Res.* **13,** 3599–3615.
8. Lopata, M. A., Cleveland, D. W., and Sollner-Webb, B. (1984) High level expression of a chloramphenicol acetyltransferase gene by DEAE-dextran mediated DNA transfection coupled with a dimethyl sulphoxide or glycerol shock treatment. *Nucleic Acids Res.* **12,** 5707–5717.
9. Hancock, J. F., Marshall, C. J., McKay, A. I., Gardner, S., Houslay, M. D., Hall, A., and Wakelam, M. J. O. (1988) Mutant but not normal p21^{ras} elevates inositol phosphate breakdown in two different cell systems. *Oncogene* **3,** 187–193.
10. Miller, J. and Germain, R. N. (1986) Efficient cell surface expression of class II MHC molecules in the absence of associated invariant chain. *J. Exp. Med.* **164,** 1478–1489.
11. Seed, B. (1987) An LFA-3 cDNA encodes a phospholipid linked membrane protein homologous to its receptor CD2. *Nature* **329,** 840–842.
12. Maniatis, T., Fritisch, E. F., and Sambrook, J. (1982) *Molecular Cloning: A Laboratory Manual* (Cold Spring Harbor Laboratory, Cold Spring Harbor, NY).

CHAPTER 15

Papillomaviruses and Assays for Transforming Genes

Karen H. Vousden

1. Introduction

1.1. History of Papillomavirus Research

Human papillomavirus (HPV) research dates back to the turn of the century, when Ciuffo demonstrated that human warts are caused by an infectious agent, only later identified as a virus *(1)*. More recently, the HPVs have generated particular interest, since they include some of the few viruses clearly involved in the development of certain human cancers *(2)*. Although most papillomaviruses give rise to benign, self-limiting proliferations, a number of human and animal papillomaviruses have also been shown to have oncogenic potential. This link between papillomaviruses and malignant disease was first recognized more than 50 years ago, when it was shown that benign lesions induced by the cottontail rabbit papillomavirus could progress to invasive carcinoma *(3)*. The change from benign lesions to malignancy and the identification of factors that contribute to this conversion has been extremely useful as a model for tumor progression. However, following the initial studies that so elegantly identified the oncogenic potential of papillomaviruses, research into these viruses slowed considerably for many years. Further investigation has been hampered by the failure to develop an in vitro culture system for viral propagation. In general, papillomaviruses are limited to a single host and show strong tropism with respect to the cell type that they

From: *Methods in Molecular Biology, Vol. 8: Practical Molecular Virology: Viral Vectors for Gene Expression*
Edited by: M. Collins

infect. Most papillomaviruses infect only cutaneous or mucosal epithelial cells at specific anatomical sites. Viral DNA replication and transcription of some viral sequences occurs in cells throughout the epithelium. However, virion production is limited to the terminally differentiated keratinocytes in the upper layers of the epithelium, probably because of the requirement of cell-type-specific host factors for transcription of viral structural proteins. Since there is no culture system for cells that exhibit this precise level of differentiation, propagation of the virus is not possible, and as a consequence almost nothing is known about the natural life-cycle of the papillomaviruses.

Two events in the 1970s led to a revival in interest in papillomaviruses. The first was the development of recombinant-DNA technology, which allowed the small papillomavirus genome to be cloned and replicated as a plasmid. In this way, the obstacle of a lack of a culture system was at least partially overcome. The second was the suggestion that certain genital HPVs play a causative role in the development of cervical carcinoma *(4)*. There is now much compelling epidemiological and experimental evidence to support this hypothesis *(5)*, making HPVs some of the most important human viruses involved in the development of cancer. Other links between HPVs and human cancer are clearly seen in a rare skin disorder, epidermodysplasia verruciformis *(6)*.

1.2. Organization of the Papillomavirus Genome

Papillomaviruses are small DNA viruses that were originally classified in the papovavirus family, although now they are considered to constitute an independent group. The 55-nm viral particle encloses a circular double stranded DNA genome of approx 8 kbp in length, which is normally maintained episomally in the nucleus of infected cells. A large number of different papillomaviruses infecting a broad range of vertebrates have been identified, including about 60 different types of HPV, which usually give rise to benign lesions or to warts. Different types of papillomaviruses are distinguished on the basis of differences in DNA sequence *(7)*, and most research has concentrated on a few papillomaviruses that have been cloned and sequenced. In particular, attention has been directed toward bovine papillomavirus type 1 (BPV-1) and human papillomavirus type 16 (HPV-16), although a number of other human and animal papillomaviruses have also been studied. BPV-1 has been used extensively as a model for the other papillomaviruses, and the transcription and function of viral gene products is well understood in this papillomavirus. HPV-16 is the human papillomavirus type most strongly associated with the development of cervial carcinoma and is a strong candidate for a human tumor virus. Sequence analysis has revealed that all papillomaviruses share the same basic genome organization and the genome of HPV-16 is shown in Fig. 1. An examination of the nucleotide sequences

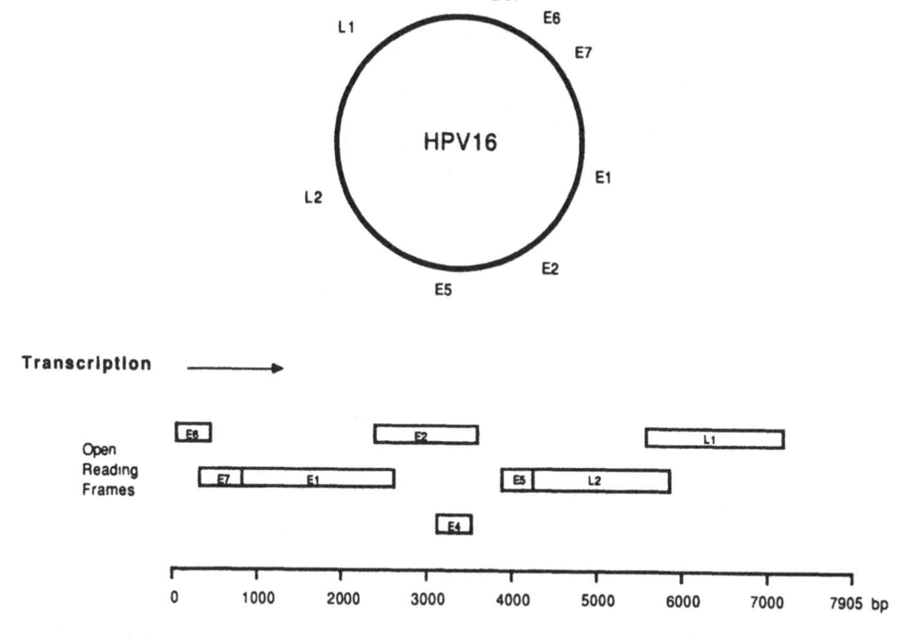

Fig. 1. The genome of HPV-16. Above, the genome is represented as a circular, double-stranded DNA molecule. This form of the viral DNA is usually maintained episomally in the nucleus of the infected cell. Below, the genome is shown linearized, with the positions of the major open reading frames and the direction of transcription. All papillomaviruses have basically the same genome organization.

reveals a number of open reading frames (ORFs) that are common to all papillomavirus types. Only one strand of the circular DNA genome is transcribed, using several alternative promoters and splicing patterns. At least one protein product encoded by each of the major ORFs has been identified. The genome has been broadly divided into three regions, the early and late ORFs and the upstream regulatory region (URR). The two large, late ORFs encode the viral structural proteins. The early ORFs encode proteins that are involved in viral transcription and replication, and in cellular transform-ation. One exception is E4, which encodes a protein that is involved in viral-particle maturation and should be classified as a late protein. The known activities of the early proteins are shown in Table 1. Some of these functions have been identified only in BPV-1, others have also been demonstrated for HPV proteins. The URR, which lies between the late and early regions, contains many of the *cis*-acting elements of transcriptional control that are responsive to both virally encoded and cellular factors *(8)*.

Table 1
Functions of Papillomavirus Early Proteins

ORF-encoding proteins	Function
E6 and E7	Transformation and immortalization of rodent and human cells in culture Transcriptional *trans*-activation Maintenance of high copy number of episomal viral DNA
E1	Maintenance of episomal replication (only shown for BPV1)
E2	*Trans*-activation and *trans*-repression of viral transcription
E5	Transformation (shown only for BPV1)

1.3. Recipient Cells Used for Transformation Studies

The epidemiological association between certain papillomaviruses and human malignancies has been complemented by studies examining the ability of the HPV DNA sequences to transform cells in culture. The transforming activities of papillomaviruses can be assayed following transfection of cloned papillomavirus DNA into suitable recipient cells. Broadly, cells used as recipients are either primary cells or cell lines that have already been established in culture. The type of recipient cell used in transfection experiments determines the type of activity that can be measured; primary cells are used to assay immortalizing activities of transfected DNA sequences, whereas established cell lines can be used to identify transforming activities. Immortalization is measured by the ability of transfected primary cells to grow continuously in culture, rather than entering senescence. Transformation is measured by a number of characteristics that are usually considered hallmarks of tumor cells, such as anchorage independence, or tumorigenicity in mice. However, the term "transformation" is often used more generally to include both transforming and immortalizing activities.

1.4. Transforming Genes of BPV-1 and HPV-16

Using the methods of calcium phosphate mediated transfection of BPV-l DNA into recipient tissue-culture cells described elsewhere in this vol, full-length BPV-l was shown to transform established cell lines *(9)*. Similarly, expression of HPV-16 sequences also transforms cells in culture, although this transformation is detected less easily *(10)*. The transforming activity of both BPV-l and various HPVs was localized to specific ORFs by the construction of plasmids that contained and expressed subgenomic regions of the

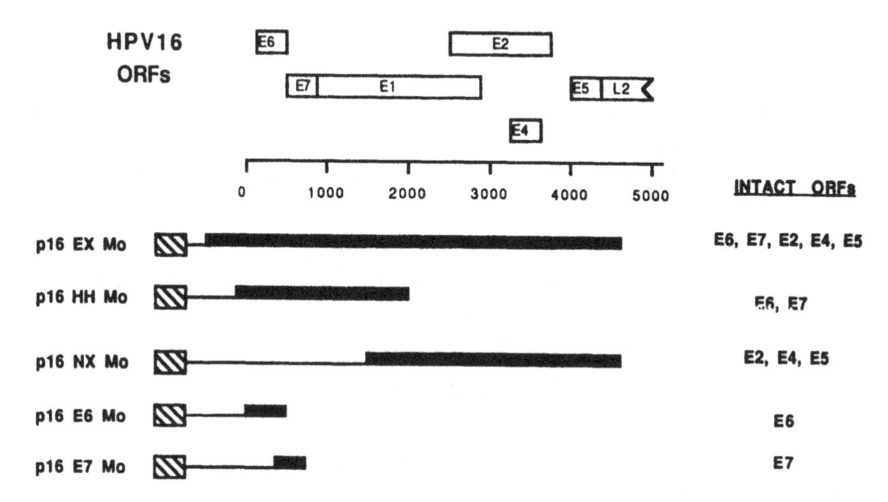

Fig. 2. Subgenomic constructs of HPV-16 used to identify transforming genes of this virus. The black boxes represent the regions of HPV-16 DNA excised from the full-length genome and placed into a plasmid vector. As indicated, each construct contains different regions of the HPV-16 genome and has the capacity to encode only a subset of the viral ORF. The hatched box represents the enhancer and promoter elements from the Molony murine leukemia virus (MoLTR). This indicates that the HPV-16 sequences were under the control of this strong heterologous promoter in all these constructs to ensure efficient expression of the viral sequences in the transfected cells.

papillomaviruses. Figure 2 shows a series of subgenomic constructs of HPV-16 that were used to identify the transforming genes of this virus in rodent cells *(10)*. For these experiments, subgenomic viral regions were linked to strong heterologous promoters, such as the Molony murine leukemia virus long terminal repeat (MoLTR), to ensure efficient expression of the viral sequences in the transfected cell.

This chapter describes assays for detecting immortalizing and transforming activities of transfected DNA sequences. The introduction and expression of DNA into recipient cells by calcium phosphate precipitation are discussed elsewhere. In this chapter a brief review of other methods of introducing exogenous DNA into cells is followed by a description of the assays most commonly used to determine transformation of cells. These include the focus-formation assay in established cells or in primary cells (in cooperation with an activated *ras* oncogene), selection of drug resistant transfected cells, determination of anchorage independence, assays for plating efficiency, growth in low serum, growth to high saturation density, reduction in population doubling time, and tumorigenicity in mice.

2. Materials

1. Routine tissue-culture equipment, including incubators, laminar-flow hoods, and tissue-culture-grade plastics.
2. Means of sterilizing bottles and solutions, i.e., autoclave and 0.22-µm filters.
3. Means of counting cells, i.e., hemocytometer or Coulter counter.
4. Recipient cells for transfection, e.g., NIH3T3 cells, rat embryo fibroblasts (REFs), or baby rat kidney cell (BRKs).
5. Tissue-culture medium: For all the above cells, use Dulbecco's Modified Eagle's Medium (DMEM) (GIBCO-BRL).
6. Calf serum to supplement medium: For NIH3T3 cells, use donor calf serum, for primary rat cells, such as REFs and BRKs, use fetal calf serum (GIBCO-BRL).
7. Dulbecco's phosphate buffered saline solution A (PBS): 0.20g/L KCl, 0.20g/L KH_2PO_4, 8.00g/L NaCl, 2.16g/L $Na_2HPO_4 \cdot 7H_2O$, pH 7.2–7.4.
8. 0.05% Trypsin/0.02% EDTA in modified Puck's Saline A (GIBCO-BRL).
9. Geneticin® (G418 sulfate) (GIBCO-BRL): Dissolve in water to 50 mg/mL microbiological potency, filter-sterilize, and store 5-mL aliquots at –20°C.
10. Hygromycin B (Calbiochem-Behring): Prepare as G418 at 10 mg/mL.
11. Formaldehyde (10%) made up in PBS.
12. Giemsa stain, 0.4% (Siga).
13. Noble agar (Difco Laboratories).
14. Stainless steel cloning cylinders, 8 mm in diameter.
15. Vacuum grease.
16. 10-cm glass Petri dish.
17. Sterile 1-mL syringes with 22-gage needles.
18. Nude mice—strain *nu/nu*, 4–8 wk old.

3. Methods

3.1. Transfection of Cells in Culture

The method most commonly used to introduce DNA into cells is that of using a calcium phosphate coprecipitate, as described in Chapter 21. However, some cells are difficult to transfect using this technique, and other methods using electroporation, liposome fusion, and polybrene or protoplast fusion provide alternatives. Electroporation and liposome fusion have frequently been used in papillomavirus studies to introduce HPV DNA into primary human keratinocytes, since these cells are induced to terminal differentiation when exposed to calcium.

3.1.1. Electroporation

Many factors affect the efficiency of transfection by electroporation—for example, temperature, current, and electroporation medium. A number of pilot experiments using various conditions need to be carried out to optimize the conditions for a particular cell type. One drawback of electroporation is that a large number of cells are needed for each experiment, so this would not be the technique of choice for cells that are difficult to grow. The following protocol suggests some alternatives.

1. Change the medium of the cells the day before electroporation.
2. Trypsinize cells and resuspend in growth medium or PBS at $2–8 \times 10^6$/mL. Aliquot 0.25–1.0 mL into electroporation cuvets and leave on ice or at room temperature.
3. Add 1–20 μg DNA in a volume of 1–50 μL and mix (*see* Note 1).
4. Electroporate at a range of voltages and resistances. Depending on the electroporation apparatus used, try ranges between 2000 v/25 μF and 250 v/960 μF (*see* Note 2).
5. Leave at room temperature for 2 min or place on ice for 15 min.
6. Add 2 mL of growth medium to the cuvet and wash everything (including the debris) into tissue culture dishes or flasks. Return the cells to the incubator and change the medium the following day.

3.1.2. Lipofection

Lipofection is best carried out using the Lipofectin™ reagent supplied by GIBCO-BRL. The method involves the formation of a lipid–DNA complex that fuses with the cell membranes. A detailed protocol is supplied with the reagent, and the method is very simple, entailing only mixing the DNA with the Lipofectin™ reagent and adding this to a dish of cells. The drawback of this method is that the reagent is very expensive.

3.2. Focal Transformation of NIH3T3 Cells

Focus formation is the most straightforward assay for transformation following transfection. Success is dependent on the recipient cell line, which must be sensitive to transformation, yet have a low incidence of spontaneous transformation. In practice this means that the recipient cells must become efficiently contact-arrested and remain as a flat monolayer despite being kept at confluency for a number of weeks.

1. Introduce exogenous DNA sequences into the cells using any of the described protocols. Calcium phosphate precipitation is a very good transfection method for NIH3T3 cells; transfect a 60-mm dish that was seeded with $2–3 \times 10^5$ cells the day before transfection.

2. Trypsinize the cells 24 h after transfection, and reseed them into medium containing only 5% serum. Take transfected cells from one 60-mm dish and divide them equally between two 100-mm dishes. This serum reduction from 10% gives a monolayer that is less dense, on which the transformed foci can be more easily detected.
3. Feed the cells twice a week in 5% serum (feeding on Monday and Friday is often convenient) (*see* Note 3). Examine for foci after 10 d, although weakly transforming sequences may not result in the formation of foci until after 3 wk or more (*see* Notes 4 and 5).
4. Foci are best counted under low power of an inverted microscope. It is convenient to use a grid to ensure that the entire plate has been examined. The lid of a 100-mm dish, etched with a razor blade with lines wide enough to fill an entire field, is convenient for this purpose. After a certain amount of practice, it becomes quite easy to see foci by holding the dish up to the light (taking care not to spill the medium) and examining by eye.
5. The number of foci seen depends on the amount of DNA used and the transforming gene under study. The efficiency of transformation is very dependent on the amount of transforming DNA used in the transfection. To a certain limit, the number of foci increases with increasing amounts of DNA, but after this point, the transforming efficiency drops with increased DNA. Normally between 0.1 and 2 µg of plasmid DNA or 5–20 µg genomic DNA are used in each transfection, but it is worth optimizing the system for each new DNA to be studied (*see* Note 6). To establish the assay system, transfections with known transforming sequences should be carried out. Cloned activated Ha-*ras* sequences should give about 3000 foci/µg transfected DNA. Full-length BPV-1 DNA gives about 300 foci/µg of DNA. It is worth spending time to make the assay as sensitive as possible, since other transforming genes can be far less active. For example, HPV-16 gives rise to only 1 focus/µg of DNA.

3.3. Focus Formation in Primary Cells in Cooperation with ras

This assay for immortalizing genes is essentially the same as the focus assay using NIH3T3 cells described above. Primary rat cells are used as recipients, usually either REFs or BRK cells. The assay depends on the cooperation between an immortalizing gene (the test sequences) and a transforming gene (activated Ha-*ras* sequences) to give continuously growing, transformed foci of cells on a background of senescing primary cells. REFs are more sensitive than BRK cells and can be more easily immortalized and transformed in this assay. They do, however, give a slight background of spontaneously immortalized cells, and immortalization of BRK cells is considered a more stringent assay. Clearly this assay can also be used to look for transforming genes that can cooperate with known immortalizing sequences, such as adenovirus Ela or *myc*.

1. Two DNAs are used in equal quantities in the transfection, a plasmid carrying activated Ha-*ras* sequences, e.g., pEJ *(11)* and the plasmid carrying the potential immortalizing sequences. Normally 1–10 µg of each DNA are used to transfect one dish of cells (*see* Note 7).
2. The cells need not be replated after the transfection and are maintained in normal growth medium (usually containing 10% fetal calf serum) for 2–3 wk. Foci are scored as described for NIH3T3 cells.

3.4. Cotransfection and Selection for Drug Resistance

Genes that do not lead to obvious morphological transformation can be assayed for other parameters of transformation, such as growth in soft agar. To improve chances of detection, the transfected cells are first selected, to make sure that only cells that have taken up DNA are examined. This is achieved by cotransfecting a plasmid encoding drug resistance with the test DNA and selecting resistant cells. This method can be used with primary or established cells. Weak morphological transformation is often more evident when the cells are not surrounded by a monolayer of untransfected cells.

1. Transfect cells with both DNA encoding drug resistance and the test DNA in a ratio of 1:10 (*see* Note 8). In established cells, such as NIH3T3, this is normally 100 ng:1µg; in primary cells for which transfection efficiencies are lower, use 1–2 µg:10–20 µg. The plasmids most commonly used are those encoding resistance to Geneticin® (G418) and hygromycin B.
2. Split the cells 24 h after transfection (*see* Note 9).
3. After a further 24 h, feed the cells with growth medium that has been supplemented with the drug, i.e., G418 or hygromycin B (*see* Note 10).
4. Feed two or three times a week with selective medium. Drug-resistant colonies should start to appear after about 1 wk.
5. When all the untransfected cells are dead and the resistant colonies have grown up, they can be isolated by ring cloning as described below.
6. A permanent record of foci or colonies can be made by fixing the cells and staining with Giemsa as described below.
7. Selection for drug resistance and focus formation may be profitably combined in one transfection assay as shown in Fig. 3. Using this proto-

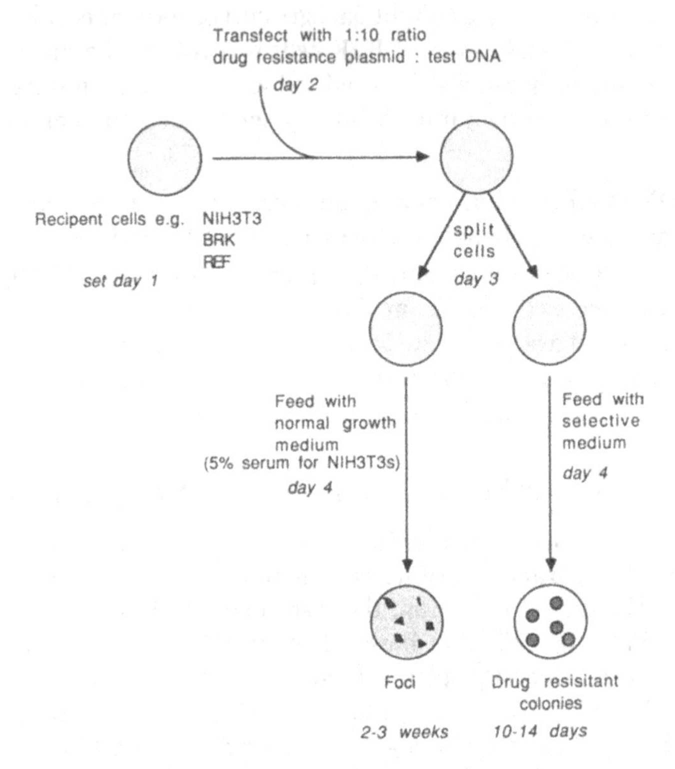

Fig. 3. Protocol for combining assays for focal transformation and selection of drug-resistant transfected colonies in one experiment.

col, an estimation of the efficiency of transfection can be determined from the number of drug-resistant colonies that arise.

3.5. Giemsa Staining Cells

Staining flasks or dishes of tissue-culture cells following a transfection experiment provides a permanent record of the experiment and makes photography and scoring of foci or colonies easier.

1. Remove medium from the cells that are to be stained and wash once in PBS.
2. Add 10% formaldehyde (about 5 mL/100 mm dish) and leave to fix for at least 30 min or overnight.
3. Pour off the formaldehyde and rinse in distilled water.
4. Add enough undiluted Giemsa stain to cover the bottom of the dish and leave for 10–20 min.

5. In order to save the stain, one may decant as much as possible back into the bottle, although eventually this will result in some dilution. If this happens, it may be necessary to leave the stain on the plates for a longer period.
6. Rinse the dish in running distilled water for a minute or until the background is clear and the colonies can be seen clearly. If the dishes are labeled on the lid, be careful to keep the correct lid and base together.
7. Invert the dish onto tissue paper to drain, then air-dry.

3.6. Isolation of Foci or Colonies by Ring Cloning

Foci or drug-resistant colonies can be isolated and expanded separately for further investigation by ring cloning.

1. Stand clean stainless steel cloning cylinders in a thin film of vacuum grease in a glass Petri dish and sterilize by autoclaving.
2. Mark the position of the foci or colonies to be isolated with a circle on the bottom of the dish, either under the microscope or, if possible, by eye (*see* Note 11). Use dishes on which the foci are well-seperated, to prevent cross-contamination (*see* Note 2).
3. Remove the medium from the dish of cells containing the colony to be isolated, wash with PBS, and then remove the PBS, leaving the dish as dry as possible.
4. Using flamed forceps, place the cloning cylinder over the focus so the greased edge sits on the dish. Lightly push the cylinder down to ensure a water-tight seal and create a small well containing the cells to be isolated.
5. Add 0.1 mL of 0.25% trypsin. The cells of the focus will become detatched after a few minutes and can be washed out of the cylinder with a sterile Pastuer pipet using medium containing 10% serum.
6. Place the cells from the focus into a well of a 24-well dish containing 1.0 mL of growth medium and expand the cells as usual.

3.7. Anchorage-Independent Growth

Tests for anchorage-independent growth involve suspending the cells in a semisolid medium and assaying colony formation. Agar or agarose may be used as the support; agar allows only highly transformed cells to grow, whereas agarose (which lacks sulfated polysaccharides) is less selective. However, the sensitivity of agarose is matched by a higher background, and untransfected NIH3T3 cells show significant growth in agarose. Agar is therefore the medium of choice for NIH3T3 cells.

1. Make up 6% agar in distilled water and autoclave. Cool to 65°.
2. Warm growth medium to 45°C. Do not leave medium at 45° for longer than necessary.
3. Make up 0.6% agar medium by mixing 9 parts of growth medium with 1 part of 6% agar, and keep at 45°C (*see* Note 13).
4. Pour agar bases by adding 5 mL 0.6% agar medium to 60-mm dishes. Allow these to set at room temperature (*see* Notes 14 and 15).
5. Trypsinize and count the cells. Take up into 1 mL growth medium enough cells for three dishes. The cell density depends on the expected transforming efficiency; normally cells are plated at about 10^4–10^5/dish (*see* Note 16).
6. Add 2 mL of 0.6% agar medium to 1 mL of cells. Plate 1 mL of the cell suspension onto each of three agar bases, working as rapidly as possible to prevent the agar setting in the pipet. Allow the tops to set by leaving the dishes in the hood for ~20 min; then return them to the incubator.
7. After 2 d, carefully add 3 mL of 0.4% agar medium to the top of each dish. Feed once a week with about 3 mL of 0.4% agar medium, or feed more frequently if the dishes appear to be drying out.
7. Colonies will be apparent after 1–3 wk. They can be counted using an inverted microscope. Normally growth is expressed as percentage colony formation compared to the number of cells seeded, adjusted to account for cell viablility if required (*see* Note 16).
8. Agar colonies can be stained using a 1 mg/mL solution of *p*-iodonitrotetrazolium violet in water. Add 2 mL to each dish and leave at room temperature overnight. After staining, decant the soft agar onto Whatman 3MM chromotography paper and allow to dry.

3.8. Plating Efficiency

Some transformed cells have a higher plating efficieny than their untransformed parents, which increases the chances of a single cell forming a colony. This method can also be used to normalize growth in agar of different cell lines as described above.

1. Trypsinize cells and count.
2. Seed 10^2, 10^3, and 10^4 cells into 100-mm Petri dishes in normal growth medium.
3. Allow colonies to grow, feeding the cells as necessary.
4. When colonies are visible (normally 7–10 d for NIH3T3 cells), fix and stain the dishes (*see* Note 17).
5. Count colonies and express plating efficiency as the percentage of cells seeded that formed a colony.

3.9. Alterations in Growth Properties

Such properties as growth in low serum, population doubling time, growth-factor independence, and growth to high saturation density can be assayed by plating the cells into 24-well dishes, adjusting the growth conditions, and counting cells at 24-h intervals (*see* Note18). A Coulter counter is very useful if many growth experiments are to be carried out.

1. Trypsinize cells and dilute to 10^3–10^5/mL in normal growth medium.
2. Plate 1 mL into each well of a 24-well dish. Normally each count is carried out in triplicate.
3. Allow the cells to attach, usually overnight, before changing growth conditions.
4. To assay population doubling and saturation density, the cells are left in normal growth medium and counted every 24 h. The growth rate of subconfluent cells can be used to estimate the population doubling time. For this assay the cells should be plated out fairly sparsely to allow enough counts to be taken before confluence is reached. To estimate saturation density, plate the cells more densely and count the number of cells in a confluent well over a number of days (*see* Note 19).
5. To assay growth in low serum, change the medium on the cells to one containing 0.1–0.5% serum (*see* Note 20). Count the cells at 24-h intervals.
6. If the cells cannot grow in low serum, the effect of adding specific growth factors can be examined. The growth factors used are very dependent on both the cell type used as a recipient and the transforming gene under examination. Cells are moved to supplemented serum-free medium and cell growth is assessed by counting at 24-h intervals.

3.10. Tumorigenicity

The ability of cells to form invasive tumors is generally accepted as the "gold standard" for oncogenic transformation. Ideally, tumorigenicity should be assessed in a syngeneic, immunocompetent host, but this is not always possible, especially when the cells to be tested are human. The genetically athymic "nude" mouse is often used as a host for tumorigenicity studies. Colonies of nude mice must be maintained in a pathogen-free environment, since they are very suceptible to infection.

1. Trypsinize the cells and count.
2. Spin down the required number of cells and resuspend in 0.2 mL of serum-free medium. Usually between 10^5 and 10^7 cells are injected at each site (*see* Note 21).
3. Inject the cells subcutaneously in up to 4 sites/mouse. Tumors may

arise at any time after 1 wk, and the mice should be examined frequently. In some cases tumors may not appear until more than 10 wk after inoculation.

4. Tumors can be harvested by dissection from the sacrificed animal. They may then be used for histological analyses, snap-frozen in liquid nitrogen for subsequent DNA and RNA analyses, established as a new cell line, or transplanted into another animal.

4. Notes

1. Linear DNA is transfected about 10 times more efficiently than circular DNA, so plasmids should be digested before electroporation.
2. If too little current is used, the DNA will not be taken up by the cells; too much current will kill all the cells. In general, aim to recover cells that are about 50% viable following electroporation.
3. When feeding the transfected cells, pipet the medium onto the cells very gently. Vigorous pipetting can disrupt the monolayer and give rise to false "foci."
4. The focus assay depends absolutely on the ability of the recipient cell line to remain as a flat monolayer during the course of the experiment, which could be 4 wk or more. Many recipient lines show some degree of spontaneous transformation, which increases as the cells are maintained as a confluent monolayer. After some practice, spontaneous foci may be distinguished by their morphological appearance. However, the morphology of foci induced by different oncogenes also varies greatly, and appearance is not an absolute guarantee that a focus is genuine. Comparison of the number of foci on the experimental dishes with the number on the negative control dishes is essential before getting excited about the discovery of a new oncogene. To reduce the background of spontaneous foci, be very careful in maintaining the recipient cell line. Keep the cells at a low density and do not allow them to become confluent.
5. One component of tissue culture that can vary widely and give different degrees of success in transfection experiments is the serum. It is worth testing a number of different batches of serum before buying a large amount. In particular, some batches of serum enhance the problem of spontaneous foci.
6. Use the same total amount of DNA for each transfection. When using plasmid DNAs it is usually necessary to make up the DNA concentration with carrier DNA; normal mouse-liver DNA is often used.
7. Control transfections should include Ha-*ras* sequences transfected alone to assess the background level of immortalization. Introduction of

Ha-*ras* will give a few foci on REFs after transfection, but BRK cells should remain negative. Good positive controls include adenovirus Ela or v-*myc* sequences cotransfected with Ha-*ras*. Test DNAs could also be transfected without Ha-*ras* sequences to assess their ability to immortalize cells alone.

8. It is useful to include a control dish of untransfected cells to ensure that all the normal cells have been killed by the drug selection.
9. If the cells are too dense before the selection is started, they will not die. It is better to split the transfection into more dishes to ensure that the cells are sufficiently sparse. For NIH3T3 cells, divide a subconfluent 60-mm dish between three 100-mm dishes. Primary cells need not be treated quite so harshly, a 1:3 split is usually sufficient. If the cells in the transfection dish are very sparse, splitting may be unnecessary.
10. The concentration of the drug depends on the recipient cell line being used. Tests should be carried out to assess toxicity and determine the lowest concentration that will kill untransfected cells within a few days. G418 at a concentration of 500 μg/mL will kill NIH3T3 cells, whereas only 300 μg/mL is needed for BRK cells, and 150 μg/mL may be sufficient for REFs. Be careful to make up the G418 solution with respect to microbiological potency (which varies from batch to batch) and not on the basis of weight. To kill NIH3T3 cells, 100 μg/mL of hygromycin B is sufficient.
11. Foci or colonies should be sufficiently large to ensure successful isolation. Colonies of <500 cells are difficult to expand, although with care and practice, smaller colonies can be isolated.
12. The cells of the focus will be contaminated with some of the untransfected background cells and possibly with transformed cells from another focus on the same dish. If it is necessary to have a clonal population of cells, they should be single-cell cloned at this stage.
13. This protocol results in the overall dilution of the growth medium to 90%. This does not seem to affect NIH3T3 cells, but for more sensitive cells it may be necessary to obtain 10× growth medium and adjust the dilution to allow for the addition of the agar.
14. Since the cells in this assay do not need to attach to the bottom of the dish, bacterial-grade plastics (which are much cheaper than tissue-culture grade) can be used.
15. Agar bases can be stored in the cold room overnight.
16. If required, a parallel estimation of plating efficiency can be carried out as described below. The number of colonies in agar can be adjusted to account for plating efficiency.

17. Do not let the colonies get too big or smaller satellite colonies will begin to appear which represent cells reseeded from an existing colony rather than an independently derived colony.
18. Cell growth can also be assessed by measuring tritiated thymidine incorporation by cells over a fixed period (normally around 1 h).
19. To get accurate cell counts, it is very important to have a single cell suspension, especially when using a Coulter counter. Particular care must be taken when counting cells that have been confluent for some time.
20. The amount of serum needed to ensure that the untransformed parental cells do not die, rather than stop dividing, is dependent on the cell lines. Some cells can tolerate medium that is completely serum-free, whereas others will need up to 0.5% serum to ensure viability.
21. Many cells used as recipients for transfection give rise to tumors after extended periods when injected at high numbers. The incidence of spontaneous tumors can be assessed using the appropriate negative controls.

Acknowledgment

I am indebted to C. J. Marshall, who taught me most of these techniques.

References

1. Ciuffo, G. (1907) Innesto positivo con filtrato di verruca volgare. *G. Ital. Mal. Venereol.* **48,** 12–17.
2. Zur Hausen, H. (1989) Papillomaviruses as carcinomaviruses. *Adv. Viral Oncol.* **8,** 1–26.
3. Rous, P. and Beard, J. W. (1935) The progression to carcinoma of virus-induced rabbit papillomas (Shope). *J. Exp. Med.* **62,** 523.
4. Zur Hausen, H. (1976) Condyloma acuminata and human genital cancer. *Cancer Res.* **36,** 794.
5. Vousden, K. H. (1989) Human papillomaviruses and cervical carcinoma. *Cancer Cells* **1,** 43–50.
6. Orth, G. (1987) Epidermodysplasia verruciformis, in *The Papovaviridae 2. The Papillomaviruses* (Salzman, N. P. and Howley, P. M., eds.), Plenum, New York, pp. 199–243.
7. Pfister, H. (1987) Papillomaviruses: General description, taxonomy and classification, in *The Papovaviridae 2. The Papillomaviruses* (Salzman, N. P. and Howley, P. M., eds.), Plenum, New York, pp. 1–38.
8. Spalholz, B. A. and Howley, P. M. (1989) Papillomavirus–host cell interactions. *Adv. Viral Oncol.* **8,** 27–53.
9. Dvoretzky, I., Shober, R., and Lowy, D. (1980) Focus assay in mouse cells for bovine papillomavirus. *Virology* **103,** 369–375.
10. Vousden, K. H., Doniger, J., DiPaolo, J. A., and Lowy, D. R. (1988) The E7 open reading frame of human papillomavirus type 16 E7 gene encodes a transforming gene. *Oncogene Res.* **3,** 167–175.
11. Shih, C. and Weinberg, R. A. (1982) Isolation of a transforming sequence from a human bladder carcinoma cell line. *Cell* **29,** 161–169.

CHAPTER 16

Herpes Simplex Virus Life Cycle and the Design of Viral Vectors

David S. Latchman

1. Introduction

Herpes simplex viruses types 1 and 2 (HSV-1 and HSV-2) are double-stranded DNA viruses with a genome size of 152 kbp. The genome consists of two unique regions, U_L (long) and U_S (short), flanked by repeated sequences (Fig. 1; for review, *see* ref. *1*). The two viruses are closely related, and both infect humans, producing mucocutaneous sores that are predominantly facial in the case of HSV-1 and genital in the case of HSV-2.

In both cases, the sores produced by these viruses represent sites of lytic infection in which the synthesis of viral polypeptides and the replication of viral DNA is followed by cell lysis and the release of progeny virions (for review, *see* ref. *2*). This lytic cycle can be prevented and the lesions abolished by treatment with acyclovir, which is activated by the viral thymidine kinase *(3)* and prevents viral DNA replication. At the time of initial infection however, the virus also establishes asymptomatic latent infections in neurons of the dorsal root ganglia, innervating the site of the lesion (reviewed in refs. *4,5*). These infections, which do not contain replicating DNA and are therefore not susceptible to acyclovir treatment, provide a reservoir from which the virus can emerge to repeatedly reinfect susceptible epithelial cells, producing recurrent sores.

The clinical importance of these viruses has led to considerable study of their life cycle, which in turn has paved the way for the creation of vectors

From: *Methods in Molecular Biology, Vol. 8: Practical Molecular Virology: Viral Vectors for Gene Expression*
Edited by: M. Collins

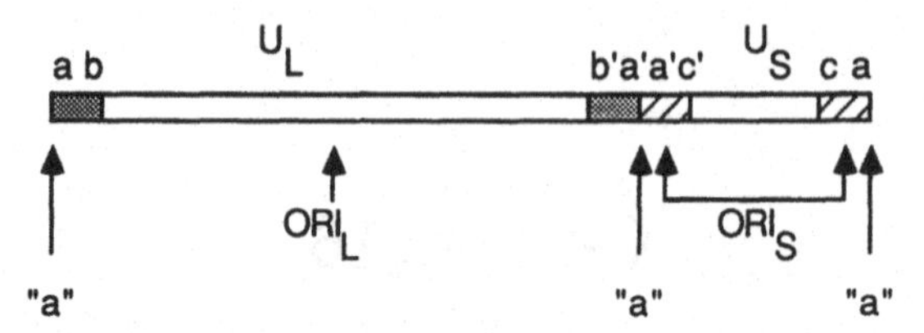

Fig. 1. Structure of the HSV genome. The two regions of unique DNA, U_L and U_S, are indicated, together with the repeated DNA elements that flank them. The positions of the replication origins, ORI_L and ORI_S, are shown, together with the three copies of the "a" sequence that mediates cleavage and packaging of viral DNA.

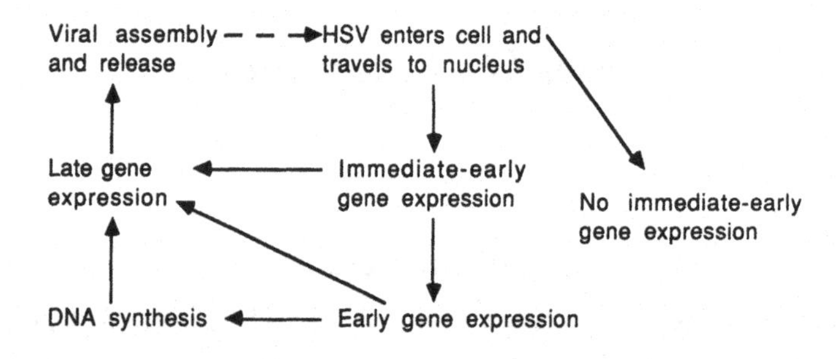

Fig. 2. Life cycle of HSV. The various events occurring during lytic infection of epithelial cells in vivo and of cultured cells in vitro are shown, together with the point at which the cycle is aborted in latent infections of neuronal cells in vivo.

based on these viruses and capable of replicating and expressing foreign genes. These aspects of the viruses are discussed in this introduction.

1.1. Life Cycle of HSV

Although in vivo HSV-1 and HSV-2 infect epithelial cells, in vitro both viruses can infect a wide range of cell types, producing a similar characteristic lytic infection that results in the death of the cell and release of virus. The events occurring in lytic infection are summarized in Fig. 2, which shows the relationship of this process to that occurring in latent infection of neuronal cells.

Initially, the virus binds to a specific receptor on the cell surface *(6)*, the virion envelope fuses with the cell membrane *(7)*, and the viral nucleocapsid enters the cell. Following the movement of the nucleocapsid into the nucleus and the release of viral DNA, the genome is transcribed to produce three classes of viral RNA and protein, namely, immediate-early, early, and late (for review, *see* ref. *2*).

The immediate-early, or α, proteins are the first to be made in the infected cell and are defined on the basis that transcription of their genes does not require prior protein synthesis *(8,9)*. This is because a 65-kD viral protein (called Vmw65) carried in the viral particle *(10)* interacts with preexisting cellular transcription factors *(11)* to promote transcription of these genes. Five proteins of this class exist and are identified by their mol wt (in kilodaltons) as Vmw175, Vmw110, Vmw68, Vmw63, and Vmw12 *(12)*.

The activity of several of these proteins is necessary for subsequent stages of viral gene expression. Most dramatically, temperature-sensitive mutations in the gene encoding Vmw175, as in the HSV-1 mutant tsK, abolish the ability to grow at the nonpermissive temperature inconsequence of a failure in the production of the early and late proteins *(13)*. Other defects in viral gene expression, of a less severe nature, have also been observed in strains carrying mutations in the genes encoding Vmw110, Vmw68, and Vmw63 *(14–16)*. The immediate-early proteins induce the synthesis of the early, or β, proteins, many of which are involved in aspects of nucleic acid metabolism necessary for viral growth and which include the viral DNA polymerase, the major DNA binding protein, and viral thymidine kinase. Some of these, such as DNA polymerase, are absolutely required for viral growth in culture; others, such as thymidine kinase, are dispensable *(17)*.

Because of the involvement of the early gene products in DNA replication, their synthesis is necessary before viral DNA replication can begin. In turn, high-level synthesis of the late, or γ, proteins is dependent on DNA replication as well as the prior synthesis of the immediate-early and early proteins *(18)*. These late proteins specify virion structural proteins, which are required for the assembly of progeny virions and include both proteins, such as the major capsid protein, and glycoproteins, such as gC.

Following entry into the cell, the linear viral genome circularizes *(19)* and replicates via a rolling circle mechanism to produce a linear concatomeric DNA (reviewed in ref. *1*). Distinct origins of replication have been defined within the viral DNA (Fig. 1). One of these, ORI_L, is found within the U_L region *(20)*, whereas two copies of the other, ORI_S, are found in the repeated elements flanking the U_S region *(21)*. Each of these sequences is sufficient to direct replication of DNA linked to it in HSV-infected cells *(22)*.

Following replication, the DNA is cleaved into unit length genomes and packaged into capsids *(23)*. Cleavage and packaging take place as a coupled process involving recognition of specific signals within viral DNA (Fig. 1), which are located within the 248-bp "a" sequences found at the ends of the viral genome and at the junction of the repeated elements flanking U_L and U_S

(24,25). Finally, the nucleocapsid leaves the cell nucleus and is transported via the endoplasmic reticulum to the cell surface.

1.2. HSV-Based Vectors

As can be seen from the previous section, HSV has many advantages as a potential vector. Some of these are

1. The virus grows very well in tissue culture. Hence, high-titer stocks of wild-type and recombinant viruses can readily be obtained.
2. The virus infects a very wide variety of cell types, allowing the gene of interest to be introduced into many different cells.
3. The virus expresses many of its genes at very high levels in the infected cell. Hence, if foreign genes introduced into the virus are linked to HSV promoters, they too will be expressed at high levels, allowing large amounts of the foreign protein to be produced.
4. The virus has a very large genome, and many of the genes within it (for example, that encoding thymidine kinase) are not required for growth in culture. Hence, large foreign genes can readily be accommodated within the virus without rendering it defective for growth.

Unfortunately, however, the large size of the virus also renders the introduction of foreign DNA difficult, since no single-restriction-enzyme cutting sites at which DNA can be inserted exist in the viral genome. Two approaches have been taken to circumvent this problem. These involve either the introduction of foreign DNA by homologous recombination to create viable recombinant viruses, or the insertion of DNA into short regions of the viral genome and subsequent amplification to produce a defective viral genome. These two approaches will be discussed in turn.

1.2.1. Viable Recombinant Viruses

The difficulty in cloning restriction fragments of foreign DNA into the viral genome has led to the use of methods involving recombination of the intact viral genome with a piece of DNA during infection as a means of introducing DNA into the virus. A number of methods for achieving such recombination with a specific nonessential region of the viral genome and for assaying the progeny have been devised (reviewed in ref. *26*).

These methods rely on the introduction into cells of full-length deproteinized viral DNA, together with a large excess of a plasmid molecule containing the foreign DNA linked to a specific region of the viral genome. Because deproteinized viral DNA is infectious *(27)*, a lytic cycle will take place in the infected cell and homologous recombination will occur at high frequency between the region of the viral genome linked to the foreign DNA and the

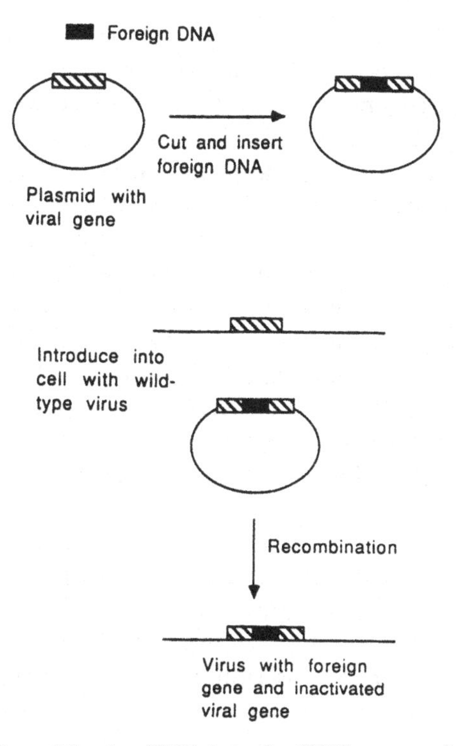

Fig. 3. Introduction of foreign DNA into the HSV genome by recombination.

corresponding region of the virus *(28)*, resulting in the insertion of the foreign DNA into a recombinant virus (Fig. 3).

In such recombination experiments, it is possible to link foreign DNA to any nonessential region of the viral DNA and obtain insertion of the foreign DNA in this region of the virus. In practice, however, it will not be possible to identify such relatively rare and possibly slowly growing recombinant viruses among the mass of nonrecombinants without employing large-scale screening procedures aimed at determining whether the foreign DNA is present in individual viruses. For this reason, most attempts to prepare recombinants of this type have involved the use of a region of viral DNA whose inactivation can readily be assayed.

The most frequently used viral gene of this type is that encoding thymidine kinase (*tk*), which, as discussed above, is not essential for viral growth. Viruses lacking this gene can readily be selected from among those that possess it by growth in the presence of deoxynucleotide analogs such as acyclovir or bromodeoxyuridine, which are phosphorylated by the intact thymidine kinase of wild-type virus resulting in their incorporation into its DNA and consequent failure of viral DNA replication. Hence, if a DNA fragment from a foreign gene is linked to a portion of the viral *tk* and introduced into cells with wild-type virus, progeny in which the viral *tk* has been inactivated by insertion of the foreign DNA will be the only viruses that can grow in the presence of acyclovir or bromodeoxyuridine and, hence, can be readily selected.

This strategy was used by Tackney et al. *(29)* to introduce into HSV the Chinese hamster gene encoding the enzyme adenine phosphoribosyl transferase (aprt). After selection of *tk⁻* virus, it was possible to determine whether *aprt* was being expressed by determining whether it could confer on infected cells that lacked this gene the ability to grow in a mixture of adenine and azaserine, which requires a functional aprt enzyme. No expression of the aprt protein was detected when the virus was allowed to carry out a full lytic cycle, although expression was observed when the recombinant virus was irradiated with UV light prior to infection, preventing it from carrying out the full lytic cycle.

This work illustrates the difficulties of expressing genes within the HSV genome. If the gene is introduced under the control of its own promoter, as in this case, although it may be expressed early in infection, such expression will be "switched off" later in infection, presumably by the same mechanisms that repress most cellular gene transcription in the infected cell *(30,31)*. Hence, this system is not ideal for applications requiring high-level expression of foreign DNA.

To circumvent this problem, several groups have introduced foreign genes into HSV under the control of HSV promoters, ensuring that such promoters will be highly active in the infected cell. By doing this, Shih et al. *(32)* obtained high-level production of the S gene product of hepatitis B virus, the protein produced being indistinguishable from that produced by the hepatitis virus itself on the basis of antigenicity, ability to form particles, and the like.

Hence, HSV vectors can be used to produce high levels of particular products and may well be of particular use in the synthesis of proteins encoded by other viruses that are dangerous or difficult to grow or those encoded by cellular genes normally expressed at a low level. The utility of these viruses as vectors is limited by two problems, however: the need to introduce the foreign DNA by recombination at a very few selectable sites in the viral genome and the fact that viral lytic infection will eventually kill the cell producing the foreign protein, limiting the amount that is actually produced.

Some progress has been made in overcoming the first of these problems. Systems have been developed that allow the ready identification of viruses carrying foreign DNA at particular sites other than *tk* within the genome. Thus, if the gene encoding a particular protein is disrupted by recombination, cells infected with the resulting recombinant virus will not react with an antibody against this protein; hence, recombinant virus can be identified. In this manner, by adapting a well-characterized test for antibody staining of viral plaques, known as the black plaque test, Enquist and colleagues have successfully monitored for the insertion of DNA into the glycoprotein III (gIII) gene of the pig herpes virus, pseudorabies virus and have expressed human immunodeficiency virus type I envelope glycoproteins as fusion proteins with gIII *(33)*. Similarly, the replacement of a particular region in the Herpes samirii genome required for the immortalization of T cells with foreign DNA results in a virus that is readily distinguishable from the wild type by its ability to produce only latent persistent infection in such cells. A virus of this type has been used to express high levels of bovine growth hormone in infected cells *(34)*.

In addition to the use of other selectable or monitorable sites in the viral genome, systems have also been derived that use genetic techniques to allow the power of *tk* selection to be used for selecting insertions at other sites within the viral genome. These methods (reviewed in ref. *26*) include the isolation of *tk*⁻ viruses by deletion of *tk* followed by its insertion into a new site in the genome by recombination. Foreign DNA linked to *tk* sequences can then be introduced into *tk* at its new site, exactly as previously described by recombination and selection for *tk*⁻ progeny.

Methods such as this, and others that are more complex, have the potential to introduce foreign DNA into any site in the genome, but have not been widely used because of their complexity. Moreover, like the simpler methods involving selectable sites, the efficiency of such methods is limited by the fact that DNA is introduced into only a small region of the viral genome. Hence, such methods fail to take advantage of the very large size of the viral genome as a means of introducing multiple copies of foreign DNA and thus achieving high expression.

This limitation, together with the fact that the vectors described so far are viable viruses that will eventually kill the cell, resulting in a cessation of foreign-protein production, has led to a search for other types of HSV based vectors that would overcome these disadvantages while maintaining the advantages of HSV vectors in terms of high expression and wide host range. Such defective viral vectors will now be discussed.

1.2.2. Defective Viral Vectors

If stocks of HSV are serially propagated by repeated infection of cells at high multiplicity, so that more than one virus infects each cell, defective viruses accumulate in the stock (reviewed in ref. *35*). The 150-kbp genome of these viruses, which can grow only in the presence of wild-type virus, consists of head-to-tail reiterations of portions of the viral genome as short as 3 kbp in length. The regions of viral DNA contained within such defective viruses al-

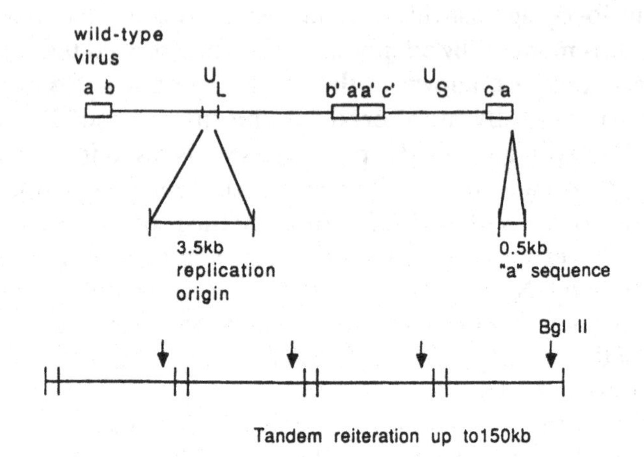

Fig. 4. Relationship of a defective viral genome isolated from the Patton strain of HSV-1 to the wild-type viral genome. The points in the intact genome from which the two portions of the defective genome derive are indicated, together with the unique BglII site present in each repeating unit (Data from Spaete and Frenkel *[21]*).

ways contain both a replication origin and a cleavage–packaging sequence (*see* Section 1.1. and Fig. 4). Hence, they contain all the *cis*-acting sequences required for replication of and packaging their DNA into virions and can be stably propagated in the presence of wild-type virus to provide the transacting viral proteins required for these processes.

The features of such defective viruses suggested that they could be used as a cloning vector for the amplification and expression of foreign DNA. Thus Spaete and Frenkel *(21)* isolated the single 3.9-kbp unit that forms the basis of one such defective virus (illustrated in Fig. 4) and introduced a portion of plasmid DNA (as a marker foreign DNA) into the unique BglII site in this basic unit. Upon introduction of this DNA into cells in the presence of wild-type virus, it was converted into head-to-tail concatamers consisting of a repeating unit containing the foreign DNA and the defective viral DNA. These were efficiently replicated and packaged into virions and, like naturally occurring defective virus, could be stably propagated through many generations.

Defective viral vectors of this type thus overcome two major problems associated with the use of wild-type HSV as a vector. First, because the basic DNA unit of such a vector is a small piece of DNA, it is likely to contain unique restriction sites (such as the BglII site in the vector discussed earlier) into which foreign DNA can be easily inserted without the need for homologous recombination. Second, because a relatively small DNA molecule containing the foreign DNA is amplified in the infected cell to a size at which it can be packaged, the foreign DNA constitutes a much higher proportion of the final viral DNA than of the wild-type vectors. The greater copy number therefore takes full advantage of the large size of HSV DNA and allows a high level of expression of the foreign DNA to be achieved.

Thus, using a defective vector of this type, it was possible to express the chicken ovalbumin gene at high levels from an HSV promoter in cells superinfected with wild-type virus *(36)*. More recently, the precise definition of the HSV sequences required for replication and packaging has allowed the construction of plasmid vectors containing just the sequences into which a piece of foreign DNA can be inserted *(22)*. Such vectors, which are amplified and packaged in infected cells exactly like the defective virus molecules, thus combine the convenience of plasmid vectors with the high levels of expression attainable in the HSV system. One such vector has successfully been used to express the chloramphenicol acetyl transferase gene at a high level under the control of several different HSV promoters *(22)*.

Vectors of this type, therefore, overcome many of the problems associated with the use of HSV as a vector and are likely to prove very useful as a means of expressing foreign proteins. However, one potential problem still remains. All the vectors described will be replicated and packaged into virus only in the presence of wild-type virus. This presents no problem when viral stocks are being prepared, since defective genomes can be maintained and propagated in the presence of wild-type virus provided infection is carried out at high multiplicity. However, when expression from the defective genome is sought, as with viable vectors, such expression will be limited by the eventual lysis of the cells by the wild-type virus.

To overcome this problem, several workers using defective vectors *(22,37)* have expressed the foreign DNA under the control of a viral immediate-early gene promoter. Because such a promoter is active very early in infection, it is possible, when protein production is desired, to use as the superinfecting virus a mutant capable of carrying out only the earliest stages of the viral lytic cycle. Hence, no cell lysis will occur, and large-scale protein production can be achieved. Indeed, if the superinfecting virus used is the HSV-1 mutant tsK, which has a temperature-sensitive form of the major immediate-early protein Vmw175 (*13*, *see* Section 1.1.), the immediate-early promoters driving the foreign gene will actually be more active at the nonpermissive temperature than in infection with wild-type virus, since wild-type Vmw175 normally represses the promoters of its own gene and other immediate-early genes *(13)*. A system of this type thus allows the high-level production of a foreign protein in a relatively simple system, without the complication of cellular lysis.

2. Materials

2.1. Medium

For routine culture, use Dulbecco's Modified Eagle's Medium supplemented with 10% newborn calf serum and 4.5 g of glucose/L. This is supplemented appropriately for particular experiments, as noted in the Methods section.

2.2. Solutions

1. 10× HEPES-buffered saline (HBS): 8.18% NaCl, 5.94% HEPES, 0.2% Na_2HPO_4. Store at 4°C in 50-mL aliquots. Prior to transfection, make a 2× HBS solution and adjust to exactly pH 7.12 with 1*M* NaOH. Filter-sterilize.
2. Phosphate-buffered saline (PBS): 12.2 g NaCl, 0.27 g KCl, 1.92 g $Na_2HPO_4 \cdot 2H_2O$, 0.34 g KH_2PO_4/2L.
3. Developing solution for black plaque assay: Dissolve 10 mg of 4-chloro-1-napthol in 1 mL of ethanol and add to 99 mL of water. Just before use, add 0.1 mL of 3% hydrogen peroxide and add the mixture to the cells.

3. Methods

3.1. Production of Recombinant Viruses

To isolate recombinant virus, wild-type HSV DNA and a recombinant plasmid or restriction fragment containing the gene of interest are introduced into cells by calcium phosphate mediated DNA transfection *(38)*.

1. Plate BHK-21 cells *(39)* at a density of 5×10^5 cells on a 90-mm plate.
2. One hour prior to transfection, remove medium and replace with 5 mL of fresh medium.
3. Make up a mixture containing 0.1 µg of purified HSV virion DNA, 0.2 µg of plasmid containing the gene of interest, and 31 µL of 2*M* $CaCl_2$ in a total volume of 0.25 mL.
4. Add the mixture dropwise to 0.25 mL of 2× HBS buffer, pH 7.12.
5. Immediately add the mixture to the medium on the cells.
6. After 4 h, remove the medium, wash the cells with PBS, and add 10 mL of fresh medium.
7. Incubate for a further 36 h and harvest virus exactly as when growing wild-type virus.

The resulting viral stock can then be assayed for the presence of recombinant viruses, as described below.

3.2. Assay of Recombinant Viruses

3.2.1. Selection of tk⁻ viruses

In cases in which *tk* has been inactivated, recombinant viruses can be detected and isolated by growth in deoxynucleotide analogs that are phosphorylated by the wild-type thymidine kinase and are hence incorporated into its DNA, blocking DNA replication. Two methods can be used to do this.

3.2.1.1. Selection in Bromodeoxyuridine

The procedure used is that described by Dubbs and Kit *(17)*.

1. Plate 2×10^6 L cells deficient in thymidine kinase (Ltk^-) in normal medium supplemented with 30 µg/mL of bromodeoxyuridine (BUdR).
2. Infect with the progeny of transfection at a multiplicity of 0.001 plaque-forming units/cell.
3. Leave for 3–4 d and harvest progeny virus.
4. Repeat steps 1–3 at least four times, using the progeny of each cycle to infect Ltk^- cells.
5. Titer the resulting tk^- virus on Ltk^- cells as described in the next chapter, using medium containing BUdR, and isolate virus from an individual plaque. This virus can be cloned by repeated replating at low dilution and picking of individual plaques.

3.2.1.2. Selection in Acyclovir

This procedure is essentially identical to that described above, with the exception that, because acyclovir, unlike BUdR, is not phosphorylated by the host-cell thymidine kinase, it is not necessary to perform these experiments using a tk^- cell line. Hence, any HSV-permissive cell, such as BHK-21, HeLa, or Vero, can be used with medium containing 0.2 m*M* acyclovir (acycloguanosine from Sigma).

3.2.2. The Black Plaque Test

When the recombination event has resulted in the inactivation of the gene encoding a protein against which an antibody is available, recombinants can be identified by the black plaque test. This method was originally described by Smith et al. *(40)* and has been modified by Holland et al. *(41)*.

1. Plate cells at a density of 2×10^6 cells on a 90-mm plate in DMEM containing 2% fetal calf serum and 0.5% carboxymethyl cellulose.
2. Infect with 1000 plaque-forming units (PFU) of the viral mixture.
3. Incubate at 37°C for 3 d.
4. Wash once with PBS.

5. Fix in 5 mL of PBS containing 0.25% glutaraldehyde for 5 min at room temperature.
6. Wash three times in PBS.
7. Incubate for 2 h at room temperature in an appropriate dilution of the antibody in PBS containing 3% bovine serum albumin (BSA).
8. Wash three times in PBS.
9. Incubate for 2 h at room temperature in a 1:2500 dilution of sheep antimouse (or antirabbit as appropriate) immunoglobulin in PBS containing 3% BSA.
10. Wash three times in PBS.
11. Add 2 mL of developing solution and incubate for 20 min.
12. Examine plaques; cells infected with wild-type virus and expressing intact viral protein on their surface will turn black, whereas cells infected with recombinant virus not expressing the protein remain colorless.

3.3. Isolation and Propagation of Defective Virus

1. Recombinant plasmid containing the defective-virus vector linked to the gene of interest is isolated by standard recombinant DNA techniques.
2. Introduce 0.25 μg of recombinant DNA into BHK-21 cells as described above (Section 3.1., steps 1–6).
3. Six hours after transfection, infect cells with wild-type HSV-1 or HSV-1 strain tsK at a multiplicity of 5 PFU/cell.
4. Incubate for a further 36 h at 37°C for wild type HSV-1 or 31°C (at permissive temperature) for tsK.
5. Harvest virus as for wild-type virus.

The resulting virus stock should contain both nondefective helper virus end defective viruses containing the concatamerized recombinant DNA. These can be maintained by passaging the mixed viral stock at a high multiplicity of infection or by coinfecting with nondefective virus. Viruses isolated and propagated with a tsK helper virus are grown at 31°C, which is the permissive temperature for the virus. This therefore allows growth of the helper virus and, hence, of the defective virus. The presence of recombinant defective virus genomes in the stock can be assayed either by isolating viral DNA from infected cells (*see* next chapter) followed by Southern blotting *(21)*, or by assaying infected cell extracts for the product of a gene carried by the defective virus, such as that encoding chloramphenicol acetyl transferase *(22)*.

To obtain high-level expression from the defective virus vector, the viral stock is used to infect cells at a multiplicity of 5 PFU/cell. Twelve hours after infection, cells are harvested and assayed for expression of the appropriate antigen. In the case of defective virus carrying an immediate-early gene promoter driving expression of the gene of interest *(22)* and using tsK helper virus, this infection is carried out at 38.5°C. This produces high-level expression from the immediate-early promoter, but prevents a full lytic cycle and cell death, since 38.5°C is the nonpermissive temperature for tsK.

4. Notes

HSV has a number of advantages as a potential viral vector, including its large size, wide host range, and the high-level expression of its proteins in lytic infection. Its use as a vector system has lagged behind the smaller DNA and RNA viruses, however, principally because of the need for complex manipulations to introduce foreign DNA into the viral genome by recombination and to subsequently isolate recombinants. With the construction of defective viral vectors that are both easy to manipulate and capable of replicating and packaging foreign DNA, it is likely that HSV-based systems will grow in importance and that the advantages of these systems particularly when high-level expression is desired, will be more widely appreciated.

Moreover, it is possible that in the future, in addition to its role as a vector for large-scale protein production in vitro, HSV may become the vector of choice for the introduction of functional genes into individuals with genetic diseases involving defects in neuronal cells. Thus, as described above (Section 1.1.), HSV can establish life-long asymptomatic infections in neuronal cells in vivo and, hence, unlike retroviral vectors, can stably deliver DNA into these nonreplicating cells.

Indeed, recent studies *(37,42)* using both replicating and defective HSV vectors have established that foreign DNA introduced in this way is stably expressed in neuronal cells. Thus, Geller and Breakfield *(37)* introduced a defective HSV-1 vector in whlch a viral immediate-early promoter drives a β-galactasidase gene into primary neuronal cells in conjunction with the HSV-1 mutant tsK (*see* Section 1.2.) and obtained stable expression of β-galactosidase. Similarly, Palella et al. *(42)* used a nondefective vector into which the human hypoxanthine-guanine phosphoribosyl transferase gene (*hgprt*) had been introduced by recombination to express the corresponding protein for at least 35 d in an infected rat neuroma. This latter example is of particular importance in that a defect in this gene is the cause of the human disease Lesch-Nyhan syndrome, which involves profound neurological symptoms and hence is likely to be corrected only by expression of the corresponding protein in neuronal cells rather than in circulating lymphocytes or other cell types.

The growth of simple techniques for the introduction of genes into HSV thus offers an exciting prospect for the use of HSV-based vectors in the future, involving both the expression of foreign proteins in culture and, ultimately, the possibility of gene therapy for the many genetic diseases involving the nervous system.

References

1. Roizman, B. (1979) The organization of herpes simplex virus genomes. *Ann. Rev. Genet.* **13,** 25–57.
2. Spear, P. G. and Roizman, B (1980) Herpes simplex viruses, in *DNA Tumour Viruses* (Tooze, J., ed.), 2nd Ed., Cold Spring Harbor Laboratory, Cold Spring Harbor, NY, pp. 615–746.
3. Elion, G. B., Furman, P. A., Fyfe, J. A., de Mirando, P., Beauchamp, L., and Shaeffer, H. J. (1977) Selectivity of action of an anti-herpetic agent 9- (2-hydroxyethoxymethyl) guanine. *Proc. Natl. Acad. Sci., USA* **74,** 5716–5720.
4. Roizman, B. and Sears, A. E. (1987) An inquiry into the mechanisms of herpes simplex virus latency. *Ann. Rev. Microbiol.* **41,** 543–571.
5. Latchman, D. S. (1990) Molecular biology of Herpes simplex virus latency. *J. of Exp. Pathol.* **71,** 133–141.
6. Vahlne, A., Svennerholm, B., and Lycke, E. (1979) Evidence of herpes simplex virus type-selective receptors on cellular plasma membranes. *J. Gen. Virol.* **44,** 217–225.
7. Morgan, C., Rose, W. M., and Mednis, B. (1968) Electron microscopy of herpes simplex virus I Entry. *J. Virol.* **2,** 507–516.
8. Honess, R. W. and Roizman, B. (1974) Regulation of herpes virus macromolecular synthesis. I. Cascade regulation of three groups of viral proteins. *J. Virol.* **14,** 8–19.
9. Watson, R. J., Preston, C. M., and Clements, J. B. (1979) Separation and characterization of herpes simplex virus type I immediate-early mRNAs. *J. Virol.* **31,** 42–52.
10. Campbell, M. E. M., Palfreyman, J. W., and Preston, C. M. (1984) Identification of Herpes simplex virus DNA sequences which encode a trans-acting polypeptide responsible for stimulation of immediate-early transcription. *J. Mol. Biol.* **180,** 1–19.
11. O'Hare, P. and Goding, C. R. (1988) Herpes simplex virus regulatory elements and the immunoglobulin octamer domain bind a common factor and are both targets for virion transactivation. *Cell* **52,** 435–445.
12. Pereira, L., Wolff, M. H., Fenwick, M., and Roizman, B. (1977) Regulation of herpes virus macromolecular synthesis V. Properties of alpha polypeptides made in HSV-1 and HSV-2 infected cells. *Virology* **77,** 733–749.
13. Preston, C. M. (1979) Control of Herpes simplex virus Type 1 mRNA synthesis in cells infected with wild-type virus or the temperature sensitive mutant tsk. *J. Virol.* **29,** 275–285.
14. Sears, A. E., Halliburton, I. W., Meignier, B., Silver, S., and Roizman, B. (1985) Herpes simplex virus type 1 mutant deleted in the a 22 gene: Growth and gene expression in permissive and restrictive cells and establishment of latency in mice. *J. Virol.* **55,** 338–346.
15. Sacks, W. R., Greene, C. C., Aschman, D. P., and Schaffer P. A. (1985) Herpes simplex virus type 1 ICP27 is an essential regulatory protein. *J. Virol.* **55,** 796–805.
16. Stow, N. D. and Stow, E. C. (1986) Isolation and characterization of a herpes simplex virus type 1 mutant containing a deletion within the gene encoding the immediate-early polypeptide Vmw110. *J. Gen. Virol.* **67,** 2571–2585.
17. Dubbs, D. R. and Kit, S. (1964) Mutant strains of herpes simplex virus deficient in thymidine kinase-inducing activity. *Virology* **22,** 493–502.

18. Holland, L. E., Anderson, K. P., Shipman, C., and Wagner, E. K. (1980) Viral DNA synthesis is required for the efficient expression of specific herpes simplex virus type 1 mRNA species. *Virology* **101**, 10–24.
19. Poffenberger, K. L. and Roizman, B. (1985) Studies on a non-inverting genome of a viable herpes simplex virus. *J. Virol.* **53**, 589–595.
20. Vlazny, D. A. and Frenkel, N. (1981) Replication of herpes simplex virus DNA: localization of replication recognition signals within defective virus genomes . *Proc. Natl. Acad. Sci., USA* **78**, 742–746.
21. Spaete, R. R. and Frenkel, N. (1982) The herpes simplex virus amplicon: A new eukaryotic defective-virus cloning-amplifying vector. *Cell* **30**, 295–304.
22. Stow, N. D., Murray, M. D., and Stow, E. L. (1986) Cis-acting signals involved in the replication and packaging of herpes simplex virus type-1 DNA. *Cancer Cells* **4**, 497–507.
23. Jacob, R. J., Morse, L. S., and Roizman, B. (1979) Anatomy of herpes simplex virus DNA XII. Accumulation of head to tail concatamers in nuclei of infected cells and their role in the generation of the four isomeric arrangements of viral DNA. *J. Virol.* **29**, 448–457.
24. Mocarski, E. S. and Roizman, B. (1982) Structure and role of the herpes simplex virus DNA termini in inversion, circularization and generation of virion DNA. *Cell* **31**, 89–97.
25. Deiss, L. P. and Frenkel, N. (1986) Herpes simplex virus amplicon: Cleavage of concatameric DNA is linked to packaging and involves amplification of the terminally reitreated a sequence. *J. Virol.* **57**, 933–941.
26. Roizman, B. and Jenkins, F. J. (1985) Genetic engineering of novel genomes of large DNA viruses. *Science* **229**, 1208–1214.
27. Sheldrick, P., Laithier, M., Lando, D., and Ryliner, M. L. (1973) Infectious DNA from herpes simplex virus: Infectivity of double stranded and single stranded molecules. *Proc. Natl. Acad. Sci., USA* **70**, 3621–3625.
28. Brown, S. M., Ritchie, D. A., and Subak-Sharpe, J. H. (1973) Genetic studies with herpes simplex virus type 1. *J. Gen. Virol.* **18**, 329–346.
29. Tackney, C., Cachianes, G., and Silverstein, S. (1984) Transduction of the Chinese hamster ovary *aprt* gene by herpes simplex virus *J. Virol.* **52**, 606–614.
30. Stenberg, R. and Pizer, L. I. (1982) Herpes simplex virus-induced changes in cellular and adenovirus RNA metabolism in an adenovirus-type 5 transformed human cell line. *J. Virol.* **42**, 474–487.
31. Kemp, L. M. and Latchman, D. S. (1988) Induction and repression of cellular gene transcription during herpes simplex virus infection are mediated by different viral immediate-early gene products. *Eur. J. Biochem.* **174**, 443–449.
32. Shih, N. T., Arsenakis, M., Tiollias, P., and Roizman, B. (1984) Expression of hepatitis B virus S gene by herpes simplex virus alpha and beta regulated gene chimaeras. *Proc. Natl. Acad. Sci., USA* **81**, 5867–5870.
33. Whealy, M. E., Baumeister, K., Robbins, A. K., and Enquist, L. W. (1988) A herpes virus vector for expression of glycosylated membrane antigens: Fusion proteins of pseudorabies virus gIII and human immunodeficiency virus type 1 envelope glycoproteins. *J. Virol.* **62**, 4185–4194.
34. Desrasiers, R. C., Kamine, J., Bakker, A., Silva, D., Woychick, R. P., Sakai, D. D., and Rottman, F. M. (1985) Synthesis of bovine growth hormone in primates by using an herpesvirus vector. *Mol. Cell. Biol.* **5**, 2796–2803.
35. Frenkel, N. (1981) Defective interfering herpesvirus, in *The Human Herpesviruses–An*

Interdisciplinary Prospective (Nahmias A. J., Dowdle, W. R., and Schinazy, R. S., eds.), Elsevier, New York, pp. 91–120.
36. Kwong, A. D. and Frenkel, N. (1985) The herpes simplex virus amplicon *IV*: Efficient expression of a chimeric chicken ovalbumin gene amplified within defective virus genomes. *Virology* **142**, 421–425.
37. Geller, A. I. and Breakfield, X. O. (1988) A defective HSV-1 vector expresses *Escherichia coli* beta-galactosidase in cultured peripheral neurons. *Science* **241**, 1667–1669.
38. Gorman, C. M. (1985) High efficiency gene transfer into mammalian cells, in *DNA Cloning*, (Glover, D. M., ed.) Vol 2, IRL, Oxford, pp. 143–190.
39. MacPherson, I. and Stoker, M. (1962) Polyoma transformation of hamster cell clones—an investigation of the genetic factors affecting cell competence. *Virology* **16**, 147–151.
40. Smith, K. O., Kennel, W. L., and Lamn, D. L. (1981) Visualization of minute centres of viral infection in unfixed cell cultures by an enzyme linked antibody assay. *J. Immnol. Methods* **40**, 294–305.
41. Holland, T. C., Sandri-Goldin R. M., Holland, S. E., Marlin, S. D., Levine, M., and Glorioso, J. C. (1983) Physical mapping of the mutation in an antigenic variant of herpes simplex virus type 1 by use of an immunoreactive plaque assay. *J. Virol.* **46**, 649–652.
42. Palella, T. D., Silverman, L. J., Schroll, C. T., Homa, F. L., Levine, M., and Kelley, W. N. (1988) Herpes simplex virus-mediated human hypoxanthine-guanine phosphoribosyl transferase gene transfer into neuronal cells. *Mol. Cell. Biol.* **8**, 457–460.

CHAPTER 17

Growth of Herpes Simplex Virus and Purification of Viral DNA

David S. Latchman and Lynn M. Kemp

1. Introduction

In order to use herpes simplex virus (HSV) as a vector for the transmission and expression of foreign genes, it is obviously necessary to be able to prepare stocks of the virus and to propagate both HSV and the recombinant viruses derived from it. Similarly, the introduction of foreign genes into the virus will require the purification of viral DNA for use as a cloning vector, and preparation and analysis of viral DNA will also be necessary in order to characterize the recombinant viruses produced during the cloning procedures. This chapter will therefore discuss in turn the procedures used in our laboratory for the growth of HSV, and for the preparation of viral DNA.

2. Materials

2.1. Medium

For routine culture of infected BHK cells, we use Dulbecco's Modified Eagle's Medium (DMEM) supplemented with 10% newborn calf serum and 4.5 g of glucose/L. For freezing virus stocks, use serum-free medium.

2.2. Solutions

1. Phosphate-buffered saline: 12.2 g NaCl, 0.27 g KCl, 1.92 g $Na_2HPO_4 \cdot 2H_2O$, and 0.34 g KH_2PO_4/ 2 L.

From: *Methods in Molecular Biology, Vol. 8: Practical Molecular Virology: Viral Vectors for Gene Expression*
Edited by: M. Collins © 1991 The Humana Press Inc., Clifton, NJ

2. PEST buffer: 0.1*M* EDTA, 50 m*M* Tris-HCl, 0.5% SDS, 0.2 mg/mL proteinase K, pH 8.0.
3. TE buffer: 10 m*M* Tris-HCl, pH 7.5, 1 m*M* EDTA.
4. Saturated sodium iodide: Dissolve sodium iodide at 25°C in 10 m*M* Tris-HCl, 10 m*M* EDTA, pH 8.0, containing 50 mg of sodium bisulfite/100 mL.

3. Methods

3.1. Preparation of Virus Stocks

3.1.1. Cells

For most strains of virus, either Hep-2 cells *(1)* or BHK-21 cells (clone 13: ref. *2*) are suitable, provided that they are mycoplasma-free. We routinely use BHK-21 cells in our laboratory. These cells are grown in DMEM supplemented with glucose and newborn calf serum.

3.1.2. Virus

In order to avoid repeated passaging of virus, which can lead to the accumulation of defective viruses of the type described in Chapter 16, it is necessary to prepare a master stock of each virus. This master stock is prepared from a single viral plaque as described in Section 3.2. and is stored at –70°C in serum-free medium. This stock is used in the same manner to prepare a submaster stock, which in turn is used to prepare the working stocks for infection of cells. All stocks are maintained in serum-free medium at –70°C, which will prevent loss of titer.

3.2. Growth of Virus

In order to prepare virus, cells are infected at a low multiplicity of infection (moi) of 0.001 plaque-forming units (PFU) per cell. This prevents the accumulation of defective viruses that occurs in stocks grown at high multiplicities. A high moi (≥1 PFU/cell) should be used only in investigations of the effect of the virus on the infected cell or when it is necessary to express high levels of protein from a recombinant virus in the infected cell. The method described below for growing HSV (modified from Watson et al. *[3]*) is used routinely in our laboratory. For viral growth on a very large scale, a similar method can be used, employing roller bottles containing 5×10^8 cells instead of the dishes normally used.

1. Grow BHK cells in supplemented DMEM. Seed 2×10^6 cells onto a 10-cm^2 dish in 10 mL of medium and leave overnight.
2. The next day, remove the medium and add 1 mL of serum-free medium containing 0.001 PFU of virus/cell. Leave for 1 h, agitating occasionally.
3. Add 10 mL of medium containing serum and leave for 2–3 d at 37°C,

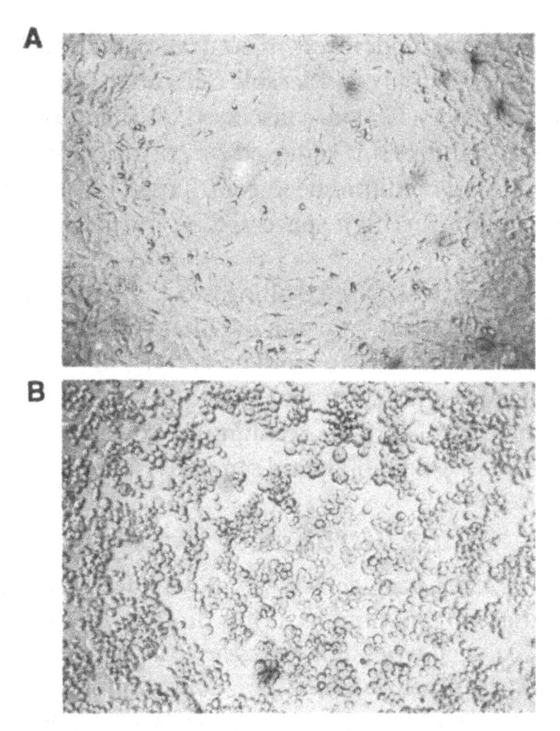

Fig. 1. (A) Uninfected BHK cells growing as a confluent monolayer, and (B) infected BHK cells showing a typical cytopathic effect.

until a cytopathic effect is seen in the cultured cells. The time for this effect to develop varies with different strains of virus and must be determined empirically for each strain used. Cells showing a typical cytopathic effect are illustrated in Fig. 1.

4. Harvest the cells by scraping into the medium using a sterile rubber policeman, and collect the cells by low-speed centrifugation (3000*g* for 10 min). These infected cells can then be resuspended in a smaller vol to concentrate the stock. The amounts of virus that are released into the medium and that remain cell-associated are dependent on the precise strain of virus and cell type used. In virus/cell combinations in which significant lysis occurs, the culture supernatant may have to be used as part of the stock. In most cases, however, the majority of virus remains cell-associated and can be harvested as described.
5. Once infected cells have been harvested, it is necessary to release the virus and obtain a cell-free preparation. In our laboratory, this is achieved by three cycles of freeze-thawing in a dry ice/ethanol bath and a 37°C

water bath. Other methods involving sonication of the cell pellet require a dedicated sonicator and considerable calibration, and are not routinely used in our laboratory. They are discussed by Killington and Powell *(4)*.

6. Following lysis, cell debris is removed by centrifugation at high speed, either in a standard medium-speed centrifuge (10,000*g* for 10 min) or, for small vol, using the high speed setting on the MSE microcentaur minicentrifuge.
7. To ensure sterility, filter the resulting stock through a Nalgene filter (diameter 0.45 μm). Take care that the filter does not become blocked, which would cause pressure to build up and the filter to rupture, resulting in a danger of exposure to the virus.
8. As with master stocks, store the resulting stock at –70°C. When needed, stocks should be thawed rapidly and repeated freeze-thaw cycles avoided.

3.3. Titration of Virus

Once the virus stock has been obtained, its infectivity must be assayed before it can be used in experiments. If it is thought that viral titer will vary considerably between cell types, it is necessary to titrate the virus in the cell type in which it will eventually be used. For routine applications however, we titrate the virus in a monolayer assay on Vero cells, which normally results in the appearance of large, easily counted plaques. Whatever cell line is employed, it is necessary that the conditions that are used allow the cells to maintain a confluent monolayer for the time taken for plaques to appear, i.e., a mimimum of 2–3 d, and up to a week for some temperature-sensitive mutants. This can be achieved by reducing the level of serum in the medium and is also critically dependent on the brand of plasticware used, which affects the ability of the cells to remain adherent for long periods. The conditions described below work well for Vero cells and for all viral strains tested.

1. Seed 10^6 Vero cells/well into 35-mm wells (use Flow, Linbro 6-well dishes) 24 h prior to use. This should give a confluent monolayer when required.
2. Remove the medium and add 0.1 mL of each virus dilution in serum-free medium to individual wells and leave for 1 h to adsorb.
3. Add 5 mL of medium containing 2% fetal calf serum and 0.5% carboxymethyl cellulose. Avoid disturbance and incubate for several days (normally 2–3 d) until discernable plaques appear.
4. When plaques are visible, remove the medium and rinse the wells gently to remove all the carboxymethyl cellulose.
5. Fix the cells briefly in methanol/glacial acetic acid (3:1).
6. Stain cells briefly with a solution of crystal violet (0.7% in 50:50

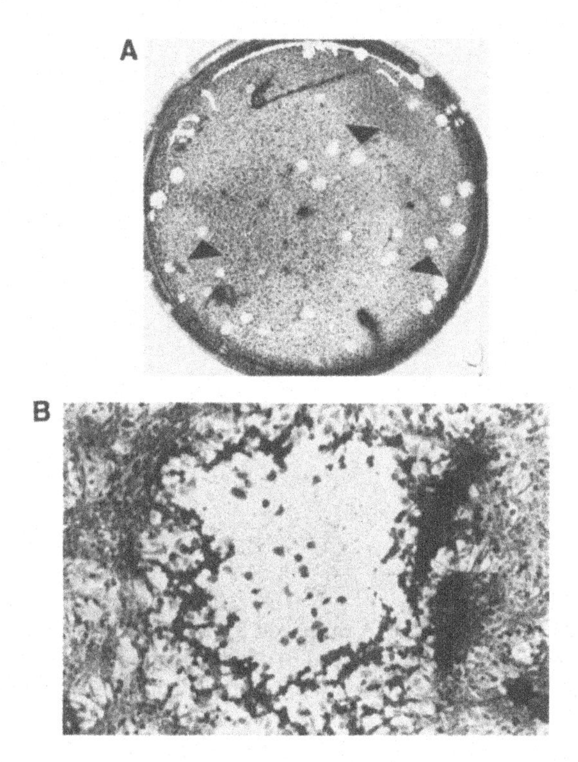

Fig. 2. (A) Plaque assay showing a stained cell monolayer punctuated by typical plaques (indicated by arrows). (B) A higher magnification of a single plaque.

methanol:water), and wash with water.

7. Plaques can be counted by eye or with the aid of a low-power objective. Typical plaques are illustrated in Fig. 2.
8. Viral titer is calculated on the basis of the number of plaques formed in the various dilutions.

In addition to the monolayer assay described above, viral titer can also be measured by a suspension method, as described by Killington and Powell (*4*).

3.4. Preparation of Viral DNA

A variety of methods are available for the preparation of viral DNA. These methods differ in the degree of purity of viral DNA that is obtained and generally increase in complexity with increasing purity. Therefore, for each individual application, it is necessary to determine the degree of purity required and, thus, the optimal method to be used. The methods available are described here in order of increasing purity of the DNA obtained.

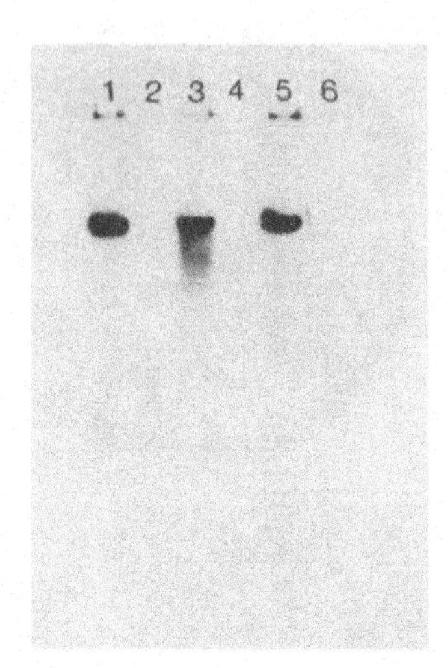

Fig. 3. Southern blot of total DNA prepared from HSV-infected cells (tracks 1, 3, and 5) or uninfected cells (tracks 2, 4, and 6) after digestion with various restriction enzymes. The blot has been hybridized with a probe from the U_L region of the viral genome.

3.5. Preparation of Total DNA from Infected Cells

For many applications, such as checking the structure of a recombinant virus, a crude DNA preparation containing cellular as well as viral DNA is quite sufficient. Thus, if total DNA is prepared from infected cells showing a cytopathic effect, and digested with appropriate restriction enzymes, viral DNA bands are readily seen following gel electrophoresis within the smear of cellular DNA. The structure and size of particular regions of the viral DNA genome can readily be determined by Southern blotting of this DNA preparation and hybridizing with appropriate probes. As a negative control, DNA can also be prepared from uninfected cells and included in the blot hybridization. A typical blot obtained in this way is illustrated in Fig. 3.

The method used in our laboratory to isolate total DNA from infected cells is given below. A similar method has been used by De Luca et al. *(5)*.

1. Infect 10^7 cells at 0.1 PFU/cell and leave at 37°C until a cytopathic effect is visible.
2. Remove medium and wash the cells in PBS.
3. Scrape the cells into the PBS and pellet by low-speed centrifugation (3,000 *g* for 10 min).

4. Incubate the cells at 37°C overnight in PEST buffer.
5. Extract the solution once with phenol and once with phenol/chloroform/isoamylalcohol (25:24:1).
6. Precipitate by adding of 2 vol of ice-cold ethanol, and leave overnight at –20°C.
7. The DNA is harvested by centrifugation (10000*g* for 20 min), washed once in 70% ethanol, dried under vacuum, and gently resuspended in an appropriate vol of TE.

The DNA is now ready for digestion with restriction enzymes or for other applications.

3.6. Purification of Viral DNA from Infected Cells

If required, viral DNA in a form that is only minimally contaminated with cellular DNA can be achieved by centrifugation of infected cell extracts on sodium iodide gradients, which results in the separation of the denser viral DNA from the less dense cellular DNA. The method used is a modification of that of Walboomers and Schegget *(6)*, who originally described this method.

1. Infect monolayers of cells at 0.1 PFU/cell and leave until the cytopathic effect has developed.
2. Remove the medium and wash cells twice in ice-cold PBS.
3. Scrape the cells into PBS and harvest by low-speed centrifugation.
4. Resuspend the pellet in TE and add SDS (to a final concentration of 0.8%).
5. Gently agitate the suspension and add proteinase K (to a final concentration of 2 mg/mL). Incubate for 3 h at 37°C.
6. Add a further aliquot of proteinase K to increase the concentration to 4 mg/mL, and incubate overnight at 37°C.
7. Add saturated sodium iodide solution to a final density of 1.525 g/mL, and ethidium bromide, to 30 µg/mL.
8. Centrifuge at 44,000 rpm in a type 65 Beckman (Palo Alto, CA) rotor or equivalent for 36 h at 20°C.
9. After centrifugation, the DNA may be seen by side illumination with long-wavelength UV light. Two bands are normally observed, and the lower (viral) band is collected by side puncture using a syringe with a 21-gage needle (*see* Fig. 4).
10. The DNA is extracted three times with isobutanol (presaturated with sodium iodide in water at a ratio of 1 mg to 1 mL) and precipitated with 2 vol of ethanol overnight at –20°C.

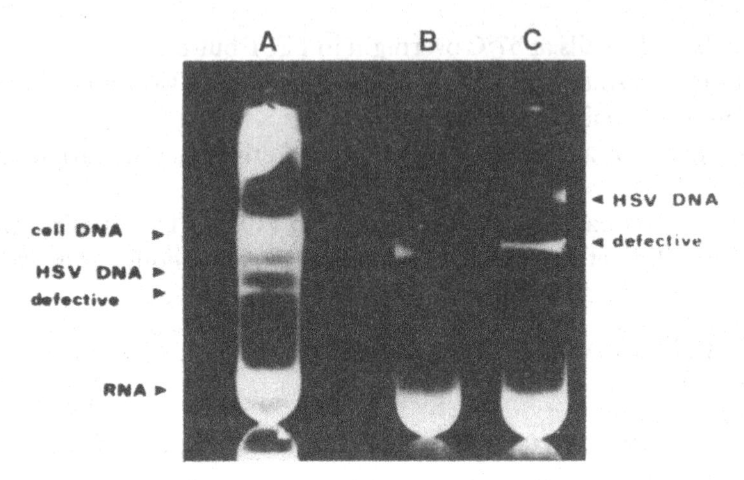

Fig. 4. Purification of viral DNA on sodium iodide gradients. Track A shows the result of the initial centrifugation; tracks B and C show the result of recentrifugation of the HSV DNA band obtained in A. Illustration provided by N. B. LaThangue.

11. The resulting precipitate is harvested, washed, and resuspended as previously described.

This method routinely yields viral DNA that is more than 95% pure, and is of sufficient purity for most applications.

3.7. Isolation of Viral DNA from Purified Virions

The DNA obtained by sodium iodide centrifugation will contain all the forms of viral DNA that are present in the cell and will be contaminated with low levels of cellular DNA. If it is desired to prepare highly purified viral DNA, representative of that which is packaged into the virion, it is necessary to purify the virions and then prepare DNA from them. In many combinations of particular viral strains, such as Kos or HFEM, and specific cell types, such as Hep-2 or Vero cells, it is possible to purify virions in reasonable quantities simply by centrifugation of the culture supernatant as described, for example, by Goldin et al. *(7)*:

1. Infect cells and incubate at 37°C until cytopathic effect is generalized.
2. Scrape the cells off into the culture medium and spin at 3000*g* for 15 min to spin out the cells.
3. Take the supernatant and centrifuge at 7500*g* for 4 h at 4°C. This is the extracellular virus pellet, which can be used for the preparation of viral DNA.

In some other virus/cell combinations, however, relatively little virus will be obtained in this way, because the majority of the virus remains cell-associated. The complex methods that are available for the purification of such cell-associated virions are described by Killington and Powell *(4)*, Spear and Roizman *(8)*, and Peden et al. *(9)*.

Once virions have been isolated by one of these means, viral DNA can be isolated by the method described by Sandri-Goldin et al. *(10)*.

1. Resuspend the virion pellet in TE buffer and lyse by addition of SDS, to a final concentration of 1%, and *N*-lauryl sarcosinate, to a final concentration of 0.5%.
2. Incubate the suspension with a final concentration of 0.5 mg/mL of ribonuclease A at 37°C for 1 h.
3. Add proteinase K to the suspension to a final concentration of 2 mg/mL, and incubate at 37°C for 12–16 h.
4. Adjust the final vol of the lysate to 25 mL and add 32.5 g of cesium chloride. The refractive index of the resulting solution should be checked and should be between 1.4005 and 1.4010.
5. Centrifuge the lysate for 20 h at 40,000 rpm in a Beckman VTi 50 rotor at 20°C.
6. After centrifugation, bottom-puncture the tubes and collect 1-mL fractions. The viral DNA peak will be present at a density between 1.72 and 1.74 (as determined by refractive index), and these fractions are pooled.
7. Repeat the centrifugation twice.
8. Dialyze the resulting viral DNA extensively against several changes of TE buffer.
9. Harvest the DNA by ethanol-precipitation at –20°C overnight and centrifugation at 20,000 rpm at 4°C for 1 h.
10. Gently resuspend the DNA in TE buffer.

4. Notes

In general, the wide host range and virulence of HSV makes it an easy virus to grow, even for those without previous experience. Similarly, although the methods available for purifying viral DNA are often difficult and complex, this can usually be overcome by choosing one of the the methods described that gives the minimal purity required for any particular purpose.

References

1. Spear, P. G. and Roizman, B. (1968) The proteins specified by herpes simplex virus. 1. Time of synthesis, transfer into nuclei and properties of proteins made in productivity infected cells. *Virology* **36,** 545–555.
2. Macpherson, I. and Stoker, M. (1962) Polyoma transformation of hamster cell clones —an investigation of genetic factors affecting cell competence. *Virology* **16,** 147–151.

3. Watson, D. A., Sheldon, W. I. H., Elliott, A., Tetsuka, T., Wildy, P., Bourgaux–Ramoisy, D., and Gold, E (1966) Virus specific antigens in mammalian cells infected with herpes simplex virus. *Immunology* **11,** 399–408.
4. Killington, R. A. and Powell, K. L. (1985) Growth, assay and purification of herpes viruses, *Virology, a Practical Approach* (Mahy, B. W. J., ed.), IRL, Oxford, UK, pp. 207–236.
5. De Luca, N. A., Courtney, M. A., and Schaffer, P. A. (1984) Temperature sensitive mutants in herpes simplex virus type I ICP4 permissive for early gene expression. *J. Virol.* **52,** 767–776.
6. Walboomers, J. M. M. and Schegget, J. T. (1976) A new method for the isolation of herpes simplex virus type 2 DNA. *Virology* **74,** 256–258.
7. Goldin, A. L., Sandri–Goldin, R. M., Levine, M., and Glorioso, J. C. (1981) Cloning of herpes simplex virus type I sequences representing the whole genome. *J. Virol.* **38,** 50–58.
8. Spear, P. G. and Roizman, B (1972) Proteins specified by herpes simplex virus. V Purification and structural proteins of the herpes virion. *J. Virol.* **9,** 143–159.
9. Peden, K., Mounts, P., and Hayward, G. S. (1982) Homology between mammalian cell DNA sequences and human herpes virus genomes detected by a hybridization procedure with high complexity probe. *Cell* **31,** 71–80.
10. Sandri-Goldin, R. M., Levine, M., and Gloriosa, J. C. (1981) Method for induction of mutations in physically defined regions of the herpes simplex virus genome. *J. Virol.* **38,** 41–49.

CHAPTER 18

Vectors Based on Epstein-Barr Virus

Paul M. Brickell and Mukesh Patel

1. Introduction

Vectors containing elements of the Epstein-Barr virus (EBV) genome are primarily used to maintain cloned DNA inserts as plasmids in mammalian cells. In addition, EBV-based vectors are proving to be valuable tools in elucidating details of EBV biology that have long eluded students of the virus. In this chapter, we will review those characteristics of EBV and its life cycle that have been used in vector construction and will describe methods that are particularly applicable to the use of EBV-based vectors.

1.1. The Life Cycle of EBV

1.1.1. General

EBV is a herpes virus comprising an icosahedral capsid within a lipid envelope. The virus particle contains a linear double-stranded genome of approx 172,000 bp, the entire sequence of which is known *(1)*. A diagram of the genomic organization of EBV, indicating those features relevant to vector construction, is presented in Fig. 1. More detailed maps have been published elsewhere *(2,3)*.

EBV naturally infects human B-lymphocytes and certain human epithelial cells, and it is the causative agent of infectious mononucleosis, the central feature of which is an EBV-induced polyclonal proliferation of B-lymphocytes. EBV is also associated with a number of human neoplasia, including Burkitt's lymphoma, which is a malignancy of B-lymphocytes, and nasopha-

From: *Methods in Molecular Biology, Vol. 8: Practical Molecular Virology: Viral Vectors for Gene Expression*
Edited by: M. Collins © 1991 The Humana Press Inc., Clifton, NJ

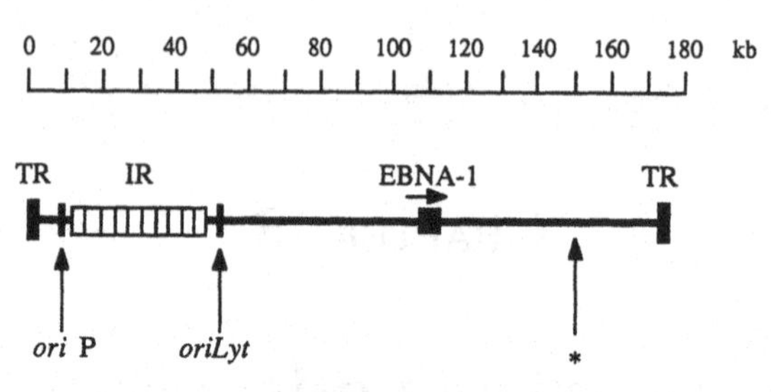

Fig. 1. Map of the genome of the B95-8 strain of EBV, showing features relevant to vector construction: The terminal repeats (TR), the internal repeats (IR), *oriP, oriLyt,* and the coding sequence of the EBNA-1 gene. Other strains of EBV have a second copy of *oriLyt* at the position indicated with an asterisk. This is deleted in the B95-8 strain.

ryngeal carcinoma *(3,4).* When normal B-lymphocytes from umbilical-cord blood or adult peripheral blood are infected with EBV in vitro, they become immortalized. Immortalized cell lines derived from such cultures share a range of phenotypic properties with antigen or mitogen-stimulated B-lymphoblasts, including the acquisition of B-lymphoblast cell-surface markers and the ability to secrete immunoglobulin *(4).* They are therefore called B-lymphoblastoid cell lines (B-LCL).

Once inside the infected cell, the virus particle is uncoated and the linear viral genome is converted into a covalently closed circle, probably as a result of base pairing between the terminal repeat sequences, shown in Fig. 1 *(5).* The genome then persists in the nucleus of the infected cell in the form of covalently closed circular supercoiled episomes with a copy number of between 5 and 500, depending on the cell line *(6).* The EBV plasmids replicate during the S phase of the cell cycle *(7).* Only rarely does the viral genome become integrated into the host-cell genome *(8).*

B-LCL cultures release infectious EBV particles in very small quantities. The majority of the infected cells remain latently infected, and only 1 in 10^4–10^6 cells switch from latent to lytic replication in each cell cycle *(9).* In latently infected cells, at least eight viral proteins are expressed from complex transcription units occupying some 100,000 bp of the viral genome. These are EBV-determined nuclear antigen-1 (EBNA-1), EBNA-2, EBNA-3, EBNA-4, EBNA-5 (also called EBNA-leader protein; EBNA-LP), the latent membrane protein (LMP), the BHRF-1 membrane protein, and the terminal protein (TP), the coding sequence of which is created by the circularization of the linear viral genome *(10).* In addition, two small RNA polymerase III transcripts, termed EBERs, are transcribed at high rates in latently infected cells. These are thought to facilitate translation of EBV mRNAs *(11).* Between them, these latent viral genes encode the functions required to establish and

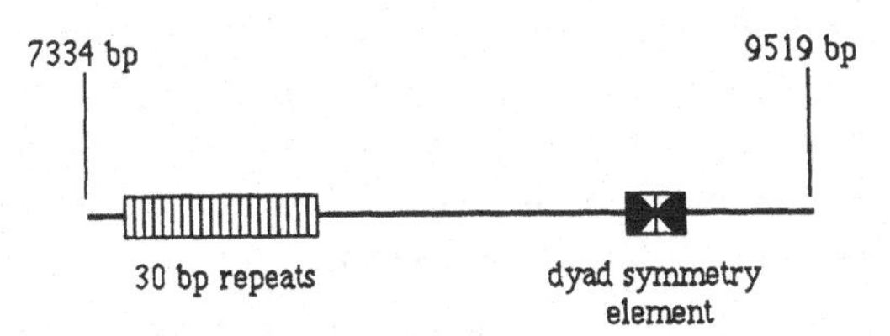

Fig. 2. Structure of *oriP*, which comprises 20 copies of a 30-bp repeat and a dyad symmetry element. Map positions on the B95-8 EBV strain genome are given in base pairs.

maintain latency, as well as those required to immortalize cells. The characteristics of the EBV latent proteins and of the genes encoding them have recently been reviewed elsewhere *(2,3)*. Their involvement in B-lymphocyte immortalization is reviewed elsewhere in this vol.

1.1.2. The Maintenance of EBV Episomes in Latently Infected Cells

Two viral functions are required to maintain EBV DNA as a plasmid. A *cis*-acting region, located within the BamHI C fragment of the EBV genome and designated *oriP* (*origin* of *plasmid* replication), permits extrachromosomal replication of linked DNA in cells that express EBNA-1 *(12)*.

As shown in Fig. 2, *oriP* contains 20 tandem copies of a 30-bp repeat, separated by approx 1000 bp from a 65-bp dyad symmetry element. This element can be drawn as a stem of 31 bp with three mismatches and a loop of three nucleotides, although there is no evidence that such a structure forms in vivo. Both the 30-bp repeat array and the dyad symmetry element are required for *oriP* to function as an origin of replication, although their relative orientations and the distance between them can be varied without loss of function *(13)*.

EBNA-1 is the only *trans*-acting factor required for replication of plasmids containing *oriP*. DNase I protection studies *(14)* have shown that EBNA-1 can bind to each of the 30-bp repeats and to four overlapping binding sites within and adjacent to the dyad symmetry region. This binding activity resides in the carboxyl-terminal third of the EBNA-1 protein *(14)*.

The behavior of EBNA-1 resembles that of SV40 T antigen, which binds to three sites within and adjacent to the dyad symmetry element of the SV40 origin of replication *(15)*. If this analogy can be pushed, it is likely that the binding of EBNA-1 to the dyad symmetry element facilitates the initiation of DNA replication within the vicinity of the dyad symmetry element. However, this has not yet been tested directly.

EBNA-1 further resembles SV40 T antigen in that binding of EBNA-1 can *trans*-activate the 30-bp repeat array of *oriP* to function as a transcriptional enhancer, both for heterologous promoters *(16)* and for at least one

EBV promoter *(17)*. Thus, the origins of plasmid replication of EBV and SV40 are both associated with transcriptional enhancer elements, as is that of bovine papilloma virus type 1 *(9)*. It is probable that, like SV40 T antigen, EBNA-1 plays a central role in coordinating viral DNA replication and gene expression, the complexities of which are just beginning to emerge *(18)*.

1.1.3. Lytic Replication of the EBV Genome

As noted above, a small fraction of latently infected B-lymphocytes spontaneously enter the lytic phase of viral replication. In these cells, most or all of the 90–100 viral genes are expressed, the viral genome is replicated and packaged, and infectious virus particles are released. Treatment of latently infected cultures with a variety of agents, including phorbol esters and sodium butyrate, can increase the proportion of cells that undergo this switch. It has also been discovered that the defective P3HR-1 strain of EBV, which cannot immortalize B-lymphocytes, is able to induce the lytic cycle upon superinfection of some EBV-infected B-lymphocyte lines. This property results from a genomic rearrangement in which the viral BZLF-1 gene, which is not normally expressed in latently infected cells, has been placed near to a promoter that is active during latency. Expression of the BZLF-1 gene alone has subsequently been shown to be sufficient to induce the lytic cycle in latently infected cells. It is believed that the BZLF-1 product may function by *trans*-activating the expression of three other viral genes (BMLF-1, BMRF-1, and BRLF-1), which themselves encode *trans*-activators of gene expression. These proteins presumably coordinate the activation of other genes necessary for viral DNA replication and virion assembly *(19)*. Perhaps the most striking feature of the EBV lytic phase identified to date is that the replication of EBV DNA during the lytic phase is mediated not by *oriP*, but by a distinct lytic origin of replication, termed *oriLyt*, which is located approx 40 kbp away from *oriP* (Fig. 1). The EBV genome is the only known eukaryotic replicon to contain two such functionally distinct origins of replication. The structure of *oriLyt* is complex and, as yet, not fully defined *(20)*.

1.2. Vectors Based on EBV

1.2.1. Vectors Based on EBV oriP *Replicons*

The most widely used methods for obtaining the stable expression of foreign genes in eukaryotic cells involve the integration of the foreign DNA into the host cell's chromosomes. This has a number of disadvantages. First, the expression of an integrated foreign gene may be suppressed or modified by the host DNA sequences flanking the integration site. Second, it is sometimes necessary to recover the introduced sequences from the host cell, and this generally involves the construction and screening of a genomic library, which is a time-consuming procedure. Considerable efforts have therefore

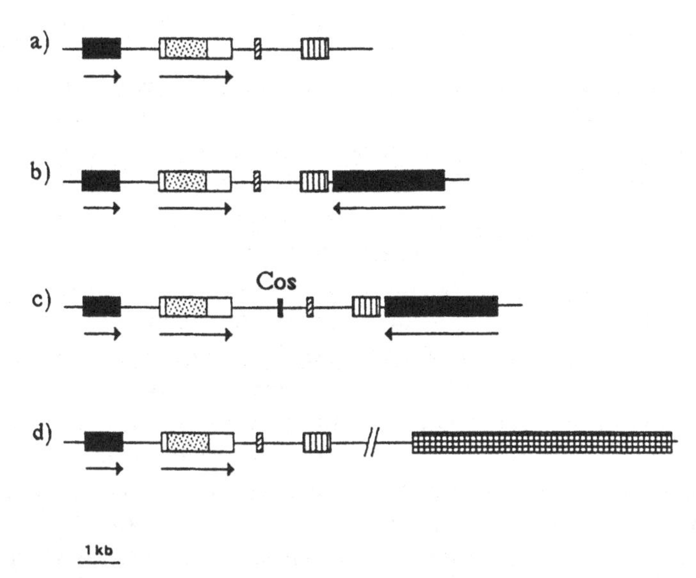

Fig. 3. (A) Struture of the EBV *oriP* vector pHEBo *(21)*. Symbols in all figures: Black box, pBR322 β-lactamase (ampicillin resistance) gene; stippled box, *E. coli hph* (hygromycin B resistance) gene; open boxes, HSV-1 thymidine kinase gene promoter sequence and polyadenylation signal; cross-hatched box, dyad symmetry element of *oriP;* vertically hatched box, 30-bp repeat element of *oriP*. The directions of transciption of the β-lactamase and *hph* genes are indicated with arrows. In all figures, the plasmid is shown as if linearized at the EcoRI site of pBR322. (B) Structure of a pHEBo derivative containing the EBNA-1 gene *(12)*. EBNA-1 sequences are represented by a black, cross-hatched box, and the direction of transcription is indicated with an arrow. (C) Structure of the cosmid vector cos202 *(22)*. The cos site derived from bacteriophage λ is indicated. (D) Structure of an *oriP*/*oriLyt* vector *(20)*. The checkered box represents a fragment of EBV containing *oriLyt*. This fragment lies approx 6 kbp away from *oriP* in the vector.

been directed toward developing vectors that are able to maintain themselves, as well as linked foreign DNA, as stable plasmids in eukaryotic cell nuclei. The first successful vectors of this type were based on BPV. However, these vectors suffered from a number of disadvantages. First, their host range was limited and, in particular, BPV plasmids were not efficiently maintained in human cells. Second, many of the vectors induced malignant transformation in the host cells, precluding their use for studying, for example, the effects of cloned oncogenes. The discovery that the EBV *oriP* region is capable of maintaining linked DNA in plasmid form in the presence of EBNA-1 led to the development of vectors based on EBV.

Sugden and colleagues *(21)* developed the shuttle vector pHEBo, depicted in Fig. 3a. This vector carries the pBR322 origin of replication and ampicillin-resistance gene to allow propagation and selection in *E. coli*. It also carries a fragment of EBV DNA containing *oriP*, which permits it to replicate as a plasmid in EBV-infected B-lymphoblasts containing EBNA-1. It was found that the traditional methods for selecting for stable transfectants were unsuitable for use with these cells *(21)*. Cells selected to be resistant to mycophenolic acid grew poorly, and G418 was toxic only at very high (and therefore expensive) concentrations. Consequently, pHEBo was constructed with the *E. coli* *hph* gene, which encodes a hygromycin B phosphotransferase. The *hph* gene was supplied with eukaryotic promoter and polyadenylation signals from the herpes simplex type 1 thymidine kinase gene. The vector was introduced into the host cells either by electroporation or by protoplast fusion, and transfectants were selected with 300 μg/mL of hygromycin B. Up to 1 in 30 of the cells plated became hygromycin-resistant, indicating that pHEBo is approx 5×10^4 times more efficient at conferring hygromycin resistance than a similar plasmid lacking *oriP* sequences. Hygromycin-resistant cells contained 1–60 copies of pHEBo maintained as an extrachromosomal plasmid, and integrated copies were never detected. The high efficiency with which pHEBo confers hygromycin resistance presumably reflects the fact that it does not require a rare integration event in order to be stably maintained, as is the case with other vectors. In support of this, pHEBo yields hygromycin-resistant cells inefficiently when transfected into cells that lack EBNA-1 and that therefore do not maintain it as a plasmid *(21)*. Using the method described in Section 3., pHEBo DNA can be readily isolated from hygromycin-resistant lymphoblasts and then shuttled back into *E. coli (21)*.

Clearly, the usefulness of pHEBo is limited to cells that express EBNA-1. In order to extend its host range, Yates and colleagues *(12)* constructed derivatives that contained the EBNA-1 gene (Fig. 3b). These plasmids were stably maintained in a range of EBV-negative cell types, including lymphoma, erythroleukemia, fibroblast, and epithelial cell lines of human, monkey, and dog origin. They were not maintained, however, in rodent cell lines.

Foreign genes cloned into such vectors can be expressed efficiently in human cells under the control of their own *(22)* or heterologous *(23–25)* promoters. This, together with the high efficiency of stable transfection and the ease with which plasmids can be recovered from transfectants, indicated that these vectors can provide an efficient route to the cloning of some eukaryotic genes. Kioussis and colleagues *(22)* constructed a human genomic library in the cosmid vector cos202, which contains *oriP*, the EBNA-1 gene, and the hygromycin-resistance gene (Fig. 3c). The library was transfected into human cells and hygromycin-resistant colonies were selected. Such colonies can be screened for the expression of a gene of interest, for example by using an antiserum against a cell-surface protein, and the cos202 clone containing the gene of interest can then be readily transferred from the positive colony to *E. coli* for further analysis.

A number of authors have used *oriP*-containing vectors to analyze the chemically induced mutagenesis of cloned genes *(26,27)*. Other vectors suffer from the disadvantage that integration into host-cell genomic DNA is in itself highly mutagenic. Vectors that contain *oriP* can be introduced into mammalian cells and maintained as plasmids without accumulating significant levels of mutation. The transfectants can be treated with mutagenic substances and the plasmids then transferred into *E. coli* for analysis of the induced mutations.

1.2.2. Vectors Based on EBV oriP/oriLyt *Replicons*

Hammerschmidt and Sugden *(28)* have recently constructed a vector containing *oriP*, *oriLyt*, and the fused terminal repeats (TR) of EBV. When introduced into cells infected with the nonimmortalizing P3HR-1 strain of EBV, this vector is maintained as a plasmid, since the cells contain EBNA-1 and the vector contains *oriP*. If the EBV lytic cycle is induced by treating the cells with 12-O-tetradecanoylphorbol-13-acetate (TPA) or by transfecting them with the EBV BZLF-1 gene, the vector DNA is replicated, since it contains *oriLyt*. The vector DNA is then packaged into EBV virions, since the fused TR act as packaging signals and the P3HR-1 virus supplies capsid proteins and other viral functions required in *trans* for packaging. These virions are infectious and thus provide a means of delivering cloned genes to human B-lymphocytes at high frequency. These cells are largely refractory to traditional methods of transfection. An *oriP*/*oriLyt*/EBV TR vector also has an advantage over related vectors that are based on herpes simplex virus type 2 replicons (*see* Chapter 16) in that the accompanying helper virus (the EBV strain of P3HR-1) is not cytocidal. Such a vector has been used to assess the role of EBV genes in the immortalization of B-lymphocytes *(28)*, as described in Chapter 19.

2. Materials

1. Lysis buffer: 10 m*M* Tris-HCl, pH 7.5, 0.6% (w/v) sodium dodecyl sulfate (SDS); 10 mM EDTA. Autoclave before use.
2. 5*M* NaCl: Autoclave before use.
3. Buffered phenol: Redistilled phenol equlibrated with 10 m*M* Tris-HCl, pH 7.5; 0.1% (w/v) 8-hydroxyquinoline.
4. Chloroform/isoamyl alcohol (24/1, v/v).
5. 3*M* sodium acetate, pH 6.5. Autoclave before use.
6. Absolute ethanol.

7. 70% (v/v) ethanol.
8. TE buffer: 10 mM Tris-HCl, pH 7.5, 10 mM EDTA. Autoclave before use.

3. Method

3.1. Isolation of Plasmid DNA from Eukaryotic Cells

This method is based on that of Hirt *(29)* and relies on the preferential precipitation of high-mol-wt chromosomal DNA in the presence of SDS and NaCl. Low-mol-wt DNA molecules, such as the EBV episomal vectors described above, remain in the supernatant.

1. Lyse cells in a centrifuge tube by resuspending them in lysis buffer at approx 10^7/mL and incubating for 10–20 min at room temperature (*see* Note 1).
2. Add 1/5 vol of 5M NaCl and mix the solution *gently* by slowly inverting the tube 10 times.
3. Place the sample at 4°C overnight.
4. Centrifuge the sample at 17,000g for 30 min at 4°C and harvest the supernatant, which should contain the plasmid DNA. Essentially all of the chromosomal DNA should be in the pellet.
5. To the supernatant, add an equal vol of buffered phenol and mix vigorously.
6. Centrifuge the sample in order to separate the phases, and harvest the aqueous (upper) phase.
7. Repeat steps 5 and 6 once.
8. To the harvested aqueous phase, add an equal vol of chloroform/isoamyl alcohol and mix vigorously.
9. Centrifuge the sample in order to separate the phases, and harvest the aqueous (upper) phase.
10. Repeat steps 8 and 9 once.
11. To the harvested aqueous phase, add 1/10 vol of 3M sodium acetate, pH 6.5, and 2.5 vol of absolute ethanol. Precipitate the DNA at –70°C for 1 h.
12. Pellet the DNA by centrifugation and wash the pellet twice in ice-cold 70% ethanol.
13. Dry the pellet briefly in a vacuum dessicator and resuspend it in TE buffer.

Plasmid DNA isolated from 1–2 × 10^6 eukaryotic cells can be used to transform 0.2 mL of competent *E. coli* by standard procedures.

4. Notes

1. During steps 1–4, it is important to avoid extensive physical shearing of the chromosomal DNA, since small fragments of chromosomal DNA would contaminate the low-mol-wt plasmid DNA. Pipeting and vigorous vortexing should therefore be avoided. Once the intact chromosomal DNA has been removed (step 4), this precaution becomes unnecessary.
2. Tubes and aqueous solutions should be autoclaved before use in order to inactivate deoxyribonucleases.

Acknowledgments

Mukesh Patel is supported by a grant from the Leukemia Research Fund, for which we are grateful.

References

1. Baer, R., Bankier, A. T., Biggin, M. D., Deininger, P. L., Farrell, P. J., Gibson, T. J., Hatfull, G., Hudson, G. S., Satchwell, S. C., Seguin, C., Tuffnell, P. S., and Barrell, B. G. (1984) DNA sequence and expression of the B95-8 Epstein-Barr virus genome. *Nature* **310,** 207–211.
2. Speck, S. H. and Strominger, J. L. (1987) Epstein-Barr virus transformation. *Prog. Nucleic Acid Res. Mol. Biol.* **34,** 189–207.
3. Dillner, J. and Kallin, B. (1988) The Epstein-Barr virus proteins. *Adv. Cancer Res.* **50,** 95–158.
4. Tosato, G. (1987) The Epstein-Barr virus and the immune system. *Adv. Cancer Res.* **49,** 75–125.
5. Dambaugh, T., Beisel, C., Hummel, M., King, W., Fennewald, S., Cheung, A., Heller, M., Raab-Traub, N., and Kieff, E. (1980) Epstein-Barr virus (B95-8) DNA VII: Molecular cloning and detailed mapping. *Proc. Natl. Acad. Sci. USA* **77,** 2999–3003.
6. Sugden, B., Phelps, M., and Domoradzki, J. (1979) Epstein-Barr virus is amplified in transformed lymphocytes. *J. Virol.* **31,** 590–595.
7. Hampar, B., Tanaka, A., Nonoyama, M., and Derge, J. G. (1974) Replication of the resident repressed Epstein-Barr virus genome during the early S phase (S-1 period) of nonproducer Raji cells. *Proc. Natl. Acad. Sci. USA* **71,** 631–633.
8. Matsuo, T., Heller, M., Petti, L., O'Shiro, E., and Kieff, E. (1984) Persistence of the entire Epstein-Barr virus genome integrated into human lymphocyte DNA. *Science* **226,** 1322–1325.
9. Mecsas, J. and Sugden, B. (1987) Replication of plasmids derived from bovine papilloma virus type I and Epstein-Barr virus in cells in culture. *Annu. Rev. Cell. Biol.* **3,** 87–108.
10. Laux, G., Perricaudet, M., and Farrell, P. J. (1988) Epstein-Barr virus gene expressed in immortalised lymphocytes is created by circularization of the viral genome. *EMBO J.* **7,** 769–774.

11. Bhat, R. and Thimmappaya, B. (1983) Two small RNAs encoded by Epstein-Barr virus can functionally substitute for the virus-associated RNAs in the lytic growth of adenovirus 5. *Proc. Natl. Acad. Sci. USA* **80**, 4789–4793.
12. Yates, J. L., Warren, N., and Sugden, B. (1985) Stable replication of plasmids derived from Epstein-Barr virus in various mammalian cells. *Nature* **313**, 812–815.
13. Reisman, D., Yates, J., and Sugden, B. (1985) A putative origin of replication of plasmids derived from Epstein-Barr virus is composed of two *cis*-acting components. *Mol. Cell. Biol.* **5**, 1822–1832.
14. Rawlins, D. R., Milman, G., Hayward, S. D., and Hayward, G. S. (1985) Sequence-specific DNA binding of the Epstein-Barr virus nuclear antigen (EBNA-1) to clustered sites in the plasmid maintenance region. *Cell* **42**, 659–668.
15. Fried, M. and Prives, C. (1986) The biology of simian virus 40 and polyomavirus, in *Cancer Cells – 4. DNA Tumor Viruses: Control of Gene Expression and Replication* (Botchan, M., Grodzicker, T., and Sharp, P. A., eds.), Cold Spring Harbor Laboratory, Cold Spring Harbor, NY, pp. 1–16.
16. Reisman, D. and Sugden, B. (1986) *Trans*-activation of an Epstein-Barr viral transcriptional enhancer by the Epstein-Barr viral nuclear antigen 1. *Mol. Cell. Biol.* **6**, 3838–3846.
17. Sugden, B. and Warren, N. (1989) A promoter of Epstein-Barr virus that can function during latent infection can be transactivated by EBNA-1, a viral protein required for viral DNA replication during latent infection. *J. Virol.* **63**, 2644–2649.
18. Jones, C. H., Hayward, S. D., and Rawlins, D. R. (1989) Interaction of the lymphocyte-derived Epstein-Barr virus nuclear antigen EBNA-1 with its DNA binding sites. *J. Virol.* **63**, 101–110.
19. Baichwal, V. R. and Sugden, B. (1988) Latency comes of age for herpesviruses. *Cell* **52**, 787–789.
20. Hammerschmidt, W. and Sugden, B. (1988) Identification and characterization of *oriLyt*, a lytic origin of DNA replication of Epstein-Barr virus. *Cell* **55**, 427–433.
21. Sugden, B., Marsh, K., and Yates, J. (1985) A vector that replicates as a plasmid and can be efficiently selected in B-lymphoblasts transformed by Epstein-Barr virus. *Mol. Cell. Biol.* **5**, 410–413.
22. Kioussis, D., Wilson, F., Daniels, C., Leveton, C., Taverne, J., and Playfair, J. H. L. (1987) Expression and rescuing of a cloned human tumour necrosis factor gene using an EBV-based shuttle cosmid vector. *EMBO J.* **6**, 355–361.
23. Young, J. M., Cheadle, C., Foulke, J. S., Drohan, W. N., and Sarver, N. (1988) Utilization of an Epstein-Barr virus replicon as a eukaryotic expression vector. *Gene* **62**, 171–185.
24. Jalanko, A., Kallio, A., and Ulmanen, I. (1988) Comparison of mammalian cell expression vectors with and without an EBV-replicon. *Arch. Virol.* **103**, 157–166.
25. Groger, R. K., Morrow, D. M., and Tykocinski, M. L. (1989) Directional antisense and sense cDNA cloning using Epstein-Barr virus episomal expression vectors. *Gene* **81**, 285–294.
26. Drinkwater, N. R. and Klinedinst, D. K. (1986) Chemically induced mutagenesis in a shuttle vector with a low-background mutant frequency. *Proc. Natl. Acad. Sci. USA* **83**, 3402–3406.

27. DuBridge, R. B., Tang, P., Hsia, H. C., Leong, P.-M., Miller, J. H., and Calos, M. P. (1987) Analysis of mutation in human cells by using an Epstein-Barr virus shuttle system. *Mol. Cell. Biol.* **7,** 379–387.
28. Hammerschmidt, W. and Sugden, B. (1989) Genetic analysis of immortalizing functions of Epstein-Barr virus in human B lymphocytes. *Nature* **340,** 393–397.
29. Hirt, B. (1967) Selective extraction of polyoma DNA from infected mouse cell cultures. *J. Mol. Biol.* **26,** 365–369.

CHAPTER 19

Immortalization of Human B-Lymphocytes by Epstein-Barr Virus

Paul M. Brickell

1. The Mechanism of B-Lymphocyte Immortalization by EBV

Epstein-Barr virus (EBV) is able to immortalize human B-lymphocytes with high efficiency. This property underlies the role of EBV in a number of human diseases. First, EBV is the causative agent of infectious mononucleosis, which is a benign proliferation of B-lymphocytes *(1)*. Second, EBV is involved in the etiology of Burkitt's lymphoma *(2)* and of invasive B-lymphocyte proliferations found in immunosuppressed individuals *(3)*.

Human B-lymphocytes immortalized by EBV, termed B-lymphoblastoid cell lines (B-LCL), have been useful research tools in a number of fields. They have facilitated the study of human B-lymphocyte biology, have provided material for the study of the structure and expression of a variety of human genes, and provide an easy means of obtaining a large number of cells from a given individual. A large number of workers have also used B-LCL for the production of human monoclonal antibodies *(4)*.

This chapter reviews recent advances in our understanding of how EBV immortalizes human B-lymphocytes, including data obtained using the *oriP/oriLyt*/EBV TR vector described in Chapter 18. Details of a method for deriving B-LCL from human peripheral blood B-lymphocytes are also given.

From: *Methods in Molecular Biology, Vol. 8: Practical Molecular Virology: Viral Vectors for Gene Expression*
Edited by: M. Collins © 1991 The Humana Press Inc., Clifton, NJ

The details of the mechanism by which EBV immortalizes human B-lymphocytes are by no means fully clear. However, a variety of evidence has implicated two viral proteins to date. These are EBV nuclear antigen 2 (EBNA-2) and latent membrane protein (LMP), which is encoded by the viral BNLF-1 gene *(5)*.

1.1. The Role of EBNA-2 in B-Lymphocyte Immortalization

The EBV strain P3HR-1, which is unable to immortalize B-lymphocytes *(6)*, has suffered a deletion that removes the entire EBNA-2 coding region and the last two exons of the gene encoding EBNA-5 (also called EBNA-leader protein [EBNA-LP]). Hammerschmidt and Sugden *(7)* recently constructed an *oriP/oriLyt/*EBV TR plasmid vector (*see* Chapter 18) containing the EBNA-2 and EBNA-LP genes. They then made derivatives of this construct in which either the EBNA-2 or the EBNA-LP gene was disrupted. These plasmids were transfected independently into a cell line containing the P3HR-1 strain of EBV. The lytic cycle was then induced, yielding stocks of infectious virus particles containing either EBNA-2^+/EBNA-LP^- plasmid DNA or EBNA-2^-/EBNA-LP^+ plasmid DNA. The EBNA-2^+/EBNA-LP^- virus stock was able to immortalize human B-lymphocytes, indicating that EBNA-2 is sufficient to induce immortalization. The EBNA-2^-/EBNA-LP^+ virus stock failed to immortalize human B-lymphocytes, indicating that EBNA-LP is not required for immortalization. When the EBNA-2^-/EBNA-LP^+ virus stock and the P3HR-1 strain were mixed and used to infect human B-lymphocytes, immortalized cells were obtained at a higher frequency than with the EBNA-2^+/EBNA-LP^- virus alone. This indicates not only that the EBNA-2^+/EBNA-LP^- virus can complement the immortalizing defect in P3HR-1 virus, but also that other viral functions supplied by P3HR-1 can cooperate with EBNA-2 in promoting efficient immortalization.

EBNA-2 can also affect the growth properties of other cell types. For example, expression of the cloned EBNA-2 gene in Rat-1 fibroblast cells leads to a decrease in their serum requirement *(8)*.

The mechanism by which EBNA-2 immortalizes cells is unknown. Transfection of the EBNA-2 gene into EBV-negative Burkitt's lymphoma cell lines leads to an increase in the levels of the cell-surface protein CD23 *(9)*. This is intriguing, since CD23 expression is also induced in B-lymphocytes infected with EBV. CD23 is the B-lymphocyte Fc ε IIa receptor, and a fragment of the protein, when cleaved and shed from the cell surface, can act as an autocrine B-cell growth factor for normal and transformed human lymphocytes *(10)*. It is possible, therefore, that CD23 is a component of a signaling pathway activated by EBNA-2. The nuclear localization of EBNA-2 raises the possibility that it is a transcription factor. However, there is no evidence for such a function as yet, and no amino acid sequence homology with other transcription factors.

1.2. The Role of Latent Membrane Protein in B-Lymphocyte Immortalization

The cloned EBV BNLF-1 gene, which encodes LMP, transforms rodent cell lines following transfection *(11)* and stimulates expression of CD23, the transferrin receptor, and the leukocyte adhesion molecules LFA-1, LFA-3, and ICAM-1 when transfected into an EBV-negative Burkitt's lymphoma cell line *(12)*. These data suggest that LMP may be involved in the immortalization of B-lymphocytes. It is interesting that CD23 expression is stimulated by both EBNA-2 and LMP, although this suggests that, rather than being a primary target of either of these proteins, it operates further downstream in the signaling pathway.

LMP consists of a short, intracytoplasmic *N*-terminal domain, a transmembrane region that spans the membrane six times; and a long, intracytoplasmic C-terminal domain. Deletion analyses have shown that only the *N*-terminal and membrane-spanning domains are required for transformation *(13)*. The structure of LMP suggests that it may be an ion channel or a cell-surface receptor related to the rhodopsin family *(5)*. However, there is no direct evidence relating to its function.

1.3. Conclusion

There is excellent evidence that EBNA-2 is centrally involved in the immortalization of B-lymphocytes by EBV, and that other viral gene products cooperate with EBNA-2 to increase the efficiency of immortalization. LMP may be one such product, and there is considerable interest in identifying others. This search will be greatly assisted by the use of *oriP/oriLyt/*EBV TR vectors containing cloned candidate EBV genes. The interaction of these viral gene products with cellular pathways of growth control will also be an important focus of future research.

2. Materials

1. RPMI 1640 tissue-culture medium containing 2 m*M* L-glutamine.
2. Fetal calf serum (FCS).
3. Ficoll-Hypaque (relative density 1.077) (e.g., Lymphoprep, Nycomed UK Ltd.).
4. Dimethylsulfoxide (DMSO).
5. Cyclosporin A solution. Dissolve cyclosporin A in absolute ethanol at 0.4 mg/mL. Store in the dark at 4°C.

3. Methods

3.1. Preparation of EBV Stocks

EBV stocks for routine immortalization of human B-lymphocytes are usually obtained from the marmoset B-lymphoblastoid cell line B95-8, which produces higher than average titers of infectious viral particles *(6)*. The EBV stock originally used to generate the B95-8 cell line came from a human patient with a form of infectious mononucleosis. Viral stocks from other lytically infected cell lines can be isolated similarly.

1. Grow B95-8 cells to high density (e.g., 10^6/mL) in RPMI 1640/10% (v/v) FCS, and leave the culture undisturbed in the tissue-culture incubator for 14 d. Cells should be grown in plastic tissue-culture flasks at 37°C in 5% CO_2 air of relative humidity >95%.
2. Pellet the cells by centrifugation and collect the supernatant, which contains the virus.
3. Filter the supernatant through a sterile filter (pore size 0.45 µm; Millipore) in order to completely remove B95-8 cells.
4. Aliquot the filtered supernatant in small cryotubes and store the sealed tubes in liquid nitrogen.
5. Thaw the aliquots immediately before use, and do not refreeze.

3.2. Preparation and Infection of B-Lymphocytes (see Note 1)

1. Collect 50 mL of human peripheral blood by venepuncture and defibrinate it by shaking in sterile glass universals, each containing 10–15 sterile glass beads. Alternatively, blood may be collected into heparin tubes. Cells from umbilical cord blood or from disrupted fetal liver or spleen may also be used.
2. Dilute the defibrinated blood with an equal volume of RPMI 1640 and layer 15-mL aliquots over 10-mL aliquots of sterile Ficoll-Hypaque in sterile plastic universals.
3. Centrifuge at 2000*g* for 30 min.
4. Carefully harvest the cells at the interface between the RPMI 1640 and Ficoll/Hypaque layers and wash them three times in RPMI 1640/3% (v/v) FCS. The cell types present in the preparation include lymphocytes, granulocytes, and monocytes.
5. Resuspend the final cell pellet in RPMI 1640/10% (v/v) FCS at approx 2×10^5 cells/mL.
6. Add 2 drops of EBV suspension/mL of cell suspension and mix gently.
7. Add 1 µL of cyclosporin A solution (final concentration 400 ng/mL) and mix gently (*see* Note 2).

8. Take a 96-well, flat-bottom microtiter plate (tissue-culture quality) and add 2 drops of the EBV/cell suspension to each well. Add 2 more drops of RPMI 1640/10% (v/v) FCS per well.
9. Incubate at 37°C in 5% CO_2 air of relative humidity >95%, feeding twice a week with RPMI 1640/10% (v/v) FCS.
10. Small clumps of cells should appear in most wells after 2–3 wk. Expand individual clones into 2-mL wells and then into tissue-culture flasks as required. Cells growing healthily in log phase may be stored at -70°C in aliquots of 5×10^6 cells in 1 mL of FCS/6% (v/v) DMSO.

4. Notes

1. Cells isolated from peripheral blood (steps 1–5) can be stored for immortalization at some future date. They should be frozen in aliquots of $7–10 \times 10^6$ cells in FCS/6% (v/v) DMSO at –70°C. They should then be thawed immediately before use, washed in RPMI 1640/3% (v/v) FCS, adjusted to a concentration of 2×10^5 cells/mL in RPMI 1640/10% (v/v) FCS, and immortalized as described in steps 6–10.
2. The purpose of the cyclosporin A is to kill any T-lymphocytes with cytotoxic activity against EBV-infected B-lymphocytes.

References

1. Henle, G. and Henle, W. (1979) The virus as the etiologic agent of infectious mononuleosis, in *The Epstein-Barr Virus* (Epstein, M. A. and Achong, B. G., eds.), Springer-Verlag, Berlin, pp. 297–230.
2. Lenoir, G. M. (1986) Role of the virus, chromosomal translocations and cellular oncogenes in the aetiology of Burkitt's lymphoma, in *The Epstein-Barr Virus: Recent Advances* (Epstein, M. A. and Achong, B. G., eds.), William Heinemann Medical Books, London, pp. 183–205.
3. Cleary, M. L., Dorfman, R. F., and Sklar, J. (1986) Failure in immunological control of the virus infection: Post-transplant lymphomas, in *The Epstein-Barr Virus: Recent Advances* (Epstein, M. A. and Achong, B. G., eds.), William Heinemann Medical Books, London, pp. 163–181.
4. Crawford, D. H. (1986) Use of the virus to prepare human-derived monoclonal antibodies, in *The Epstein-Barr Virus: Recent Advances* (Epstein, M. A. and Achong, B. G., eds.), William Heinemann Medical Books, London, pp. 251–269.
5. Sugden, B. (1989) An intricate route to immortality. *Cell* **57,** 5–7.
6. Miller, G. and Lipman, M. (1973) Release of infectious Epstein-Barr virus by transformed marmoset leucocytes. *Proc. Natl. Acad. Sci. USA* **70,** 190–194.
7. Hammerschmidt, W. and Sugden, B. (1989) Genetic analysis of immortalizing functions of Epstein-Barr virus in human B-lymphocytes. *Nature* **340,** 393–397.
8. Dambaugh, T., Wang, F., Hennessy, K., Woodland, E., Rickinson, A., and Kieff, E. (1986) Expression of the Epstein-Barr virus nuclear protein 2 in rodent cells. *J. Virol.* **59,** 453–462.

9. Wang, F., Gregory, C. D., Rowe, M., Rickinson, A. B., Wang, D., Birkenbach, M., Kikutani, H., Kishimoto, T., and Kieff, E. (1987) Epstein-Barr virus nuclear antigen 2 specifically induces expression of the B-cell activation antigen CD23. *Proc. Natl. Acad. Sci. USA* **84,** 3452–3456.
10. Swendeman, S. and Thorley-Lawson, D. A. (1987) The activation antigen BLAST-2, when shed, is an autocrine BCGF for normal and transformed B-cells. *EMBO J.* **6,** 1637–1642.
11. Wang, D., Liebowitz, D., and Kieff, E. (1985) An EBV membrane protein expressed in immortalized lymphocytes transforms established rodent cells. *Cell* **43,** 831–840.
12. Wang, D., Liebowitz, D., Wang, F., Gregory, C. D., Rickinson, A. B., Larson, R., Springer, T., and Kieff, E. (1988) Epstein-Barr virus latent infection membrane protein alters the human B-lymphocyte phenotype: Deletion of the amino terminus abolishes activity. *J. Virol.* **62,** 4173–4184.
13. Baichwal, V. R. and Sugden, B. (1989) The multiple membrane-spanning segments of the BNLF-l oncogene from the Epstein-Barr virus are required for transformation. *Oncogene* **4,** 67–74.

CHAPTER 20

Vaccinia Virus as an Expression Vector

Antonio Talavera and Javier M. Rodriguez

1. Introduction

Vaccinia virus (Vv) is a member of the genus Orthopoxvirus, one of seven genera included in the family Poxviridae. Most of these viruses infect vertebrates (Orthopoxvirus, Avipoxvirus, Capripoxvirus, Leporipoxvirus, Suipoxvirus, and Parapoxvirus), but one genus, Entomopoxvirus, infects insects. It is interesting to note that the Fibroma and Mixoma viruses of the leporipoxvirus genus cause tumors in their hosts (rabbits), these being the only tumorigenic viruses in the family *(1,2)*.

The most important species within the Orthopoxviruses is, for clinical and historical reasons, variola virus, the causative agent of smallpox. Vaccinia virus follows in importance because of two different, although related, facts: First, Vv has been instrumental in the eradication of smallpox through the procedure known as vaccination, which, with some antecedents in the Far East, was started in the Western world by Edward Jenner at the end of the eighteenth century. The success of this therapy was based on the ability of the antibodies elicited by vaccinia virus inoculation to react with and neutralize the variola virions, combined with the fact that there exists no animal that can be infected by variola virus. Hence, once eradicated from humankind, no reservoir is left in nature for this virus.

Second, vaccinia virus, the origin of which remains unknown, is the most widely studied member of the poxvirus family, for not only is it generally

From: *Methods in Molecular Biology, Vol. 8: Practical Molecular Virology: Viral Vectors for Gene Expression*
Edited by: M. Collins © 1991 The Humana Press Inc., Clifton, NJ

harmless to humans, causing only local and short-lived symptoms in the inoculation sites (exceptions to this rule are avoidable by means of attenuation), but it also exhibits a broad host range, making it possible for the virus to be grown in cultured cells from diverse origins.

The fact that vaccinia has been the first virus used as a weapon against infection by another microorganism, together with the ample knowledge of its biology accumulated in the last two decades, led two groups of investigators to search for methods that allowed the use of vaccinia virus as an infective expression vector *(3,4)*. This, in turn, would lead to the use of Vv as an agent against different microorganisms, as a live (as opposed to attenuated, inactivated, or subunit) vaccine. A recent review by Piccini and Paoletti *(5)* lists a series of foreign genes that have been expressed in recombinant vaccinia virus, among which one can find genes that code for viral and malarial antigens. That list could be considerably enlarged at present, since reports of new expressed genes in Vv recombinants are continuously appearing. The most relevant of these genes are herpes simplex virus glycoprotein D *(6)*, the N gene of human respiratory syncitial virus *(7)*, surface glycoproteins of human parainfluenza virus type 3 *(8)*, Hantaan virus M genome segment *(9)*, and the *gag-pol* precursor protein of the human immunodeficiency virus (HIV) type 1 *(10)*, among others.

In most recombinant vaccinia constructions, the effect sought has been the stimulation of the production of antibodies against the expressed novel protein. In general, antibody stimulation will depend on the infection of the host animal, which leads to disruption of the infected cells and release of the expressed antigen so that it can be recognized by the host animal's immune system. It is interesting to note, in respect to the mode of presentation of the antigen, that, at least in two reported cases, the foreign expressed protein becomes part of the recombinant viral particle *(11,12)*, where it is incorporated during the recombinant virus preparation (*see* Chapter 21).

The possibility that this system offers for the proper processing, modification, and cellular location of the antigen makes this type of vaccine appropriate to elicit cellular immunity also. This has been demonstrated in some cases *(13–16)*.

Other uses of vaccinia virus recombinants have been the study of the processing of isolated viral gene products *(17,18)* as well as the production of clinically useful gene products, such as blood clotting factor IX, needed for the treatment of hemophilia *(19)*.

1.1. Structural Features of Vaccinia Virus

Vaccinia virus particles are of an irregular oblong ("brick") shape with dimensions of about 300 × 230 × 180 nm *(1)*. The intracellular particles (which can be purified from disrupted infected cells) are composed of a core, or nucleoid, that contains the genetic material (DNA), which is associated with proteins and surrounded by a membrane, or core envelope. The nucleoid is, in turn, surrounded by the outer lipid coat. Between these two components there are two lateral bodies of unknown function. The extracellular virus particles are surrounded by another layer, the envelope, that is acquired as the particles bud through the cell membrane *(2,20)*.

The genetic material of vaccinia virus is a double-stranded linear DNA molecule of about 180,000 bp *(21)*. The two strands of the molecule are connected by terminal loops (hairpins) containing a few unpaired bases *(22–24)*, so the molecule can be considered as a continuous single-stranded circle formed by two semicircles complementary to each other. The role of the hairpins in the replication of the viral DNA is fully discussed in refs. *24* and *25*. Another interesting feature of the DNA molecule is that the 10,000 bp on the extreme left end are inversely repeated at the other end *(26,27)*. The inverted repeats contain, in their turn, two series of 13 and 17 tandem repeats of the same 70-bp sequence, flanked and separated by unique sequences. These tandem repeats have been implicated in the formation of circular DNA structures during replication *(27)*.

1.2. Life Cycle of Vaccinia Virus

A feature that distinguishes the poxvirus family from other DNA viruses is the fact that their replication takes place entirely in the cytoplasm of the host cell. The assembly of the progeny virions is located in specific cytoplasmic places called "factories," which can be separated from other cellular structures in the form of "virosomes" *(28)*. The cytoplasmic mode of replication implies that the machinery necessary for RNA synthesis and modification has to be carried by the incoming particle in order for the early viral genes to be expressed. Consequently, among the active enzymes found in proteins included in the capsid, there are an RNA polymerase *(29,30)*, capping and methylating enzymes *(31,32)*, and a polyA polymerase *(33)*. These enzymes, upon uncoating of the virus, start the transcription of a series of early genes, among which are found those responsible for DNA synthesis, such as DNA polymerase *(34,35)*, ligase *(36)*, thymidine kinase *(36–38)*, and thymidilate synthetase *(39)*. Once the DNA synthesis has taken place, the mode of transcription changes to a discontinuous one *(40,41)* to transcribe the so-called late genes.

The life cycle of vaccinia virus can be thus divided into several steps: penetration and uncoating (30 min), early transcription (1–2 h), DNA synthesis (2–4 h), late transcription and assembly (4–6 h), and release (budding) of the virions. During the late transcription period, many, although not all, early genes cease to be transcribed. It is during this phase of the cycle that the structural viral proteins are expressed.

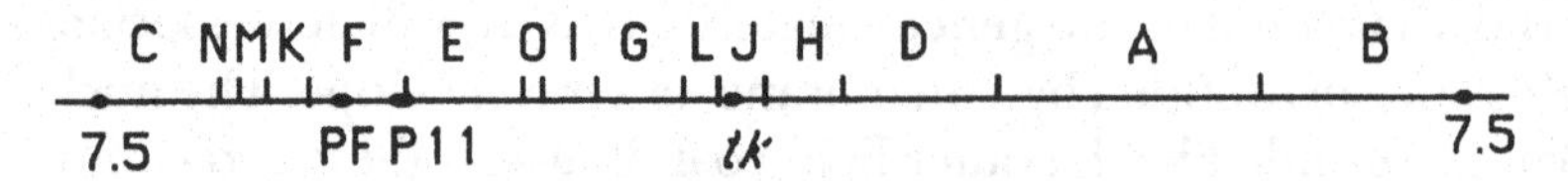

Fig. 1. HindIII map of vaccina virus DNA. The locations of the *tk* gene and the promoters referred to in the text are shown as bold lines. (Not to scale.)

An early effect of vaccinia infection on the host cell is the inhibition (shut off) of cellular protein and DNA synthesis *(20)*. The infection also causes cytopathic effects that culminate in the death and eventual lysis of the infected cell; this allows the formation of virus-containing plaques in monolayers of cells that are infected with vaccinia, and it is a very useful method for titrating and cloning the virus as well as for isolating virus variants that are the product of mutation or artificial manipulation of the vaccinia genome, as discussed in Chapter 21.

1.3. Vaccinia Virus Genes Relevant to Its Use as a Vector

Vaccinia virus DNA has a coding capacity for more than 200 gene products *(42)*. Such a genetic complexity implies that some viral genes have cellular counterparts, so that their function can be dispensed with. The best example is the gene that codes for the enzyme thymidine kinase (TK). This enzyme phosphorylates the thymidine present in the medium external to the cells. The resulting thymidine monophosphate (TMP) can be uptaken by the cell, where it is converted into thymidine triphosphate (TTP) and finally incorporated into DNA. Normal cells possess their own thymidine kinase gene (*tk* gene), so that even virus variants in which the *tk* gene has been mutated or otherwise made inactive can successfully infect normal cells. The viral *tk* gene has been mapped at about the middle part of the vaccinia genome (fragment HindIII J, refs. *36,43*) (Fig. 1), and constitutes an ideal site to insert foreign genes, since

1. Its interruption is not lethal for the virus.
2. Its inactivation by insertion makes it possible to select the TK^- recombinants among the majority of nonrecombinant TK^+ virus (*see* Chapter 21).

Other locations in the Vv genome that can be interrupted with no loss of viability are

1. A BamHI site located in fragment HindIII F. This site was the first to be used to insert a foreign gene into the vaccinia genome *(3)*.
2. Any place in a 9000-bp sequence near the left end of the vaccinia genome that appears to be spontaneously deleted in some viral variants and, consequently, carries nonessential genes *(44)*.

A characteristic of vaccinia virus genes is that their promoters do not show the features (TATA and CAT boxes) typical of eukaryotic RNA polymerase II transcribed genes *(45,46)*, although they exhibit specific features that are different in early and late genes. (For a discussion on the structure of vaccinia promoters, *see* refs. *46,47*.) An important consequence of this difference is that vaccinia promoters cannot be recognized by the cell's polymerases *(48)*, nor are eukaryotic gene promoters recognized by the vaccinia transcribing machinery. This means that any gene introduced into the vaccinia genome is not expected to be expressed in cultured cells or animals infected with the corresponding recombinant virus, unless it is properly driven by a Vv promoter. Several promoters have been used for this purpose. The first of them was the one corresponding to a gene located in the long terminal repetition, which codes for a 7.5-kDa product (7.5 gene) *(49)*. This promoter was originally considered an early one, but, by in vitro mutagenesis, it was later shown to present tandemly located early and late regulatory regions *(50)*. As a consequence, the 7.5 gene, as well as any other gene located under the control of this promoter, can be expressed throughout the whole infective cycle. Other promoters that have been commonly used are the F promoter (PF), located upstream of the aforementioned BamHI site of the HindIII F fragment *(3)*, and the promoter that controls a late gene coding for an 11-kDa protein (11 gene, ref. *38*) located in the junction of fragments HindIII F and E *(51)*. The locations of these promoters in the vaccinia genome are shown in Fig. 1.

1.4. General Procedures to Prepare Vv Recombinants

For the insertion of foreign genes into the Vv genome and their expression in recombinant-infected cells, one has to take into account several factors:

1. Because of the complexity of the vaccinia genome, no unique restriction sites are easily found that are located in nonessential regions of the genome. Even if finding such sites were possible, two additional problems would arise
   a. The difficulties inherent to the handling of the 180,000-bp DNA molecule and
   b. The impossibility of establishing an infectious cycle with the naked DNA, because of the absence of the virion-borne enzymes necessary for the first steps of the vaccinia cycle.
2. As discussed above, the virus-coded RNA polymerase does not recognize promoters other than those corresponding to vaccinia and related viruses. In consequence, no expression is to be expected from exogenous genes that include their own promoters.

To overcome these problems, the procedure followed consists of two steps:

1. Construction of an intermediate recombinant plasmid in which the foreign gene is inserted in the middle of a Vv sequence and placed under the control of a Vv promoter;
2. Introduction of this recombinant, or insertion vector, into Vv-infected cells where, as will be discussed in Chapter 21, the foreign gene is inserted into the whole Vv genome via homologous recombination.

1.5. Insertion Vectors

In general, insertion vectors have been constructed by using a bacterial plasmid into which a dispensible vaccinia DNA region has been inserted. Panicali and Paoletti's plasmid pDP3 *(3)* consisted of plasmid pBR322 *(52)* into which the HindIII F fragment of vaccinia DNA had been inserted. The BamHI site present inside the vaccinia moiety served to clone the herpes simplex *tk* gene. After isolating the vaccinia recombinants, only those in which the *tk* gene had been cloned in one of the two possible orientations exhibited herpes TK expression, thus suggesting that the transcription of the foreign gene took place from a promoter located in the vaccinia region.

A group of insertion vectors have been subsequently developed, all of which are based on the Vv *tk* gene. This is a nonessential gene as long as the host cell can provide its own TK activity; in addition, it is easy to select for viruses in which the *tk* gene has been interrupted (*see* Chapter 21). The prototype of these vectors is plasmid pGS20 *(53)* (Fig. 2a), which combines the bacterial plasmid pUC9 *(54)*, the vaccinia fragment HindIII J (containing the *tk* gene) and the P7.5 promoter inserted into the unique EcoRI site present in the *tk* gene. Two unique SmaI and BamHI sites introduced downstream of the promoter can be used for inserting foreign DNA sequences. The utility of this vector was originally tested by cloning and expressing the bacterial gene that codes for the enzyme chloramphenicol acetyltransferase (CAT) (*cat* gene). One limitation of pGS20 is the scarcity of suitable restriction sites for the insertion of foreign genes. To overcome this inconvenience, Boyle et al. *(55)* constructed a series of insertion vectors that are similar to pGS20, but contain a multiple cloning site following either of the vaccinia promoters PF and P7.5.

An insertion vector that includes the features of pGS20 plus an additional selection method, based on the properties of bacterial β-galactosidase, is the plasmid pSC11 *(56)* (Fig. 2b). This vector contains, in the middle of the viral *tk* gene, the promoter P11 followed by the *E. coli* β-galactosidase gene and, in the opposite direction, the promoter P7. 5 followed by a cloning site similar to that of pGS20. In this instance, the presence of β-galactosidase in the

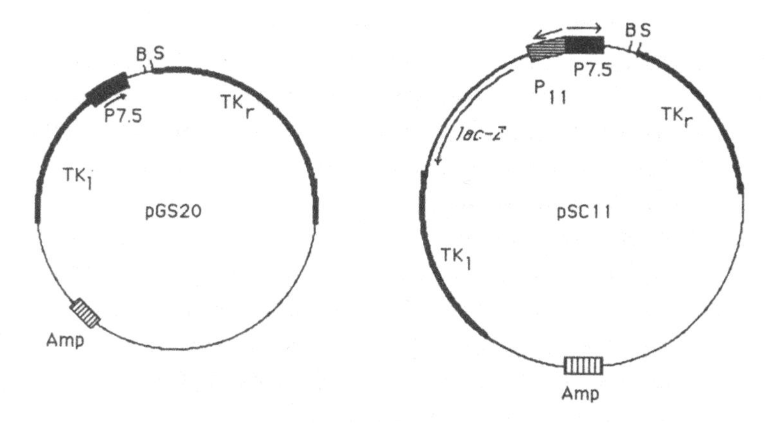

Fig. 2. Plasmids pGS20(a) and p SC11(b). TK_l and TK_r refer, respectively, to the left and right halves of the vaccinia *tk* gene (thick line), B and S represent BamHI and SmaI sites, respectively. Amp stands for the ampicillin resistance gene. The thin lines correspond to bacterial plasmid sequences. The medium-sized line corresponds to the *E. coli* β-galactosidase gene.

recombinants results in blue-stained plaques in the presence of the dye X-gal. The details of the selection of recombinant virus using these two types of insertion vectors are given in Chapter 21.

2. Materials

1. Enzymes
   a. Restriction endonuclease SmaI (Boehringer).
   b. Calf intestine phosphatase (Boehringer), 1U/μL.
   c. DNA ligase (Boehringer), 1U/μL.
   d. RNase A, 5 mg/mL (DNase–free): Dissolve 25 mg of RNase A (Sigma) in 5 mL of sterile water. Make 50-μL aliquots and store at –20°C. Before using each aliquot, heat it at 90°C for 10 min.
2. Buffers
   a. 10× Sma buffer: 0.1*M* Tris–HCl, pH 8, 0.2*M* KCl, 0.1*M* $MgCl_2$, 0.01*M* dithiothreitol (DTT).
   b. TE: 10 m*M* Tris-HCl, pH 8, 1 m*M* EDTA.
   c. 10× Ligase buffer: 0.5*M* Tris–HCl, pH 7.4, 0.1*M* Mg Cl_2.
3. Solutions
   a. 10 m*M* ATP (10× ATP).
   b. 10 m*M* DTT (10× DTT).
   c. 1 mg/mL bovine serum albumin (10× BSA).
   d. 0.1*M* $MgCl_2$.

e. 0.1*M* $CaCl_2$.
f. Solution I: 25 m*M* Tris-HCl, pH 8, 10 m*M* EDTA, 25 m*M* sucrose.
g. Solution II: 0. 2*N* NaOH, 1% sodium dodecyl sulfate [SDS]. Make up volume with H_2O before adding the SDS to the NaOH.
h. 3*M* potassium acetate, pH 4.8.
i. 3*M* sodium acetate, pH 5.2.

4. Bacteria: *E. coli* JM101.
5. Phenol equilibrated with 1*M* Tris-HCl, pH 8: Mix 500 mL of liquified phenol with 250 mL of 1*M* Tris-HCl, pH 8. When the phenol is settled, withdraw the aqueous (upper) phase. Repeat twice, leaving the aqueous phase in place the third time. Add 0.5 g of hydroxyquinoline. Shake and store at 4°C, protected from light, for no more than 1 mo.
6. Geneclean® kit (Bio 101, La Jolla, CA).

3. Methods

The following protocol is intended for the construction of pGS20 or pSC11 insertion vectors. Other similar plasmids may be used, taking into account the different cloning sites available in each instance.

3.1. Preparation of the Vector

1. Cut ~20 µg of pGS20 or pSC11 with SmaI in enzyme buffer with 20 U of enzyme for 1–2 h at 37°C.
2. Heat at 70°C for 10 min to inactivate and denature the restriction enzyme. Add 3 µL of calf intestine phosphatase (3 U). Incubate for 20 min at 37°C.
3. Load onto a 0.5% agarose gel in TAE buffer with 0.5 µg/mL ethidium bromide. Run at 5 V/cm for 1–3 h in parallel with a suitable marker (e.g., 0.5 µg BstE II–cut λ DNA).
4. Cut out the band and extract the DNA with Geneclean®, following the manufacturer's recommended procedures. Resuspend the DNA in 10–20 µL of TE.

3.2. Preparation of the Insert

1. The gene to be inserted and expressed is separated from surrounding sequences with an appropriate restriction enzyme in such a way that the initial ATG codon is near the end of the future insert (*see* Note 2). If necessary, blunt the ends of the insert by treatment with T_4 DNA polymerase.
2. Load onto an agarose gel as in Section 3.1., step 3, and purify the DNA with Geneclean®. Determine the concentration of vector and insert by

running an aliquot of each in agarose in parallel with a known amount of DNA of a similar size.

3.3. Ligation of Vector and Insert

1. Mix, in an Eppendorf tube labeled "L," 0.5 µg of dephosphorylated vector with 0.2–0.5 µg of purified fragment.
2. Add 1 µL of 10× ligase buffer, 1 µL of 10× ATP, 1 µL of 10× DTT, 1 µL of 10× BSA, H_2O up to 9 µL. Add 1 µL of DNA ligase (1 U).
3. Prepare similar mixtures, one of them lacking both ligase and insert, and the other with ligase but no insert, in tubes labeled "NL" ("No Ligase") and "IL" ("Insertless") (*see below*).
4. Incubate the three tubes overnight at 15°C.

3.4. Transformation of *E. coli* JM 101

3.4.1. Preparation of Competent Bacteria

1. Grow the bacteria in 50 mL of LB medium to an optical density of 0.4 at 600 nm.
2. Centrifuge at 4000 rpm for 10 min in the Sorvall S534 rotor or similar.
3. Resuspend the pellet thoroughly in 25 mL of ice-cold 0. 1*M* $MgCl_2$. Keep on ice for 10 min.
4. Centrifuge at 4000 rpm at 4°C for 10 min.
5. Resuspend the pellet in 25 mL of ice-cold 0.1*M* $CaCl_2$. Keep on ice for 20 min.
6. Centrifuge at 4000 rpm at 4°C for 10 min. Resuspend the pellet in 4.3 mL of cold 0.1*M* $CaCl_2$.
7. Add 0.7 mL of sterile glycerol. Mix well. Make 500-µL aliquots in sterile Eppendorf tubes. Freeze immediately in liquid nitrogen. Keep the tubes in liquid nitrogen or at –70°C.

3.4.2. Transformation of Competent Bacteria

1. Thaw three 500-µL aliquots of competent bacteria by placing them on ice. Label the tubes "L," "NL," and "IL."
2. As soon as the contents of the tubes are thawed, add to each tube 5 µL of the corresponding ligation mixture described in Section 3.3. Keep on ice for 20 min.
3. Heat-shock at 42°C for 2 min.
4. Add 1 mL of LB prewarmed at 37°C, and incubate at this temperature for 60 min.
5. Centrifuge in an Eppendorf centrifuge for 5 min. Discard the supernatant and resuspend the cells in 200 µL of LB.
6. Spread the contents of each tube onto 12 ampicillin–LB–agar plates, as

follows: 2 plates, 50 µL; 3 plates, 20 µL; 3 plates, 10 µL; 2 plates, 5 µL; and 2 plates, 1 µL. Take into account the amount of vector in each plate (2.5 ng of vector per µL spread).

7. Incubate overnight at 37°C.
8. Count the colonies and calculate the number of colonies/ng that were obtained in each instance. The number of colonies/ng in "L" plates should be at least fivefold higher than that corresponding to "IL" plates; otherwise, the chance of isolating recombinants will be poor. The number of colonies/ng in "IL" plates should be no more than twice that obtained in "NL" plates, indicating a good level of dephosphorylation.

3.5. Analysis of Transformants

The following procedure allows the analysis of multiple transformants in one day. It is convenient to make the preparations in multiples of 12 on account of the design of the Eppendorf minicentrifuges.

1. Infect, with individual colonies from the "L" set of colonies, 5 mL of ampicillin–LB in a 10-mL tube. Incubate overnight at 37°C.
2. Put 1.5 mL from each culture into an Eppendorf tube and spin for 5 min. Discard the supernatant, add 1.5 mL from the corresponding culture, and repeat centrifugation. If necessary, the procedure may be repeated once more. Keep the remainder of the culture at 4°C for possible future use.
3. Discard the last supernatant; add 100 µL of Solution I. Shake to resuspend the bacterial pellet. Keep at room temperature for 5 min.
4. Add 200 µL of Solution II. Mix by inversion, and put on ice for 2 min. The bacteria will lyse, giving the solution a viscous aspect.
5. Add 150 µL of 3*M* potassium acetate, pH 4.8. Invert and keep on ice for 10 min.
6. Transfer the supernatant to a new Eppendorf tube containing 300 µL of isopropyl alcohol.
7. Centrifuge for 5 min and dry the pellet under vacuum for 10 min at room temperature.
8. Resuspend the pellet in 400 µL of TE. Add RNase (DNase-free) at a final concentration of 10 µg/mL. Incubate at 37°C for 30 min.
9. Extract with 400 µL of phenol–chloroform–isoamyl alcohol (25:24:1). Stir and centrifuge for 1 min. Discard the organic (lower) phase. Extract with chloroform–isoamyl alcohol (24:1). Again stir and centrifuge for 1 min, and discard the organic (lower) phase.
10. Heat at 70°C for 10 min. This will inactivate phenol-resistant nucleases that could degrade the DNA preparation. Chill on ice.

11. Add 40 μL of 3*M* sodium acetate, pH 5.2, and 1 mL of absolute ethanol at –20°C. Keep on ice for 30 min.
12. Centrifuge for 20–30 min. Discard the supernatant and add 0.5 mL of 70% ethanol at –20°C. Centrifuge for 5 min and discard the supernatant. Keep the tubes upside down and wipe the edges with a sterile napkin.
13. Dry the pellet under vacuum until no more liquid can be seen. Do NOT overdry.
14. Resuspend pellets in 30 μL of TE overnight.
15. Analyze size and orientation of the insert by cutting 2–5 μL of DNA with restriction enzymes. Run in a 0.5% agarose gel along with size markers.

4. Notes

1. Single-stranded vectors: A different type of vector that allows cell transfection with single-stranded DNA has been devised by Wilson et al. *(57)* in an attempt to improve the yield of recombinant viruses. This vector is based on the single-stranded DNA bacteriophage M13. The recombinant phages obtained in this system contain single-stranded DNA that is easily extracted and purified before its use for cell transfection. Two analogous intermediate recombinants were constructed that contained the bacterial *cat* gene inserted in the Vv TK region, the only difference between them being the phage or plasmid backbone. A comparison of the recombinant yields in these instances showed that the single-stranded vector was fourfold more efficient in recombinant production. Although this difference has not been explained in molecular terms, this system offers the additional advantages of easy cloning, sequencing, and site-specific in vitro mutagenesis.
2. About the preparation of the insert: Vaccinia virus late promoters are characterized by the short distance between the starting point of transcription and the first ATG in the messenger RNA. In the case of pGS20 the BAMHI and SmaI insertion sites are located 30 bp downstream from the start site of transcription of the p7.5 promoter *(58)*. The sequence inserted will be transcribed as a continuation of the RNA initiated 30 bp upstream, and the first AUG to be encountered in the resulting RNA will act as the translation starting point. In consequence, it is convenient that the genuine starting ATG triplet of the gene to be expressed be located as near as possible to the insertion site, avoiding the presence of unwanted ATG triplets that could be present in a gene not properly trimmed.

References

1. Muller, G. and Williamson, J. D. (1987) Poxviridae, in *Animal Virus Structure* (Nermut, M. V. and Steven, A. C., eds.), Elsevier, Amsterdam, pp. 421–433.
2. Ginsberg, H. S. (1990) Poxviruses, in *Microbiology* 4th Ed. (Davis, B. D., Dulbecco, R., Eisen, H. N., and Ginsberg, H. S., eds.), Lippincott, Philadelphia, PA, pp. 947–959.
3. Panicali, D. and Paoletti, E. (1982) Construction of poxviruses as cloning vectors: Insertion of the thymidine kinase gene from herpes simplex virus into the DNA of infectious Vaccinia virus. *Proc. Natl. Acad. Sci. USA* **79**, 4927–4931.
4. Mackett, M., Smith, G. L., and Moss, B. (1982) Vaccinia virus: A selectable eukaryotic cloning and expression vector. *Proc. Natl. Acad. Sci. USA* **79**, 7415–7419.
5. Piccini, A. and Paoletti, E. (1988) Vaccinia: Virus, vector, vaccine. *Adv. Virus Res.* **34**, 43–64.
6. Wachsman, M., Aurelian, L., Smith, C. C., Lipinskas, B. R., Perkus, M. E., and Paoletti, E. (1987) Protection of guinea pigs from primary and recurrent herpes simplex virus (HSV) type 2 cutaneous disease with Vaccinia virus recombinants expressing HSV glycoprotein D. *J. Infect. Dis.* **155**, 1188–1197.
7. King, A. M. Q., Stott, E. J., Langer, S. J., Young, K. K.–Y., Ball, L. A., and Wertz, G. W. (1987) Recombinant Vaccinia viruses carrying the N gene of human respiratory syncitial virus: Studies of gene expression in cell culture and immune response in mice. *J. Virol.* **61**, 2885–2890.
8. Spriggs, M. K., Collins, P. L., Tierney, E., London, W. T., and Murphy, B. R. (1988) Immunization with Vaccinia virus recombinants that express the surface glycoproteins of human parainfluenza virus type 3 (PIV3) protects Patas monkeys against PIV3 infection. *J. Virol.* **62**, 1293–1296.
9. Pensiero, M. N., Jennings, G. B., Schmaljohn, C. S., and Hay, J. (1988) Expression of the Hantan virus M genome segment by using a Vaccinia virus recombinant. *J. Virol.* **62**, 696–702.
10. Karacostas, V., Nagashima, K., Gonda, M. A., and Moss, M. (1989) Human immunodeficiency virus-like particles produced by a Vaccinia virus expression vector. *Proc. Natl. Acad. Sci. USA* **86**, 8964–8967.
11. Franke, C. A. and Hruby, D. E. (1987) Association of nonviral proteins with recombinant vaccinia viruses. *Arch. Virol.* **94**, 347–351.
12. Huang, C., Samsonoff, W. A., and Grzelechi, A. (1988) Vaccinia virus recombinants expressing an 11-kilodalton–β galactosidase fusion protein incorporate active β galactosidase in virus particles. *J. Virol.* **62**, 3855–3861
13. Benninck, J. R., Yewdell, J. W., Smith, G. L., Moller, C., and Moss, B. (1984) Recombinant Vaccinia virus primes and stimulates influenza haemagglutinin-specific cytotoxic T cells. *Nature* **311**, 578,579.
14. Yewdell, J. W., Benninck, J. R., Smith, G. L., and Moss, B. (1988) Influenza A virus nucleoprotein is a major target antigen for cross-reactive anti-influenza A virus cytotoxic T lymphocytes. *Proc. Natl. Acad. Sci. USA* **82**, 1785–1789.
15. Jonjic, S., del Val, M., Keil, G. M., Reddehase, M. J., and Koszinowski, U. H. (1988) A nonstructural viral protein expressed by a recombinant Vaccinia virus protects against lethal cytomegalovirus infection. *J. Virol.* **62**, 1653–1658.
16. McLaughlin-Taylor, E., Willey, D. E., Cantin, E. M., Eberle, R., Moss, B., and Openshaw, H. (1988) A recombinant Vaccinia virus expressing Herpes simplex virus type I glycoprotein B induces cytotoxic T lymphocytes in mice. *J. Gen. Virol.* **69**, 1731–1734.

17. Flexner, C., Broiles, S. S., Earl, P., Chakrabarti, S., and Moss, B. (1988) Characterization of human immunodeficiency virus gag/pol gene products expressed by recombinant Vaccinia viruses. *Virology* **166**, 339–349.
18. Edwards, R. H., Selby, M. J., Mobley, W. C., Weinrich, S. L., Hruby, D. E., and Rutter, W. J. (1988) Processing and secretion of nerve growth factor: Expression in mammalian cells with a Vaccinia virus vector. *Mol. Cell. Biol.* **8**, 2456–2464.
19. de la Salle, H., Altenburger, W., Elkaim, R., Dott, K., Dieterle, A., Drillien, R., Cazenave, J.-P., Tolstoshev, P., and Lecocq, J.-P. (1985) Acitive gammacarboxyleted human factor IX expressed using recombinant DNA techniques. *Nature* **316**, 268–270.
20. Moss, B. (1985) Replication of poxviruses, in *Virology* (Fields, B. N., Chanock, R. M., and Roizman, B., eds.), Raven, New York, pp. 685–703.
21. Wittek, R., Menna, A., Schumperli, D., Stoffel, S., Muller, H. K., and Wiler, R. (1977) Hind III and Sst I restriction sites mapped on rabbit poxvirus and Vaccinia virus DNA. *J. Virol.* **23**, 669–678.
22. Geshelin, P. and Berns, K. I. (1974) Characterization and localization of the naturally occurying crosslinks in Vaccinia virus DNA. *J. Mol. Biol.* **88**, 785–796.
23. Baroudy, B. M., Venkatesan, S., and Moss, B. (1982) Incomplete base-paired flip-flop terminal loops link the two DNA strands of the Vaccinia virus genome into an uninterrupted polynucleotide chain. *Cell* **28**, 315–324.
24. Baroudy, B. M., Venkatesan, S., and Moss, B. (1983) Structure and replication of Vaccinia virus telomeres, in *Symposium on Quantitative Biology*, vol. 47, Cold Spring Harbor Laboratory, Cold Spring Harbor, NY, pp. 723–729.
25. Moyer, R. W. and Graves, R. L. (1981) The mechanism of cytoplasmic orthopoxvirus DNA replication. *Cell* **27**, 391–401.
26. Garon, C. F., Barbosa, E., and Moss, B. (1978) Visualization of an inverted repetition in Vaccinia virus DNA. *Proc. Natl. Acad. Sci. USA* **75**, 4863–4867.
27. Wittek, R. and Moss, B. (1980) Tandem repeats within the inverted terminal repetitions of Vaccinia virus DNA. *Cell* **21**, 277–284.
28. Tomley, F. M. and Cooper, R. J. (1984) The DNA polymerase activity of Vaccinia virus "Virosomes": Solubilization and properties. *J. Gen. Virol.* **65**, 825–829.
29. Munyon, W., Paoletti, E., and Grace, J. T., Jr. (1967) RNA polymerase activity in purified infectious Vaccinia virus. *Proc. Natl. Acad. Sci. USA* **58**, 2280–2287.
30. Baroudy, B. M. and Moss, B. (1980) Purification and characterization of a DNA dependent RNA polymerase from Vaccinia virus. *J. Biol. Chem.* **255**, 4372–4380.
31. Monroy, G., Spencer, H., and Hurwitz, J. (1978) Purification of mRNA guanylyl transferase from Vaccinia virions. *J. Biol. Chem.* **253**, 4481–4489.
32. Barbosa, E. and Moss, B. (1977) mRNA (nucleoside–2'–)–methyl transferase from Vaccinia virus. *J. Biol. Chem.* **253**, 7692–7697.
33. Moss, B., Rosenblum, E. N., and Gershowitz, A. (1975) Characterization of polyadenylate polymerase from Vaccinia virions. *J. Biol Chem.* **250**, 4722–4729.
34. Traktman, P., Sheidar, P., Condit, R. C., and Roberts, B. E. (1984) Transcriptional mapping of the DNA polymerase gene of Vaccinia virus. *J. Virol.* **49**, 125–131.
35. Earl, P. L., Jones, E. V., and Moss, B. (1986) Homology between DNA polymerases of poxviruses, herpesviruses and adenoviruses: Nucleotide sequence of the Vaccinia virus DNA polymerase gene. *Proc. Natl. Acad. Sci. USA* **83**, 3659–3663.
36. Weir, J. P., Bajszar, G., and Moss, B. (1982) Mapping of the Vaccinia virus Thymidine kinase gene by marker rescue and cell-free translation of selected mRNA. *Proc. Natl. Acad. Sci. USA* **79**, 1210–1214.

37. Weir, J. P. and Moss, B. (1983) Nucleotide sequence of the Vaccinia virus thymidine kinase gene and the nature of spontaneous frameshift mutations. *J. Virol.* **46,** 530–537.
38. Hruby, D. E., Maki, R. A., Miller, D. B., and Ball, L. A. (1983) Fine structure analysis and nuleotide sequence of the Vaccinia virus thymidine kinase gene. *Proc. Natl. Acad. Sci. USA* **80,** 3411–3415.
39. Smith, G. L., deCarlos, A., and San Chan, Y. (1989) Vaccinia virus encodes a thymidilate kinase gene: Sequence and transcriptional mapping. *Nucleic Acids Res.* **17,** 7581–7590.
40. Bertholet, C., Van Heir, E., ten Heggeler–Bordier, B., and Wittek, R. (1987) Vaccinia virus produces late mRNA by discontinuous synthesis. *Cell* **50,** 153–162.
41. Schwer, B. and Stunnenberg, H. G. (1988) Vaccinia virus late transcripts generated *in vitro* have a poly(A) head. *EMBO J.* **7,** 1183–1190.
42. Wittek, R., Richner, B., and Hiller, G. (1984) Mapping of the genes coding for the two major Vaccinia virus core polypeptides. *Nucleic Acids Res.* **12,** 4835–4847.
43. De Filippes, F. M. (1982) Retriction enzyme mapping of Vaccinia virus DNA. *J. Virol.* **43,** 136–149.
44. Moss, B., Winters, E., and Cooper, J. A. (1981) Deletion of a 9000 base-pair segment of the Vaccinia virus genome that encodes nonessential polypepotides. *J. Virol.* **40,** 387–395.
45. Weir, J. P. and Moss, B. (1987) Determination of the promoter region of an early Vaccinia virus gene encoding Thymidine kinase. *Virology* **158,** 206–210.
46. Davison, A. J. and Moss, B. (1990) Structure of Vaccinia virus early promoters. *J. Mol. Biol.* **210,** 749–769.
47. Davison, A. J. and Moss, B. (1990) Structure of Vaccinia virus late promoters. *J. Mol. Biol.* **210,** 771–784.
48. Pellicer, A. and Esteban, M. (1982) Gene transfer, stability, and biochemical properties of animal cells transformed with Vaccinia DNA. *Virology* **122,** 363–380.
49. Venkatesan, S. and Moss, B. (1981) In vitro transcription of the inverted terminal repetition of the Vaccinia virus genome: Correspondence of initiation and cap sites. *J. Virol.* **37,** 738–747.
50. Cochran, M. A., Puckett, C., and Moss, B. (1985) *In vitro* mutagenesis of the promoter region for a Vaccinia virus gene. Evidence for tandem early and late regulatory signals. *J. Virol.* **54,** 30–37.
51. Bertholet, C., Drillien, R., and Wittek, R. (1985) One hundred base pairs 5' flanking sequence of a Vaccinia virus late gene are sufficient to temporally regulate transcription. *Proc. Natl. Acad. Sci. USA* **82,** 2096–2100.
52. Bolivar, F., Rodriguez, R. L., Greene, P. J., Betlach, M. C., Meyneker H. L., and Boyer, H. W. (1977) Construction and characterization of new cloning vehicles. II: A multipurpose cloning system. *Gene* **2,** 95–113.
53. Mackett, M., Smith, G. I., and Moss, B. (1984) General method for production and selection of infectious Vaccinia virus recombinants expressing foreign genes. *J. Virol.* **49,** 857–864.
54. Vieira, J. and Messing, J. (1982) The pUC plasmids, an M13 mp7-derived system for insertion mutagenesis and sequencing with synthetic universal primers. *Gene* **19,** 259–268.
55. Boyle, D. B., Coupar, B. E. H., and Both, G. W. (1985) Multiple-cloning-site for the rapid construction of recombinant poxviruses. *Gene* **35,** 169–177.

56. Chakrabarti, S., Brechling, K., and Moss, B. (1985) Vaccinia virus expression vector: Coexpression of β-galactosidase provides visual screening of recombinant virus plaques. *Mol. Cell. Biol.* **5**, 3403–3409.
57. Wilson, E. M., Hodges, W. M., and Hruby, D. E. (1986) Construction of recombinant Vaccinia virus strains using single-stranded DNA insertion vectors. *Gene* **49**, 207–213.
58. Moss, B., Smith, G. L., and Mackett, M. (1983) Use of Vaccinia virus as an infectious molecular cloning and expression vector, in *Gene Amplification and Analysis* (Papas, T. K., Rosenberg, M. and Chirikjian, J. G., eds.), Elsevier, New York, pp. 201–213.

CHAPTER 21

Isolation and Handling of Recombinant Vaccinia Viruses

Antonio Talavera and Javier M. Rodriguez

1. Introduction

In Chapter 20, a general procedure was described for the construction of the intermediate vectors necessary for the insertion of foreign DNA sequences into the vaccinia virus (Vv) genome. The principles and basic methodology for the isolation of recombinant vaccinia viruses will be discussed in this chapter.

1.1. In Vivo Homologous Recombination

It has been known for a long time that wild-type viruses could be recovered in the progeny of cells doubly infected with pairs of vaccinia mutants *(1,2)*. This phenomenon was later shown to be responsible for the recovery, in the viral progeny, of genetic markers carried by DNA fragments that were introduced into vaccinia infected cells by the calcium phosphate coprecipitation method. This technique, known as "marker rescue," was first demonstrated by Nakano et al. *(3)* using a variant of vaccinia that contained a deletion in the genome, and a purified DNA fragment from wild-type vaccinia that spanned the deleted region. Virus was recovered whose DNA contained, in the proper place, the previously deleted sequence, acquired by double recombination at the homologous regions flanking the deletion, which were present in the fragment as well as in the viral genome. Subsequent marker rescue experiments led to the mapping of mutations in genes of unknown

From: *Methods in Molecular Biology, Vol. 8: Practical Molecular Virology: Viral Vectors for Gene Expression*
Edited by: M. Collins © 1991 The Humana Press Inc., Clifton, NJ

function (temperature-sensitive mutants) *(4,5)* or genes of definite function, such as the one responsible for rifampicin resistance *(6,7)*, or the ones encoding the viral thymidine kinase *(8)* or DNA polymerase *(9)*.

The occurrence of homologous recombination in vaccinia opened the possibility of inserting into the viral genome any foreign DNA previously cloned in insertion vectors. The procedure, which will be detailed in the Methods section, consists (in summary), in the introduction of the insertion vector into Vv-infected cells, followed by the selection of recombinant viruses among the viral progeny. The principles of this selection are discussed in the following section.

1.2. Selection of Recombinants

The insertion of the exogenous DNA sequences via in vitro homologous recombination is a very infrequent event. It is necessary to select, among the progeny virus, those rare individuals in which the insertion has taken place. The methods of selection are different, depending on the inserted (and expressed) genes as well as on the insertion site. The first recombinants reported *(10–12)* were those in which the expressed gene was the herpes *tk*. In one instance *(10)*, Syrian hamster (BHK 21) or African green monkey (CV1) cells were infected with wild-type vaccinia virus and immediately transfected with a calcium phosphate coprecipitate of vaccinia DNA and the insertion vector. Progeny virus was recovered by rupturing the cells 24 h after infection. The recombinant virus was screened by two methods:

1. Plating the progeny onto cell monolayers to obtain visible plaques; virus was lifted onto nitrocellulose membranes that were hybridized to a ^{32}P-labeled DNA probe containing insert sequences, followed by autoradiography;
2. Plating the virus and adding 5 (^{125}I) iododeoxycicidine, which can be phosphorylated by the herpes TK but not by the vaccinia or cellular TKs. Plaques in which the radioactive precursor had been incorporated in viral DNA were identified by lifting the virus onto nitrocellulose and autoradiography.

The first of these methods is of general application, but implies no previous enrichment of recombinants, so that it is necessary to screen a large number of plaques. In fact, Panicali and Paoletti *(10)* reported that only 0.4% of the plaques contained the herpes *tk* gene, as judged by *in situ* hybridization. Besides, as the proccess of lifting the virus destroys the cell monolayer, it was necessary to make and preserve replica filters in order to recover the recombinant viruses after their identification. The second procedure, apart from presenting the same disadvantages, is applicable only in this particular case.

A more useful screening procedure has been described for insertions into the vaccinia *tk* gene and takes advantage of its insertional inactivation *(11)*. This method, suitable for pGS20 and related insertion vectors, will be described in detail in the Methods section. In summary, the progeny virus is plated onto human TK^- 143B cells in the presence of 5-bromodeoxyuridine (BUdR). This thymidine analog can be phosphorylated by thymide kinase and incorporated into the cellular or viral DNA but, once there, it is toxic for the cells. Thus, TK^- cells can survive in the presence of BUdR, but those cells infected by nonrecombinant, TK^+ virus suicide before giving rise to progeny virus. The ones infected with recombinant TK^- virus are able to produce progeny viruses and start the formation of a plaque.

This method is, however, not 100% selective, since it does not distinguish between the TK^- recombinants sought for and the TK^- spontaneous mutants present in the infectious virus used in the construction of recombinants. An alternative selection method that is recombinant-specific has been described by Chakrabarti et al. *(13)*, and is based on the simultaneous insertion into the vaccinia *tk* gene of the required gene and the *E. coli* β-galactosidase, or *lacZ* gene, under the control of two different vaccinia promoters. Plaques with recombinant viruses containing both the required and the reporter gene can be visualized by adding 5-bromo-4-chloro-indolyl β-D-galactoside (X-gal) in the BUdR selective step. This colorless compound turns blue when hydrolized by the β-galactosidase, so the required mutant plaques can be identified by their blue color. The details of this selection procedure are given in the Methods section. A similar procedure has been described by Panicali et al. *(14)*.

2. Materials

1. Cell lines: CV1; 143B (TK^-) *(15)*.
2. Virus: Wild-type vaccinia virus (WR strain).
3. Media: Agar noble, 1.4% in H_2O; low-melting temperature (LMT) agarose, 2% in H_2O; 1.4% agar-neutral red; 2% LMT agarose-neutral red.
4. Buffers:
   a. HP: 0.1% glucose, 0.14M_4 NaCl, 5 m*M* KCl, 1 m*M* Na_2PO_4H, 20 m*M* HEPES, pH 7.1;
   b. 0.1*M* NaPO4, pH 8.5;
   c. 10 × TNE: 100 m*M* NaCl, 100 m*M* EDTA, 100 m*M* Tris-HCl, pH 8;
   d. TE: 10 m*M* Tris-HCl, pH 8, 1 m*M* EDTA.
5. Solutions:
   a. 1*M* $CaCl_2$;
   b. 2.5 mg/mL BUdR in H_2O, sterilized by filtration through a 0.45-μm pore filter;

c. 50 × Denhardt's solution (1% Ficoll™, 1% polyvinyl-pyrrolidone, 1% bovine serum albumin);
d. 10% sodium dodecyl sulfate (SDS), sterilized by filtration as above.

6. 5-Bromo-4-chloro-indolyl β-D-galactoside (X-gal).

3. Methods

3.1. Introduction of the Insertion Plasmid into Vaccinia-Infected Cells

Day 0.

1. Seed CV1 cells in Dulbecco's Modified Eagle's Medium (DMEM) +10% Fetal Calf Serum (FCS) in 50-cm² flasks or M-24 wells at 50,000 cells/cm². Incubate overnight at 37°C in a 10% CO_2 atmosphere.

Day 1.

2. Infect cells with Vv at a multiplicity of infection (moi) of 0.01–0.05 in 0.5 mL of medium+FCS. Give a preadsorption time of 2 h at 37°C. Meanwhile, prepare, for each flask to be transfected, a miture consisting of 1 mL of HP buffer containing 1 µg of insertion vector, 1 µg of vaccinia DNA (this increases the frequency of homologous recombination), and 1 µg of carrier (salmon sperm or λ DNA). Add to the mixture (dropwise and stirring between drops) 125 µL of 1*M* $CaCl_2$. Leave at least 45 min at room temperature. A cloudy aspect will develop, indicating precipitation of calcium phosphate, with which the DNA coprecipitates.
3. After the adsorption period, take the medium from the cells and add to each flask 0.8 mL of the corresponding precipitate.
4. Incubate 30 min at 37°C.
5. Add 8 mL of DMEM+FCS and incubate at 37°C for 3.5 h.
6. Wash once with serum-free medium, add 8 mL of complete medium, and incubate for 48 h (or until the cytopathic effect is total).

Day 3.

7. Remove and save the medium.
8. Detach cells by means of a rubber spatula and centrifuge the cells in the trypsin for 10 min at 1000 rpm.
9. Resuspend the cell pellet in the corresponding medium and sonicate to release the intracellular viral particles. Alternatively, freeze and thaw three times.

3.2. Selection of Recombinants

3.2.1. BUdR Selection

1. Make serial dilutions (10^{-1}–10^{-5}) of the viral preparations in DMEM containing 2% FCS. Plate on semiconfluent monolayers of 143B (TK^-) cells (*see* Section 3.4.1.) in 35-mm plates, adding 25 µg/mL BUdR (1:100 dilution from stock) to the top agar layer in the 10^{-1}–10^{-3} dilutions. Dilutions 10^{-4} and 10^{-5} are plated in the absence of BUdR to have a measure of the viral titer and the proportion of TK^- viruses.
2. Incubate for 2 d at 37°C. At this point, some plaques are visible under the microscope. Stain overnight with agar-neutral red. Count the plaques in the absence (total virus) and in the presence of BUdR (TK^- mutants *and* recombinants). In this case, the monolayer of the plates infected with the lowest dilutions may appear completely destroyed because of the high moi.
3. Pick the virus-containing agar from plaques in the BUdR plates by using Pasteur pipets. Resuspend the agar in 200 µL of medium with 2% FCS.
4. Infect 2-cm wells containing monolayers of CV1 cells with 50 µL of each plaque resuspension. After 2 h of adsorption, add 1 mL of complete medium. Incubate for 2–3 d, or until a complete cytopathic effect (cpe) is seen.
5. Collect the medium and cells in numbered Eppendorf tubes. Freeze and thaw three times. Make dot blots onto a nitrocellulose sheet with 200–500 µL of each viral suspension.
6. Denature the DNA on the nitrocellulose by placing the sheet on a filter paper soaked with 0.5*M* NaOH for 5 min. Blot the nitrocellulose dry on a filter paper. Repeat on a filter soaked in 1*M* Tris-HCl, pH 7.5. Repeat this step. Blot and place on a filter soaked with 2× SSC. Dry and bake at 80°C for 2 h.
7. Prehybridize by placing the nitrocellulose in a sealed plastic bag with 5–10 mL of 5× Denhardt's solution, 2× SSC, and 0.1 mg/mL of denatured carrier DNA (salmon sperm). Incubate at 65°C for 4 h.
8. Meanwhile, label with ^{32}P the purified insert used to make the insertion vector (*see* Chapter 20). Denature the probe by boiling it in 500 µL TE, 1% SDS, for 5 min. Place on ice for 5 min.
9. Open a corner of the plastic bag and introduce the denatured probe. Reseal and incubate overnight at 65°C.
10. Open the bag, dispose of the liquid, and wash the nitrocellulose twice at room temperature with 200 mL of 0.1 × SSC, 1% SDS. Then repeat three times with 200 mL of the same buffer at 65°C, for 30–60 min each time.
11. Blot the nitrocellulose on filter paper. Wrap in Saran Wrap™ and auto-

radiograph for 24 h. Identify the positives and use the contents of the corresponding Eppendorf tubes to expand the virus.

3.2.2. BUdR–X-gal Selection

1. Plate the virus as in Section 3.2.1. Incubate for 2 d.
2. Add, to the plates containing BUdR, a layer of low-melting-temperature (LMT) agarose-neutral red with an amount of X-gal, such that the final concentration in the total vol of agarose is 300 µg/mL. The X-gal stock should be prepared immediately before use at 150 mg/mL (500×) in dimethyl sulfoxide (DMSO). Let the agarose settle 10 min at room temperature.
3. Incubate for 2–6 h at 37°C (a longer time will result in virus inactivation). A blue color will appear in the plaques produced by recombinant virus. To enhance the color, leave the plates for 30 min at room temperature.
4. Pick blue plaques with Pasteur pipets and resuspend the agarose in 200 µL of medium with 2% FCS. Plate 1, 10, and 100 µL of the resuspensions in the same conditions. At this stage, more than 90% of the plaques should be blue; otherwise, give one more purification step.

3.3. Expansion and Analysis of Recombinants

1. Prepare one 60-mm plate with semiconfluent 143B cells. Inoculate each plate with 1/2 vol of the resuspension of a purified plaque. After a 2-h adsorption period, add 9.5 mL of complete medium plus 25 µg/mL of BUdR. Incubate at 37°C until the cpe is complete. Collect the cells and medium. Freeze and thaw three times. Centrifuge at 5000 rpm for 5 min to remove the cell debris. Titrate the supernatant.
2. Purify virus and DNA (*see* Sections 3. and 4.). Digest 1 µg of DNA with HindIII. Subject the digest to 0.5 % agarose electrophoresis in parallel with a HindIII digest of wild-type Vv DNA, and also size markers. Transfer the DNA to nitrocellulose by the method of Southern *(16)*.

Prehybridize as in the previous section, and hybridize with the same probe used for the selection of recombinants (probe A) or the whole insertion vector labeled with ^{32}P (probe B). The results expected are

- Probe A: No hybridization in the Vv DNA lane; hybridization of as many bands of the recombinant lane as there are HindIII sites in the insert, plus one. The sum of the sizes of the labeled bands should equal the sum of Vv DNA fragment HindIII J (4.8 kbp) plus the size of the insert.
- Probe B: Hybridization with fragment HindIII J in the Vv DNA lane.

Slight hybridization in the same lane with fragments B and C, as a result of the presence in the probe of the P7.5 promoter, also present at the end fragments of the DNA (*see* Chapter 20, Fig. 1). Disappearance of band HindIII J in the recombinant lane, replaced by the same bands as with the probe A.

3.4. General Procedures

3.4.1. Titration of Vaccinia Virus

1. Make serial dilutions of the virus suspension in 1 mL of DMEM+2% FCS.
2. Remove the medium from 35-mm plates (or M-24 wells) containing monolayers of CV1 cells. (VERO or BSC1 cells can also be used.)
3. Put 0.25 mL (35-mm plates) or 100 µL (M-24 wells) from each dilution onto duplicate monolayers. Incubate at 37°C for 2 h, shaking every 30 min (adsorption time).
4. Meanwhile, melt by boiling in a water bath, 2% LMT agarose. Keep at 50°C. Mix equal amounts of agar and doubly concentrated DMEM (2× DMEM), add 2% of the total vol of FCS, and keep at 45–50°C .
5. Remove the inocula from the plates (optional) and overlay the infected monolayers with the agarose mixture (3 mL for 35-mm plates, 1 mL for M-24 wells).
6. Let the agarose harden at room temperature for 10–15 min. Incubate at 37°C in a 10% CO_2 atmosphere. After 2 d, plaques should be visible under the microscope in the plates with high virus concentration.
7. Melt agar-neutral red. Mix with an equal amount of 2× DMEM without adding serum. Add 1 mL (35-mm plates) or 0.5 mL (M-24 wells) over the bottom agar. Let the overlay settle and incubate at 37°C overnight to allow diffusion of the dye to the cell monolayer.
8. Count the plaques and calculate the concentration of the original virus suspension as follows:
   a. For 35-mm plates, $T = N/0.25 \times D$, i.e. $T = 4 \times N/D$, where T is the titer in plaque forming units per mililiter (PFU/mL), N is the number of plaques in a plate, and D is the dilution corresponding to that particular plate.
   b. For M-24 wells, $T = 10 \times N \times D$.

The optimum amount of plates is 50–200 (depending on plaque size) for 35-mm plates, and 20–50 for M-24 wells.

3.4.2. Virus Purification

1. Infect monolayers of host cells in 100-mm plates or roller bottles at a multiplicity of infection (moi) between 0.1 and 1 PFU/cell. The vol of

the inocula should be 1 mL for plates or 5 mL for bottles. After 2 h at 37°C, add DMEM–2% FCS to bring the vol to 10 mL for plates or 50 mL for bottles (it is not necessary to withdraw the inocula). Incubate at 37°C until the cpe is total (cells are rounded and half-detached).

2. Remove and save the medium. Detach the still-attached cells by means of a rubber spatula. Add the cell suspension to the saved medium.
3. Centrifuge for 10 min at 5000 rpm at 4°C in a Sorvall GS3 rotor or equivalent. The supernatant contains the virus particles released from the cells (extracellular virus, EV). Most of the viral particles, however, are not released (intracellular virus, IV) and must be extracted from the cell pellet.

3.4.2.1. Purification of the IV

1. Resuspend the pellet in 10 mL of 0.01*M* phosphate buffer, pH 8.5.
2. Disrupt the cells with a Dounce homogenizer 10 times *(17,18)*.
3. Centrifuge 3 min at 750*g*. Discard the pellet (nuclear fraction).
4. Add 0.1 vol of 10 × PBS to the supernatant (cytoplasmic fraction).
5. Add purified trypsin, to a final concentration of 0.25 mg/mL. Incubate for 30 min at 37°C. This step loosens virus clumps without inactivating the virus infectivity *(19)*. Centrifuge for 10 min at 2000 rpm.
6. Layer the supernatant on a 6-mL cushion of 36% sucrose *(17)* in PBS, in a 17-mL tube of a Sorvall AH627 rotor. Centrifuge 90 min at 20,000 rpm, 4°C.
7. Discard the supernatant. Resuspend the viral pellet formed in the bottom of the tube in 1 mL of complete PBS (the pellet is easily dispersed by cavitation). Separate 10 µL for titration. Store the viral preparation in 50–100-µL aliquots at –70°C.

3.4.2.2. Purification of the EV

1. Centrifuge the medium at 8000 rpm, 4°C for 6 h (or overnight).
2. Resuspend the viral pellet in 10 mL of complete PBS at 4°C overnight and by cavitation.
3. Continue as in Step 5 above.

3.4.3. DNA Purification

1. Put 0.4 mL of purified virus in an Eppendorf tube. Add 62 µL of 10 × TNE, 62 µL of 10% SDS, and 100 µL of a stock solution of 5 mg/mL autodigested proteinase K. Incubate overnight at 37°C.
2. Extract once with phenol:chloroform:isoamyl alcohol (25:24:1). Centrifuge. Separate the organic (bottom) phase. Extract the aqueous phase with chloroform:isoamyl alcohol (24:1). Again centrifuge and separate the organic phase. Add 1/10 vol of 3*M* sodium acetate, pH 5.2 + 2 vol absolute ethanol at –20°C. Keep on ice for 30 min.

3. Centrifuge for 30 min. Decant the supernatant, and wash the pellet with 0.5 mL of 70% ethanol at –20° C. Centrifuge for 5 min. Decant the ethanol and blot the edge of the tube on filter paper, keeping the tube inverted. Air-dry at room temperature until no more liquid can be seen. DO NOT overdry. Dissolve the DNA pellet in 50–100 µL of TE. Determine the concentration by absorbance at 260–280 nm.

4. Notes

4.1. Other Selection Methods

The BUdR selection method is applicable only when the insertion site is the vaccinia *tk* gene. Two similar selection methods have been developed that do not depend on the TK phenotype of the infected cell. The first one, described by Franke et al. *(20)*, is based on the resistance to the antibiotic G418 conferred by the bacterial enzyme neomycin phophotransferase. Plasmids containing the *neo* gene driven by promoter P7.5, and the reporter bacterial *cat* gene driven by the Vv promoter PF *(21)*, were constructed in such a way that both genes were inside the Vv *tk* gene and were in the four possible relative orientations *(22)*. All recombinant viruses that were selected with G418 expressed a high level of CAT, suggesting no interference between neighboring genes, which could impair the coexpression of the selective and reporter genes. The *cat* gene used in this study can be replaced by any other gene.

Another antibiotic–enzyme system was simultaneously described by two independent groups *(23,24)*. It is based on the sensitivity of Vv growth to mycophenolic acid. This compound inhibits the cellular enzyme inosine monophosphate dehydrogenase, interfering with the synthesis of purine nucleotides. Eukaryotic cells that express the bacterial xanthine guanine phosphoribosyltransferase (the product of the *E. coli* gene *gpt*) are able to uptake xanthine and hypoxanthine from the culture medium as an alternative source of purine nucleotides. Insertion vectors were constructed that contained the *gpt* gene, preceded by the promoter P7.5 in the Vv *tk* gene. In one case *(24)*, a second Vv sequence, containing the vaccinia promoter P11 and the proximal part of the vaccinia gene 11 (including the initiation ATG triplet) was added upstream of the P7.5 promoter, after addition of none, one, or two G residues after the ATG codon to allow the expression of any open reading frame inserted in the multicloning site that followed. The vaccinia recombinants obtained could be selected in the presence of mycophenolic acid.

It is important to note that, although in both selection methods the recombinant viruses were prepared in such a way that the insert interrupted the vaccinia *tk* gene, no BUdR selection was used, indicating that these methods could be easily applied to the construction of vaccinia recombinants carrying the inserted genes in dispensible regions other than that of the *tk* gene.

4.2. Improvement of the Expression of Foreign Genes

Although for most purposes the expression of foreign gene that can be attained by using the "normal" Vv promoters (P7.5, P11, and PF) is sufficiently high, different methods have been used to optimize it. The most direct has been the use, in the insertion vectors, of a strong cowpox virus (CPv) promoter that directs the transcription of a 160-KDa protein that is the major component of the A-type cytoplasmic inclusions caused by CPv infection and reaches, at late times after infection, 4% of the total protein content in the cell *(25)*.

Another, more indirect, method made use of the prokaryotic RNA polymerase induced in *E. coli* on infection with the bacteriophage T7. This enzyme is the product of the T7 gene 1 and is very efficient (compared with other prokaryotic RNA polymerases) and specific for T7 promoters *(26)*. The feasibility of applying these characteristics to the expression of genes in Vv recombinants was investigated by Fuerst et al. *(27)* by inserting the T7 gene 1 into vaccinia under the promoter P7.5. This recombinant virus was used to infect cells that were also transfected with a plasmid containing the reporter *cat* gene flanked by a T7 promoter and a T7 transcription terminator. Assays of transient expression of the *cat* gene in the infected and transfected cells showed that only cells in which the T7 RNA polymerase gene was present were able to synthesize CAT. The same virus construct was subsequently used to infect cells that were simultaneously infected with other Vv recombinants containing, under the T7 promoter P10 control, genes of diverse origin, such as the one coding for the hepatitis B virus surface antigen (HBsAg), the *E. coli lacZ* gene, or the human immunodeficiency virus (HIV) envelope proteins *(28)*. In all cases studied, a higher level of the corresponding gene product was obtained in the double-infection system than was obtained in the parallel cases in which a Vv recombinant bearing the same gene under the control of a Vv promoter was used. This system, however, presented two interesting and different problems: On one hand, the protein levels obtained were not as high as predicted, taking into account the aforementioned efficiency of the T7 RNA polymerase. On the other hand, the obvious simplification of the system—a double Vv recombinant containing both the T7 RNA polymerase and the gene to be expressed—was not possible, as no double recombinants could be selected, apparently in consequence of their low viability. A more thorough study of the system *(29)* led to the conclusion that most of the mRNA translated in the mammalian cells by the T7 RNA polymerase remained uncapped, as a result of the stem-loop secondary structure at the 5' ends of the T7 transcripts, thus explaining the discrepancy, previously observed, between the high level of mRNA and the relatively low level of translation.

4.3. Attenuation of Vaccinia Recombinants

Although Vv has been widely used as a safe vaccine against smallpox for nearly two centuries, occasional complications other than the benign local pustules have appeared, such as encephalitis *(30)*, especially when the vaccine is administered to immunosupressed or immunocompromised individuals. This fact, albeit rare, has impeded the licensing of recombinant Vv as vaccination agents *(31)*. To avoid this inconvenience, attenuated vaccinia strains or mutants have been isolated. Among the first attenuated vaccinia viruses reported was a recombinant in which the viral *tk* gene had been used as the insertion site *(32)*. It appeared that the TK^- phenotype itself was responsible for the decreased virulence.

Other ways suggested to solve this problem have included the use for the preparation of recombinants of either vaccinia strains that showed spontaneous attenuation *(31)* or mutants prepared for such purpose. In this manner, Rodriguez et al. *(33)* used Vv mutant 48.7, which contains an 8-mDa deletion near the left side of the genome *(34)*, as well as point mutations in the gene encoding for an envelope protein *(35)*.

4.4. Practical Advice on Cell Seeding

Usually, when cells are seeded in small plates or wells, a vortex effect causes the cells to accumulate at the center of the plate, causing the subsequent monolayer to be uneven, which makes it difficult to count virus plaques. This effect cannot be avoided, as generally thought, by bumping the plates in two orthogonal directions (which usually makes vortexing worse). The best remedy is to avoid moving the plates for 20–30 min once seeded; operating as follows for 35-mm plates,

1. Calculate the total vol of medium and the total amount of cells needed. Prepare a suspension of cells in such a vol. Shake.
2. Arrange plates on trays inside the laminar-flow hood. Putting all other trays aside, place the first tray in the center of the hood.
3. Take 10 mL of the cell suspension. Open the first plate and put in 2 mL. Repeat for four more plates. Shake the cell suspension and proceed with five more plates. Fill the plates, letting the liquid spread slowly.
4. When the first tray is completely full, place the second tray on top, avoiding brusque movements. Fill all the plates of the second tray.
5. Stacks of four or five trays can be managed easily. If more trays are to

be prepared, start making the first stack toward one side of the hood, so there is enough room to start the second stack without moving the first one.

6. When finished, leave all the trays in the hood for 20–30 min. After that time, the cells will be deposited on the bottom of the plates, and the slight movement of the medium caused by transporting the trays to the incubator is not strong enough to displace the cells.
7. This procedure can easily be adapted to multiwell plates.

References

1. Ensinger, M. J. (1982) Isolation and genetic characterization of temperature-sensitive mutants of vaccinia virus WR. *J. Virol.* **43,** 778–790.
2. Drillien, R., Spehner, D., and Kirn, A. (1982) Complementation and genetic linkage between vaccinia virus temperature-sensitive mutants. *Virology* **119,** 372–381.
3. Nakano, E., Panicali, D., and Paoletti, E. (1982) Molecular genetics of vaccinia virus: Demonstration of marker rescue. *Proc. Natl. Acad. Sci. USA* **79,** 1593–1596.
4. Condit, R. C., Motyczka, A., and Spizz, G. (1983) Isolation, characterization, and physical mapping of temperature-sensitive mutants of vaccinia virus. *Virology* **128,** 429–443.
5. Drillien, R. and Spehner, D. (1983) Physical mapping of vaccinia virus temperature-sensitive mutations. *Virology* **131,** 385–393.
6. Tartaglia, J. and Paoletti, E. (1985) Physical mapping and DNA sequence analysis of the rifampicin resistance locus in vaccinia virus. *Virology* **147,** 394–404.
7. Baldick, C. J. and Moss, B. (1987) Resistance of vaccinia virus to rifampicin conferred by a single nucleotide substitution near the predicted NH_2 terminus of a gene encoding an M_r 62,000 polypeptide. *Virology* **156,** 138–145.
8. Hruby, D. E. and Ball, L. A. (1982) Mapping and identification of the vaccinia virus thymidine kinase gene. *J. Virol.* **43,** 403–409.
9. Jones, E. V. and Moss, B. (1984) Mapping of the vaccinia virus DNA polymerase gene by marker rescue and cell-free translation of selected RNA. *J. Virol.* **49,** 72–77.
10. Panicali, D. and Paoletti, E. (1982) Construction of poxviruses as cloning vectors: Insertion of the thymidine kinase from herpes simplex virus into the DNA of infectious vaccinia virus. *Proc. Natl. Acad. Sci. USA* **79,** 4927–4931.
11. Mackett, M., Smith, G. I., and Moss, B. (1982) Vaccinia virus, a selectable eukaryotic cloning and expression vector. *Proc. Natl. Acad. Sci. USA* **79,** 7415–7419.
12. Mackett, M., Smith, G. I., and Moss, B. (1984) General method for production and selection of infectious vaccinia virus recombinants expressing foreign genes. *J. Virol.* **49,** 857–864.
13. Chakrabarti, S., Brechling, K., and Moss, B. (1985) Vaccinia virus expression vector: Coexpression of β-galactosidase provides visual screening of recombinant virus plaques. *Mol. Cell. Biol.* **5,** 3403–3409.
14. Panicali, D., Grzelecki, A., and Huang, C. (1986) Vaccinia virus vectors utilizing the β-galactosidase assay for rapid selection of recombinant viruses and measurement of gene expression. *Gene* **47,** 93–99.

15. Rhim, J. S., Cho, H. Y., and Huebner, R. J. (1975) Nonproducer human cells induced by murine sarcoma virus. *Int. J. Cancer* **15,** 23–29.
16. Southern, E. (1975) Detection of specific sequences among DNA fragments seperated by gel electrophoresis. *J. Mol. Biol.* **98,** 503–517.
17. Joklik, W. K. (1962) The preparation and characterization of highly purified radioactively labeled poxvirus. *Biochim. Biophys. Acta* **61,** 292–302.
18. DeFilippes, F. M. (1976) Restriction enzyme digests of rapidly renaturing fragments of vaccinia virus DNA. *J. Virol.* **17,** 227–238.
19. Plantrose, D. N., Nishimura, C., and Salzman, N. P. (1962) The purification of vaccinia virus from cell cultures. *Virology* **18,** 294–301.
20. Franke, C. A., Rice, C. M., Strauss, J. H., and Hruby, D. E. (1985) Neomycin resistance as a dominant selectable marker for selection and isolation of vaccinia virus recombinants. *Mol. Cell. Biol.* **5,** 1918–1924.
21. Bertholet, C., Drillien, R., and Wittek, R. (1985) One hundred base pairs flanking sequence of a vaccinia virus late gene are sufficient to temporarily regulate transcription. *Proc. Natl. Acad. Sci. USA* **82,** 2096–2100.
22. Franke, C. A. and Hruby, D. E. (1988) Use of the gene encoding neomycin phosphotransferase II to convect linked markers into the vaccinia virus genome. *Nucleic Acids Res.* **16,** 1634.
23. Boyle, D. B. and Coupar, B. E. H. (1988) A dominant selectable marker for the construction of recombinant poxviruses. *Gene* **65,** 123–128.
24. Falkner, F. G. and Moss, B. (1988) *Escherichia coli gpt* gene provides dominant selection for vaccinia virus open reading frame expression vectors. *J. Virol.* **62,** 1849–1854.
25. Patel, D. D., Ray, C. A., Drucker, R. P., and Pickup, D. J. (1988) A poxvirus-derived vector that directs high levels of expressed and cloned genes in mammalian cells. *Proc. Natl. Acad. Sci. USA* 9931–9935.
26. Studier, F. W. and Moffat, B. A. (1986) Use of bacteriophage T7 RNA polymerase to direct selective high-level expression of cloned genes. *J. Mol. Biol.* **189,** 113–130.
27. Fuerst, T. R., Niles, E. G., Studier, F. W., and Moss, B. (1986) Eukaryotic transient-expression system based on recombinant vaccinia virus that synthesizes bacteriophage T7 RNA polymerase. *Proc. Natl. Acad. Sci. USA* **83,** 8122–8126.
28. Fuerst, T. R., Earl, P. L., and Moss, B. (1987) Use of a hybrid vaccinia virus T7 RNA polymerase system for expression of target genes. *Mol. Cell. Biol.* **7,** 2538–2544.
29. Fuerst, T. R. and Moss, B. (1989) Structure and stability of mRNA synthesized by vaccinia virus-encoded bacteriophage T7 RNA polymerase in mammalian cells. Importance of the 5' untranslated leader. *J. Mol. Biol.* **206,** 333–348.
30. Lane, J. M., Ruben, F. L., Neff, J. M., and Millar, J. D. (1969) Complications in smallpox vaccination. 1968. National surveillance in the United States. *N. Engl. J. Med.* **21,** 1201–1208.
31. Shida, H., Hinuma, Y., Hatanaka, M., Morita, M., Kidokoro, M., Suzuki, K., Maruyama, T., Takahashi-Nishimari, F., Sugimotoo, M., Kitamura, A., Miyazama, T., and Hayami, M. (1988) Effects and virulences of recombinant vaccinia viruses derived form attenuated strains that express the human T-cell leukemia virus type I envelope gene. *J. Virol.* **62,** 4474–4480.
32. Buller, R. M. L., Smith, G. L., Cremer, K., Notkins, A. L., and Moss, B. (1985) Decreased virulence of recombinant vaccinia virus expression vectors is associated with a thymidine kinase-negative phenotype. *Nature* **317,** 813–815.

33. Rodriguez, D., Rodriguez, J. R., Rodriguez, J. F., Trauber, D., and Esteban, M. (1989) Highly attenuated vaccinia virus mutants for the generation of safe recombinant viruses. *Proc. Natl. Acad. Sci. USA* **86,** 1287–1291.
34. Dallo, S. and Esteban, M. (1987) Isolation and characterization of attenuated mutants of vaccinia virus. *Virology* **159,** 408–422.
35. Dallo, S., Rodriguez, J. F., and Esteban, M. (1987) A 14 K envelope protein of vaccinia virus with an important role in virus-host cell interactions is altered during virus persistence and determines the plaque size phenotype of the virus. *Virology* **159,** 423–432.

CHAPTER 22

Design and Construction of Poliovirus Epitope Expression Vectors

Karen L. Burke, Jeffrey W. Almond, and David J. Evans

1. Introduction

We have developed the very safe and efficacious live-attenuated Sabin 1 poliovirus vaccine strain as a vehicle for the presentation of defined epitopes from foreign pathogens *(1–3)*. Precise modification of the poliovirus capsid is made possible by the application of recombinant DNA technology to cloned cDNA copies of the viral genome, which are infectious for mammalian cells in culture *(4)*.

Poliovirus occurs as three distinct serotypes 1, 2, and 3. Each consists of a single-stranded, positive-sense RNA genome of approx 7450 bases, with a genome-linked VPg protein at the 5' end and a polyA tract at the 3' end, which is enclosed within an icosahedral capsid. Poliovirus has become one of the best characterized animal viruses: The genome organization has been determined *(5)*, the three-dimensional structure solved *(6, 7)* and the regions of the capsid responsible for the induction of virus-neutralizing antibodies mapped (reviewed in ref. *8*).

We *(9)* and others *(10,11)* have reported the successful construction of intertypic poliovirus chimeras by exchange of one of the antigenic sites of type 1 poliovirus by the corresponding region of serotypes 2 or 3. The resulting chimeras exhibited composite antigenicity and immunogenicity, being

From: *Methods in Molecular Biology, Vol. 8: Practical Molecular Virology: Viral Vectors for Gene Expression*
Edited by: M. Collins © 1991 The Humana Press Inc., Clifton, NJ

able to induce an immune response against the parental serotype 1 and either type 2 or type 3, depending on the origin of the introduced sequences. These results led us to explore the interesting possibility of expressing antigenic determinants from other pathogens on the surface of the poliovirus capsid.

2. Poliovirus Structure

The poliovirus capsid is composed of 60 copies each of four structural proteins, VP1–4. The proteins VP1–3 have a similar core structure characterized by an eight-stranded antiparallel β-pleated sheet flanked by two short α helices *(6,7)*. The individual proteins are differentiated by their size, their amino and carboxy termini, and their connecting loops, which form the antigenic sites of the virion. The protein VP4 is somewhat smaller and, being entirely internal, does not contribute to the antigenic structure of the virus.

Four independent antigenic sites have been identified on the poliovirion (reviewed in ref. *8*). Of these, sites 2, 3, and 4 are complex, being composed of residues from more than one capsid protein. In contrast, antigenic site 1 comprises a linear epitope of 12 amino acids, which forms a distinct surface protrusion surrounding the fivefold axis of the icosahedral virus particle. The linear nature of this epitope made it the candidate of choice in the construction of the intertypic poliovirus chimeras mentioned previously. The construction of these chimeras involved modification of a subgenomic fragment of the cloned genome, followed by its subsequent incorporation into a full-length genomic clone.

Because of both the nature of the flanking sequences and the restriction sites present in the pBR322 vector, the region of cDNA encoding antigenic site 1 of Sabin 1 is most readily subcloned, mutated, and rebuilt on a 3.6-kb partial KpnI restriction fragment (nucleotides 66–3660; *12,13*). This imposes severe limitations on the rapid and extensive modification of antigenic site 1, and it was therefore decided to construct a cassette vector to facilitate this process. In its most basic form, a cassette vector contains introduced restriction sites flanking the region of interest that are unique to the entire construct, and that minimally modify the coding region of the vector.

3. Antigenic Site 1 Mutagenesis Cassette Vector pCAS1

Taking advantage of the degenerate nature of the triplet amino acid code, the nucleotide sequence of the Sabin 1 cDNA was examined for locations at which restriction sites could be introduced with minimum modification of the encoded amino acid sequence. (The University of Wisconsin

Genetics Computing Group [UWGCG] suite of sequence analysis programs *(14)* can be used to expedite this process.) The cDNA sequence surrounding site 1 allowed the introduction of unique SalI and DraI restriction sites at nucleotides 2753 and 2783, respectively, which was achieved using standard gapped duplex mutagenesis *(15)* of the 3.6-kb KpnI fragment previously described. Although the introduction of the DraI site resulted in the replacement of residue 102 of VP1 (aspartic acid) with phenylalanine, this was not considered disadvantageous, since the coding change was situated at the 5' side of the blunt restriction enzyme site, and so, would be removed upon DraI digestion prior to replacement of antigenic site 1.

The modified 3.6-kb KpnI partial fragment was introduced into pT7Sabin 1, an infectious clone of Sabin 1, onto which a T7 promoter had been engineered at the 5' end of the poliovirus sequence (K. Burke et al., unpublished data). The presence of the bacteriophage T7 DNA-dependent RNA polymerase promoter allows the synthesis in vitro of infectious poliovirus RNA from which virus can be recovered following transfection of suitable cells (*16*; *see* Chapter 23). The modified full-length clone was subsequently transferred to the pBR322-derived vector pFB1(2) (Pharmacia) from which the three DraI sites had been removed, by complete DraI digestion, and religated in the presence of an EcoRI linker.

The resulting cassette vector (designated pCAS1) therefore consists of a full length Sabin 1 cDNA under the control of a T7 promoter containing unique restriction sites flanking the region encoding antigenic site 1 (Fig. 1a). Transfection of subconfluent Hep2c monolayers with RNA generated from pCAS1 resulted in the production of a cytopathic effect within 3 d, thereby demonstrating that the single amino acid change (Asp_{102}–Phe) within antigenic site 1 had no apparent effect on virus viability. The full construction details of pCAS1 have been reported recently *(1)*.

4. Construction of Recombinant Poliovirus cDNA

The choice and design of the sequences to replace antigenic site 1 are beyond the scope of this chapter. However, there appears to be considerable flexibility in this region of the poliovirus capsid, since we have replaced the nine amino acids (-NSASTKNKD-) of site 1 with up to 21 amino acids and still retained virus viability. We have also deleted the majority of the site by replacement with the sequence -NGG- between the flanking vector Asp_{93} and Lys_{103} residues (J. Meredith and D. Evans, unpublished data). Not all recombinant cDNA constructs have allowed the recovery of viable poliovirus antigen chimeras, and we are currently trying to identify the stage(s) in the virus life cycle (e.g., assembly, receptor binding, or uncoating) at which the block occurs.

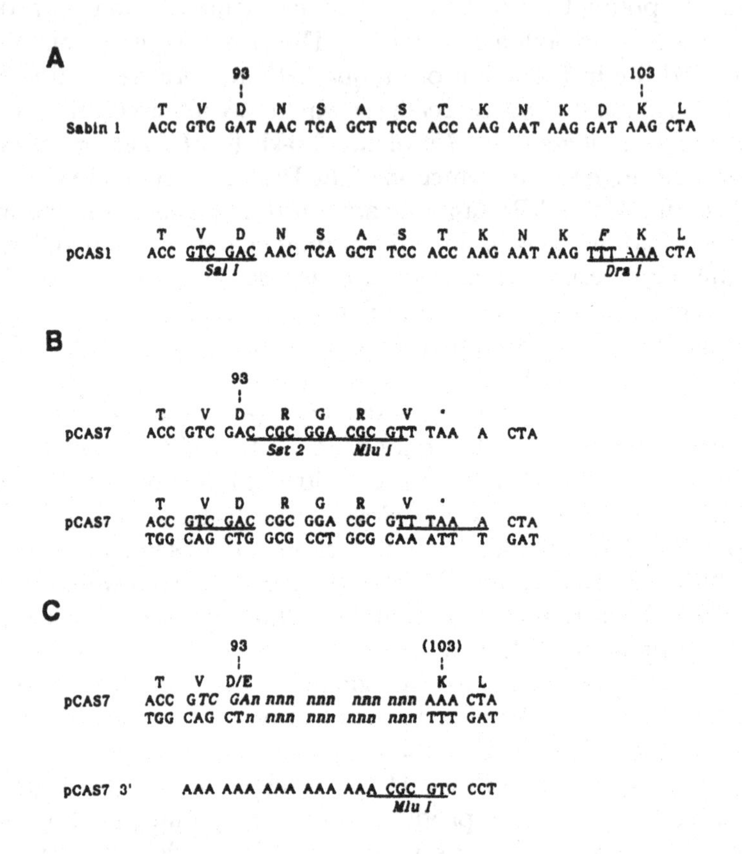

Fig. 1(A) Construction of the poliovirus cassette vector pCAS1. Unique SalI and DraI restriction sites were introduced into a Sabin 1 cDNA, occupying the nucleotides encoding residues 92/93 and 102/103 of VP1. The resulting change of Asp_{102} to Phe is indicated in italics. (B) Construction of the improved cassette vector pCAS7. The sequences encoding antigenic site 1 of Sabin 1 were removed and the unique SalI and DraI restriction sites separated by a linker that introduces a frameshift into VP1, together with a premature termination codon following VP1 96. Two additional restriction sites (SstII and MluI) were also introduced to facilitate screening recombinants. The * indicates the introduced termination codon. (C) Use of the pCAS7 vector. The vector is cleaved with SalI and DraI (underlined) and ligated with annealed complementary oligonucleotides encoding sequences of choice (indicated in italics). Since only the four nucleotides 5'-TCGA-3' are required for ligation with cut SalI vector, the second triplet of the SalI site can be used to encode an Asp (-GAT- or -GAC-, as in the parental virus) or a Glu (-GAA- or -GAG-) residue. The introduced MluI site at the 3' end of the poly-A tail allows synthesis of T7 runoff RNA without the retention of additional nonvirus sequences.

The vector pCAS1 is designed in such a way that recombinant poliovirus cDNAs can be rapidly generated, the entire process being readily accomplished in 2–3 d. Generally, two complementary oligonucleotides are synthesized, the coding strand being four nucleotides longer as a result of the presence of the 5'-TCGA-3' sequence compatible with the overhang that remains following SalI digestion of the vector.

Vector DNA is prepared by digestion with the restriction endonucleases Sal1 and Dra1, and purified on 1% agarose gels using standard techniques *(17)*. Approximately 100 pmol of each of the two complementary oligonucleotides are heated at 100°C for 3 min and slowly cooled to room temperature to anneal prior to ligation with 50–100 ng of the prepared vector. We generally design the insert oligonucleotides in such a manner that their ligation with the vector destroys the Dra1 site. Therefore, digestion of the ligated DNA with Dra1 prior to transformation of competent *E. coli* minimizes the recovery of unmodified vector. Screening of transformants for the insertion of oligonucleotides is usually carried out by digestion of plasmid DNA with Sal1 and Dra1. In addition, incorporation of restriction endonuclease sites into the insert sequence can also aid screening *(2)*. The recombinant DNA construct is subsequently used as a template for the production in vitro of RNA, which is used for the transfection of permissive cells, as outlined in Chapter 23.

5. Improved Cassette Vectors

As described in Section 3., infectious poliovirus can be recovered from pCAS1 even before antigenic site 1 is modified. We have recently improved the cassette system so that the vector itself is nonviable as a result of the introduction of both a frameshift and a termination codon within the region that should be replaced (Fig. 1b; J. Meredith and D. Evans, unpublished results). A number of other modifications have been made to the original pCAS1 that simplify the construction of cDNA recombinants and poliovirus antigen chimeras. The pBR322-derived origin of plasmid replication has been retained, but the tetracycline antibiotic selection has been replaced with kanamycin, thereby allowing *E. coli* strains harboring recombinants to grow more rapidly. The GC "tails" separating the poliovirus polyA tail and the plasmid vector sequences have been removed and replaced with a MluI restriction site. Recombinant cDNA constructs can therefore be linearized with MluI prior to producing the T7 RNA runoff transcript. Since MluI recognizes the palindrome A'CGCGT, cleavage with this enzyme results in a polyA tail with no additional sequences. It has recently been reported that such a modification increases the infectivity of runoff RNA produced from such cDNA constructs over those that retain nonvirus sequences at the 3' end *(18)*. The improved modified vector is designated pCAS7, and its use is shown in Fig. 1c.

6. Further Developments

Although most of our attention has focused on antigenic site 1, it is possible that other regions of the poliovirus capsid may be sufficiently flexible to allow the incorporation of foreign amino acid sequences. We have recently constructed a cassette vector, pCAS2b, that facilitates the exchange of a surface exposed region of capsid protein VP2 (K. L. Burke, unpublished data). The engineering of unique HpaI and SalI restriction endonuclease sites at positions 1429 and 1478, respectively, of Sabin 1 cDNA allow substitution of the region encoding amino acids VP2 160–176. Although part of a complex antigenic site, there is evidence suggesting that the region VP2 164–175 contains an epitope that may have some linear characteristics contributing to the conformational antigenic site 2 *(8)*. We are currently assessing the suitability of this region for the expression of heterologous antigenic domains.

The efficient construction of cassette vectors for other antigenic sites may be more rapidly achieved using a modification of the polymerase chain reaction (PCR; *19*). We have demonstrated that cassette vectors based on a modified "nonviable" cDNA construct are preferable to those from which unaltered vector virus can be recovered. The development of PCR for gene splicing by overlap extension *(20)* allows the rapid introduction of two unique restriction sites flanking a region of choice. The intervening sequences are determined by the choice of PCR primers and can be designed to include termination codons and/or frameshifts to ensure that the cassette vector is nonviable.

7. Concluding Remarks

We have described here a simple and rapid method for the construction of recombinant poliovirus cDNAs. Using this approach, we have been able to generate a large number of viable Sabin 1-derived chimeric viruses expressing on their surface known or predicted antigenic domains from a wide variety of sources (*1–3*, unpublished data).

Poliovirus epitope expression vectors based on site 1 allow the presentation of a defined antigenic determinant in a known immunogenic location. The epitope presented by the chimeric virus particle is constrained at the amino and carboxyl termini, which may qualitatively, and beneficially, affect the induced immune response. Furthermore, the poliovirus capsid contains a number of other surface-projecting loops, suggesting that there may be potential for the presentation of multiple and/or conformational epitopes.

There is no well-documented case of poliomyelitis resulting from vaccination with the Sabin 1 strain of poliovirus, implying that it is one of the safest vaccines currently in use. We are currently determining whether poliovirus antigen chimeras may have a role, not only as epitope expression systems, but also in future vaccine development.

Acknowledgment

We would like to thank Janet Meredith for expert assistance.

References

1. Burke, K. L., Evans, D. J., Jenkins, O., Meredith, J., D'Souza, E. D. S., and Almond, J. W. (1989) A cassette vector for the construction of antigen chimaeras of poliovirus. *J. Gen. Virol.* **70,** 2475–2479.
2. Evans, D. J., McKeating, J., Meredith, J., Burke, K. L., Katrak, K., John, A., Ferguson, M., Minor, P. D., Weiss, R. A., and Almond, J. W. (1989) An engineered poliovirus elicits broadly reactive HIV-1 neutralizing antibodies. *Nature* **339,** 385–388.
3. Jenkins, O., Cason, J., Burke, K. L., Lunny, D., Gillen, A., Patel, D., McCance, D., and Almond, J. W. (1990) An antigen chimaera of poliovirus induces antibodies against Human Papillomavirus type 16. *J. Virol.* **64,** 1201–1206.
4. Racaniello, V. R. and Baltimore, D. (1989) Cloned poliovirus complementary DNA is infectious in mammalian cell. *Science* **214,** 916–919.
5. Kitamura, N., Semler, B. L., Rothberg, P. G., Larsen, G. R., Adler, C. J., Dorner, A. J., Emini, E. A., Hanecack, R., Lee, J. J., van der Werf, S., Anderson, C. W., and Wimmer, E. (1981) Primary structure, gene organization, polypeptide expression of poliovirus RNA. *Nature* **219,** 547–553.
6. Hogle, J. M., Chow, M., and Filman, D. J. (1985) The three-dimensional structure of poliovirus at 2.9A resolution. *Science* **229,** 1358–1365.
7. Filman, D. J., Syed, R., Chow, M., Macadam, A. J., Minor, P. D., and Hogle, J. M. (1989) Structural factors that control conformational transitions and serotype specificity in type 3 poliovirus. *EMBO J.* **8,**1567–1579.
8. Minor, P. D. (1989) Picornavirus antigenicity, in *Current Topics in Immunology and Microbiology* (in press).
9. Burke, K. L., Dunn, G., Ferguson, M., Minor, P. D., and Almond, J. W. (1988) Antigen chimaeras of poliovirus as potential new vaccines. *Nature* **332,** 81,82.
10. Murray, M. G., Kuhn, R. J., Arita, M., Kawamura, N., Nomoto, A., and Wimmer, E. (1988) Poliovirus type 1/type 3 antigenic hybrid virus constructed in vitro elicits type 1 and type 3 neutralizing antibodies in rabbits and monkeys. *Proc. Natl. Acad. Sci. USA* **85,** 3202–3207.
11. Martin, A., Wychowksi, C., Couderc, C., Crainic, R., Hogle, J. M., and Girard, M. (1988) Engineering a poliovirus type 2 antigenic site on a type 1 capsid results in a chimaeric virus which is neurovirulent for mice. *EMBO J.* **7,** 2839–2847.
12. Nomoto, A., Omata, T., Toyada, H., Kuge, S., Horie, H., Katoaka, Y., Genba, Y., Nakano, Y., and Imura, N. (1982) Complete nucleotide sequence of the attenuated poliovirus Sabin 1 strain genome. *Proc. Natl. Acad. Sci. USA* **79,** 5793–5797.

13. Stanway, G., Hughes, P., Westrop, G. D., Evans, D. M. A., Dunn, G., Minor, P. D., Schild, G.C., and Almond, J. W. (1986) Construction of poliovirus intertypic recombinants by use of cDNA. *J. Virol.* **57,** 1187–1190.
14. Devereux, J., Haeberli, P., and Smithies, O. (1984) A comprehensive set of sequence analysis programs for the VAX. *Nucleic Acids Res.* **12,** 387–395.
15. Kramer, W., Drutsa, V., Jansen, H., Kramer, B., Pflugfelder, M., and Fritz, H. (1984) The gapped duplex DNA approach to oligonucleotide-directed mutation construction. *Nucleic Acids Res.* **12,** 9441–9456.
16. van der Werf, S., Bradley, J., Wimmer, E., Studier, F. W., and Dunn, J. J. (1986) Synthesis of infectious poliovirus RNA by purified T7 RNA polymerase. *Proc. Natl. Acad. Sci. USA* **83,** 2330–2334.
17. Maniatis, T., Fritsch, E. F., and Sambrook, J. (1982) *Molecular cloning: A Laboratory Manual.* Cold Spring Harbor Laboratory, Cold Spring Harbor, NY.
18. Sarnow, P. (1989) Role of 3'-end sequences in infectivity of poliovirus transcripts made in vitro. *J. Virol.* **63,** 467–470.
19. Saiki, R. K., Gelfand, D. H., Stoffel, S., Scharf, S. J., Higuchi, R., Horn, G. T., Mullis, K. B., and Erlich, H. A. (1988) Primer-directed enzymatic amplification of DNA with a thermostable DNA polymerase. *Science* **239,** 487–491.
20. Horton, R. M., Hunt, H. D., Ho, S. N., Pullen, J. K., and Pease, L. R. (1989) Engineering hybrid genes without the use of restriction enzymes: Gene splicing by overlap extension. *Gene* **77,** 61–68.

CHAPTER 23

Growth and Characterization of Poliovirus Antigen Chimeras

David J. Evans and Philip D. Minor

1. Introduction

Chapter 22 outlines the construction of engineered, full-length poliovirus cDNAs in which the region encoding a well-characterized antigenic site has been replaced by sequences of choice. This chapter briefly describes the methods used to generate, maintain, and characterize infectious chimeric viruses. These techniques include several developed during the course of poliovirus study that have recently been published *(1)*. This overview describes modifications to these techniques where relevant, but concentrates on methods not detailed by Minor *(1)*. The production and analysis of poliovirus antigen chimeras can be readily subdivided into three discrete areas.

1. The generation of infectious virus particles from a suitably modified cDNA;
2. The routine growth and maintenance of the antigen chimera; and
3. The characterization of the chimeric virus in terms of its genetic and antigenic structure.

These stages are prerequisites for the study of the immunogenicity of chimeric viruses, which, although outside the scope of this chapter, involves techniques mentioned in Section 3.

From: *Methods in Molecular Biology, Vol. 8:*
Practical Molecular Virology: Viral Vectors for Gene Expression
Edited by: M. Collins © 1991 The Humana Press Inc., Clifton, NJ

2. Materials

1. 10× Concentrated T7 reaction buffer: 400 m*M* Tris-HCl (pH 8), 80 m*M* $MgCl_2$, 500 m*M* NaCl.
2. 10× Concentrated *Taq* polymerase buffer: 100 m*M* Tris-HCl (pH 8.3), 500 m*M* KCl, 15 m*M* $MgCl_2$, 0.01% (w/v) gelatin.
3. Hep2C cells, maintained in Eagle's Minimal Essential Medium (EMEM) and supplemented with 5% newborn calf serum (NCS), 2.2 g/L Na_2CO_3, 120 mg/L penicillin, and 100 mg/L streptomycin.

3. Methods

3.1. T7 Reaction and Transfection

The cassette vector pCAS1 is designed in such a manner that, following linearization, infectious RNA can be produced in vitro using the DNA-dependent RNA polymerase T7 as previously described *(2)*. Briefly, suitably engineered pCAS1-derived cDNAs are linearized with NaeI (or other restriction endonucleases that cleave vector sequences alone and produce either 5' overhanging or blunt ends), phenol-extracted, precipitated, and resuspended in sterile water treated with diethyl pyrocarbonate (DPC). Approximately 100 ng of linearized DNA is used as a template for the T7 reaction. DNA produced by standard alkaline lysis protocols *(3)* is of sufficient quality to act as a template and generally produces more reproducible results than DNA purified by methods involving the use of cesium chloride.

1. Add the following to a microcentrifuge tube and incubate at 37°C for 30 min: 5.0 µL DNA (100 ng), 2.5 µL T7 reaction buffer, 7.5 µL 100 m*M* DTT, 2.5 µL 20 m*M* spermidine, 2.5 µL 5 m*M* rNTPs, 0.5 µL RNAguard™ (Pharmacia), 5.0 µL DPC-treated H_2O, 0.5 µL T7 RNA polymerase (50 U; Stratagene, La Jolla, CA).
2. Prior to DEAE-dextran transfection of subconfluent Hep2c monolayers *(2)*, visualize the products of the T7 reaction by agarose gel electrophoresis. RNA transcripts produced from NaeI-linearized pCAS1 migrate at approx 2.2 kb. Limited degradation products are often visible, but extensive smearing suggests contamination with RNases.
3. Following transfection, incubate cells at 34°C until a cytopathic effect (cpe) becomes apparent. This usually takes 3–4 d, but may take as little as 36 h or as much as 7 d.

It is advisable to perform parallel positive and negative controls (e.g., a known viable chimera and a nontransfected cell sheet) to allow the progress of the test chimera to be monitored. In certain cases, transfection yields no detectable cpe before destruction of the cell monolayer occurs as a result of natural cell aging. This may be a consequence of either the nonviability of virus encoded by a specific cDNA or the slow or retarded production of infectious particles from certain cDNA constructs. In the latter case, passage of cell-free supernatant to a fresh monolayer should produce a cpe. A number of constructs generated in this laboratory have failed to yield viable virus on transfection. The reasons for this are as yet unclear; however, preliminary analysis suggests that the charge distribution of residues occupying the first four positions of the engineered antigenic site 1 may be critical for viability (J. Thornton and H. Stirk, personal communication).

3.2. Routine Growth of Antigen Chimeras

Initial transfection of a suitable susceptible cell line, such as Hep2c, with infectious RNA produced in vitro from a recombinant cDNA generates a master stock of chimeric virus. This stock is usually of low titer, and is unsuitable for further analysis; therefore, it should be stored frozen, aliquoted at –70°C, and used to raise submaster stocks for further characterization.

1. Prior to aliquoting and freezing, freeze-thaw the tissue-culture supernatant three times and spin at 3000 rpm to remove cell debris.
2. Raise virus in 75-cm^2 flasks by the addition of 0.1 mL of stock to a PBS-washed cell monolayer and incubate at 34°C to allow virus to adsorb.
3. Add 5–10 mL of medium and continued incubation at 34°C until a full cpe develops.
4. The production of [^{35}S]methionine-labeled poliovirus has been described in detail elsewhere, as have the $TCID_{50}$ and plaque assays used to detemine the titer of recovered virus *(1)*.

3.3. Characterization of Poliovirus Antigen Chimeras

Poliovirus chimeras should be shown to be neutralized by pooled human serum containing poliovirus antibodies before preparing large, high-titer pools or purifying significant amounts of virus. We also routinely determine whether infection of susceptible cell lines with a chimeric virus is blocked by a monoclonal antibody to the poliovirus receptor. These steps ensure that the virus retains sufficient poliovirus character to be handled under containment conditions appropriate to poliovirus. Before detailed studies of the immunogenicity of a particular construct are initiated, a series of precautions should be taken to ensure that no spurious deletions or modifications have occurred during the transfection process. These include the reaction of the chimera with appropriate monoclonal or polyclonal antibodies, and the sequence determination of a region spanning the modified antigenic site.

3.4. Sequence Analysis of Recovered Chimeric Virus

The nucleic acid sequence of the recovered chimeric poliovirus can be determined in two ways, either by directly sequencing the purified viral RNA or by amplification of the region of interest using the polymerase chain reaction (PCR; *see* ref. *4*). We have successfully used the technique of Rico-Hesse et al. *(5)* for the purification and extraction of RNA from virions, and sequenced the RNA using the dideoxy-chain-termination method and ^{35}S *(6)*. This method may yield unsatisfactory results if the chimera grows only to low titer or if the preparation is contaminated with RNAses. We therefore routinely characterize recombinant viral genomes by PCR amplification of a 300-bp region spanning antigenic site 1, coupled with sequencing directly from the amplified product. This method requires an extremely small amount of starting material and takes only 3 d to complete.

3.5. Viral RNA Extraction, Reverse Transcription, and PCR Amplification

The primers used are positive sense 5'-GGGGCCACAAATCCAC-TAGTCCC-3' and negative sense 5'-GCCCATTGTTAGTCTCAGTG-3', which are complementary to, respectively, nucleotides 2627–2649 and 2924–2905 of Sabin 1 *(7)*.

1. Mix 25 µL of vanadyl ribonucleoside complexes (VRC; from Sigma) with 250 µL of cell-free tissue-culture supernatant.
2. Add 14 µL of 10% SDS and incubate on ice for 10 min.
3. Add 20 µL of proteinase K (20 mg/mL) and incubate at 37°C for at least 20 min.
4. Phenol/chloroform the extract twice, taking care to avoid interfacial material. Precipitate nucleic acids by the addition of 4 µL of 2 *M* NaAc and 750 µL of ice-cold ethanol, and leave at –20°C for at least 5 min prior to centrifigation at 12,000*g*.
5. Resuspend the pellet (which is black as a result of the presence of VRC) in 40 µL of TE.
6. Add the following to a 500-µL microcentrifuge tube and incubate at 37°C for at least 5 min: 2.0 µL RNA (from Step 5), 1.0 µL negative-sense primer (100 ng/µL), 2.0 µL MMLV RT buffer (5× concentrate), 2.5 µL 2 m*M* dNTPs, 1.5 µL H_2O, and 1.0 µL MMLV reverse transcriptase (BRL).
7. Use the entire reaction amount from Step 6 for PCR amplification. Add the following to the same tube: 10 µL of 2 m*M* dNTPs, 10 µL of 10× *Taq* polymerase buffer (as specified by the supplier), 500 ng of positive-sense primer, 500 ng of negative-sense primer, H_2O (to a total volume of 100 µL), and 2 U of Cetus "AmpliTaq™" cloned *Taq* polymerase. Cover the

reaction with 100 μL of mineral oil and amplify by completing 25–35 cycles at 95°C for 1 min, 55°C for 2 min, and 72°C for 3 min.

3.6. Purification, Cloning, and Sequencing of PCR Fragments

One-fifth (20 μL) of the PCR reaction is electrophoresed on a 1% agarose gel, and the 300-bp fragment corresponding to the amplification product extracted by elution, either onto 3M filter paper *(8)* or using Geneclean™ (Bio-101, La Jolla, CA), the recovered DNA finally being resuspended in 10 μL of TE. Considerable problems have been encountered in the direct (blunt-end) cloning of such amplification products into suitable vectors, such as M13- or pUC-derivatives. "False whites" are often obtained, necessitating the screening of large numbers of recombinants. This is best overcome by including suitable restriction sites into the 5' end of the amplification primers to allow efficient subcloning (*see* Section 4.).

Unless cloned PCR product is required for subsequent analysis, it is usually sufficient to sequence directly for the gel-purified fragment using the method described by Winship *(9)*.

1. Mix approx 60 ng of gel-purified PCR product with 100 ng of positive-sense PCR primer in 40 m*M* Tris-Cl (pH 7.5), 25 m*M* $MgCl_2$, 50 m*M* NaCl, and 10% DMSO.
2. Boil for 3 min and snap-cool in a dry ice/ethanol bath.
3. Add 4 μL of labeling mix (0.025*M* DTT, 10 μCi ^{35}S dATP SA 1200 Ci/mmol, 2 U Sequenase™).
4. Add 2.5-μL aliquots to reaction tubes containing 2 μL of 80 μ*M* dCTP, dGTP, and dTTp; 50 m*M* NaCl; 10% DMSO supplemented with 0.08 μ*M* ddATP (A reaction); 8 μ*M* ddCTP (C); 8 μ*M* ddGTP (G); or 8 μ*M* ddTTP (T).
5. Incubate for 5 min at 37°C.
6. Add 2 μL of 0.25 m*M* dATP, dCTP, dGTP, dTTP, 50 m*M* NaCl; and 10% DMSO chase, and incubate for a further 5 min at 37°C. Following the addition of 4 μL of loading dye, reactions are denatured at 75°C for 10 min before electrophoresis.

3.7. Antigenic Characterization of Chimeric Viruses

The problems associated with the use of ELISA and other conventional immunoadsorbent assays for the detection of poliovirus antigens have been documented previously *(1)*. In practice, two types of assay are routinely used: single radial immunodiffusion, or "antigen blocking" *(10)*, and neutralization of virus infectivity. Antigen blocking assays involve the use of

[^{35}S]methionine-labeled virus preparations, and yield results in 1–2 d. The precise techniques involved in the application of this method have been described in detail *(1)*. Assays for virus neutralization are performed in sterile 96-well microtiter plates using a standard 2-well assay. Antiserum is serially diluted in twofold steps in EMEM containing 5% NCS and 50-µL aliquots incubated with an equal volume of virus (containing 100 $TCID_{50}$) for 2 h at 34°C. Remaining virus infectivity is determined by the addition of 100 µL of Hep2c cell suspension (approx 10^4 cells/well) and incubation at 34°C for a further 5 d prior to staining with naphthalene black. Wells are scored for the presence or absence of cpe. A virus titration control should be included in parallel with the neutralization assay.

3.8. Concluding Remarks

Poliovirus antigen chimeras represent a novel approach to foreign epitope presentation that has been successfully applied to a wide range of pathogens *(11)*, including human immunodeficiency virus *(12)* and human papillomavirus *(13)*. The construction and characterization of such chimeras involve the application of a series of techniques widely used in molecular virology, and so could be applied to antigenicity studies on any protein of interest. The immunogenicity of poliovirus antigen chimeras has only been tested on parenteral administration to rodents. Future studies are designed to investigate the immune response on oral administration to primates. These experiments will determine whether chimeras derived from the Sabin 1 strain of poliovirus may have potential in the development of novel, live-attenuated vaccines.

4. Notes

PCR amplification is a uniquely sensitive technique and, as such, is prone to serious problems associated with DNA contamination of buffers, reagents, and primers. It is advisable to prepare a "supermix" when more than one amplification is being performed, with extreme care being taken to avoid cross-contamination. The reader is advised to read the commentary by Kwok and Higuchi *(14)* for other safeguards, and to perform suitable parallel controls (positive and negative).

The primers described in Section 3.5. amplify a 300-bp fragment containing unique *SphI* and *XbaI* restriction sites, thereby facilitating the direct cloning into a vector such as MBmp18 or puc18.

References

1. Minor, P. D. (1985) Growth, assay and purification of picornaviruses, in *Virology: A Practical Approach* (IRL Press, Oxford, UK), pp. 25–41.

2. van der Werf, S., Bradley, J., Wimmer, E., Studier, F. W., and Dunn, J. J. (1986) Synthesis of infectious poliovirus RNA purified T7 RNA polymerase. *Proc. Natl. Acad. Sci. USA* **83,** 2330–2334.
3. Birnboim, H. C. and Doly, J. (1979) A rapid alkaline extraction procedure for screening recombinant plasmid DNA. *Nucleic Acids Res.* **7,** 1513–1523.
4. Saiki, R. K., Gelfand, D. H., Stoffel, S., Scharf, S. J., Higuchi, R., Horn, G. T., Mullis, K. B., and Erlich, H. A. (1988) Primer-directed enzymatic amplification of DNA with a thermostable DNA polymerase. *Science* **239,** 487–491.
5. Rico-Hesse, R., Pallansch, M. A., Nottay, B. K., and Kew, O. M. (1987) Geographic distribution of wild poliovirus type 1 genotypes. *Virology* **160,** 311–322.
6. Sanger, F., Nicklen, S., and Coulsen, A. R. (1977) DNA sequencing with chain-terminating inhibitors. *Proc. Natl. Acad. Sci. USA* **74,** 5463–5467.
7. Nomoto, A., Omata, T., Toyada, H., Kuge, S., Horie, H., Katoaka, Y., Genba, Y., Nakano, Y., and Imura, N. (1982) Complete nucleotide sequence of the attenuated poliovirus Sabin 1 strain genome. *Proc. Natl. Acad. Sci. USA* **79,** 5793–5797.
8. Maniatis, T., Fritsch, E. F., and Sambrook, J. (1982) *Molecular Cloning: A Laboratory Manual.* Cold Spring Harbor Laboratory, Cold Spring Harbor, NY.
9. Winship, P. (1989) An improved method for directly sequencing PCR amplified material using dimethyl sulphoxide. *Nucleic Acids Res.* **17,** 1266.
10. Ferguson, M., Minor, P. D., Magrath, D. I., Qui, Y-H., Spitz, M., and Schild, G. C. (1984) Neutralization epitopes on poliovirus type 3 particles: An analysis using monoclonal antibodies. *J. Gen. Virol.* **65,** 197–201.
11. Burke, K. L., Evans, D. J., Jenkins, O., D'Souza, E., and Almond, J. W. (1989) A cassette vector for the construction of antigen chimaeras of poliovirus. *J. Gen. Virol.* **70,** 2475–2479.
12. Evans, D. J., McKeating, J., Meredith, J., Burke, K. L., Katrak, K., John, A., Ferguson, M., Minor, P. D., Weiss, R. A., and Almond, J. W. (1989) An engineered poliovirus elicits broadly reactive HIV-1 neutralizing antibodies. *Nature* **339,** 385–388.
13. Jenkins, O., Cason, J., Burke, K. L., Lunny, D., Gillen, A., Patel, D., McCance, D., and Almond, J. W. (1989) An antigen chimaera of poliovirus induces antibodies against Human Papillomavirus type 16. *J. Virol.* **64,** 1201–1206.
14. Kwok, S. and Higuchi, R. (1989) Avoiding false positives with PCR. *Nature* **339,** 237,238.

Chapter 24

Expression Vectors for the Construction of Hybrid Ty-VLPs

Sally E. Adams, S. Mark, H. Richardson, Susan M. Kingsman, and Alan J. Kingsman

1. Introduction

The synthesis of recombinant proteins or protein domains in microbial, insect, or mammalian systems is now commonplace in molecular biology laboratories. The gene or gene fragment encoding the protein of interest is inserted into a specialized expression vector, flanked by efficient transcription and translation control sequences. The expression vector is then inserted into recipient cells and expression of the protein induced. The expressed protein then has to be purified from other cellular or medium components. Purification can be facilitated by expressing the recombinant protein as a fusion with a carrier protein that assembles into particulate structures. This approach has been developed using a protein encoded by the yeast retrotransposon Ty, which self-assembles into virus-like particles (VLPs) *(1,2)*. Additional protein coding sequences can be fused to the carrier protein gene and expressed in yeast to produce hybrid Ty-VLPs *(3,4)*. The physical characteristics of the VLPs have been exploited to produce a rapid purification procedure that is essentially generic for any hybrid construction. Hybrid VLPs can be used in many laboratory applications (*see* elsewhere in this vol), including the production of polyclonal and monoclonal antibodies, struc-

From: ***Methods in Molecular Biology, Vol. 8: Practical Molecular Virology: Viral Vectors for Gene Expression***
Edited by: M. Collins © 1991 The Humana Press Inc., Clifton, NJ

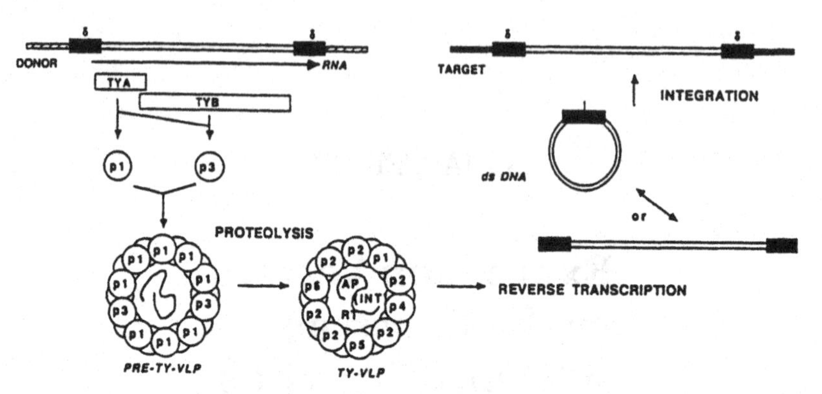

Fig. 1. The Ty transposition cycle. A Ty element is shown at a chromosomal "donor" site, with flanking sequences shown as striped boxes. Flanking yeast sequences at the "target" site are shown as hatched boxes. Direct terminal-repeat sequences, or deltas (δ), are shown as shaded boxes. AP, INT, and RT correspond to proteins with homologies to acid proteases, integrases, and reverse transcriptase, respectively.

ture/function analyses, the detection of important antigenic determinants, and epitope mapping of monoclonal antibodies.

1.1. The Yeast Retrotransposon Ty

The yeast retrotransposon Ty is a member of a class of eukaryotic DNA elements that move from one genomic location to another via an RNA intermediate and a reverse transcription reaction *(1,5,6)*. The haploid genome of laboratory strains of yeast contains 30–35 copies of Ty that are found at variable chromosome locations *(7)*. Over the past few years, a detailed analysis of the genetic organization and expression strategies of Ty elements has revealed many similarities with other retrotransposons, such as the copia-like sequences of *Drosophila* and retroviral proviruses *(1,5,8–10)*.

The majority of Ty elements are 5.9 kb in length, which can be subdivided into a unique 5.2-kb internal region flanked by 340-bp δ sequences that are analogous to retroviral LTRs (long terminal repeats) (Fig. 1). The major Ty RNA is a 5.7-kb transcript that begins in the 5' δ and ends in the 3' δ such that it is terminally redundant *(11,12)*. This RNA is both the major message and the intermediate in Ty transposition. The 5.7-kb Ty RNA is divided into two open reading frames, *TYA* and *TYB*, that are analogous to retroviral *gag* and *pol* genes, respectively *(8,10)*. *TYA* is expressed by simple translation to produce a 50-kDa protein called p1 *(9)*, which is a polyprotein precursor that is subsequently processed to obtain a 45-kDa protein called p2 and minor products p4, p5, and p6 *(1)*. *TYB* is expressed as a fusion with the *TYA* gene to produce a 190-kDa *TYA:TYB* fusion protein called p3 *(8,10,13)*. The 5' end of

TYB overlaps the 3' end of *TYA*, with *TYB* being in the +1 reading phase with respect to *TYA*. The production of p3 therefore requires a ribosomal frameshift event within the overlap region *(13)*. In addition, p3 is proteolytically processed and contains a retroviral-like protease, reverse transcriptase, and integrase *(1,9,10)* (Fig. 1).

The transposition cycle of Ty involves the packaging of Ty RNA into pre-Ty-VLPs composed of p1 and p3. This is followed by proteolytic maturation, resulting in part from the action of the *TYB*-encoded protease (Fig. 1) *(1)*. Mature Ty-VLPs are approx 50 nm in diameter and are composed mainly of p2, the p1 cleavage product. In addition, the VLPs contain reverse transcriptase, a tRNA primer, and probably integrase *(1,5)*. These components enable the VLPs to carry out an endogenous reverse transcription to produce a double-stranded DNA copy that integrates into the genome, and it is possible that this is their role in vivo.

1.2. The Formation of Hybrid Ty-VLPs

Although the major Ty protein found in mature VLPs is p2, the cleavage event that generates p2 from p1 is not necessary for particle formation *(1)*. This observation suggests that it may be possible to construct Ty-VLPs from p1 fusion proteins and thereby create particulate, polyvalent derivatives of, potentially, any protein or protein domain. This idea has been tested by producing p1 fusion proteins comprising most of p1 and all or parts of human α-interferon *(14)* or various viral proteins, including those encoded by human immunodeficiency virus-1, influenza, feline leukemia, and bovine papilloma viruses (*3,4,14,15;* unpublished data). In all cases the fusion proteins retained the ability to assemble into hybrid Ty-VLPs that could be readily purified, and they elicited strong immune responses against the added antigen.

In order to express Ty fusion proteins in yeast, specialized expression vectors have been designed and constructed (Fig. 2). The *TYA* gene has been trimmed at the 3' end in such a way that it encodes the first 381 codons of p1. At codon 381 is a BamHI site for the insertion of additional protein coding sequences, followed by termination codons in all three reading frames and a transcription terminator *(3)*. Expression of the truncated *TYA* gene (*TYA*[d]) is driven from the efficient yeast phosphoglycerate kinase (PGK) promoter *(16)*. This expression cassette containing the *TYA*(d) gene flanked by PGK promoter and terminator sequences has been inserted into an *E. coli*/yeast shuttle vector to create plasmid pMA5620 *(3)*. This contains DNA origins of replication that allow propagation in both types of cells, an ampicillin-resistance gene for selection in *E. coli*, and a *LEU2* gene for selection in yeast strains auxotrophic for leucine production. Plasmid copy number within the yeast cell is maintained at 100–200 copies/cell. Plasmids pMA5621 and pMA5622 are derivatives of pMA5620 in which the BamHI expression sites are in the +1

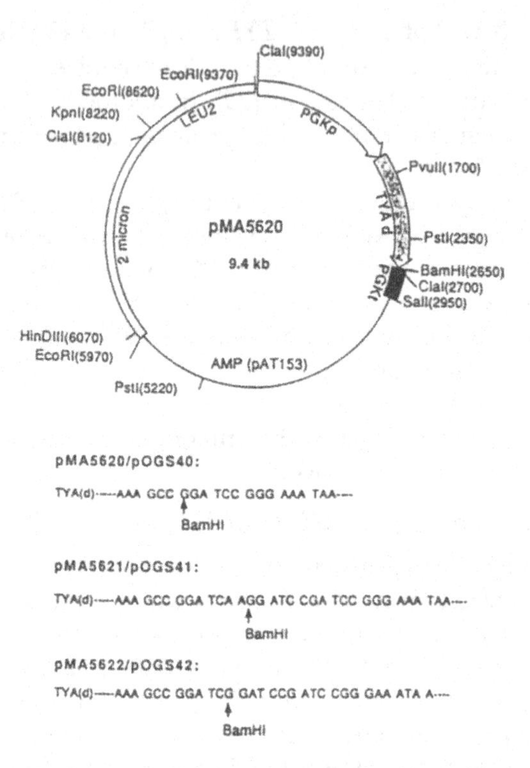

Fig. 2. Structure of the VLP expression vector pMA5620. Plasmid pMA5620 contains the yeast PGK promoter (PGKp; open arrow), the first 381 codons of the *TYA* gene (*TYA*[d]; lightly shaded arrow), a PGK terminator region (PGKt; darkly shaded box), selection and replication modules for yeast (2μ; *LEU2*; open box) and *E. coli* (thin line). The sequence around the unique BamHI site and reading phase in pMA5620 and derivatives is shown below the map.

and +2 reading frames with respect to the BamHI site of pMA5620 (Fig. 2). This series of vectors therefore allows any coding sequence to be expressed in frame with the truncated *TYA* gene.

The use of the PGK promoter results in high-level constitutive expression. In some instances, large amounts of a particular product may be toxic to the cells. An alternative expression strategy is therefore to grow the transformed cells to a high cell density before "switching on" production of the protein. Plasmids pOGS40, pOGS41, and pOGS42 are derivatives of pMA5620, pMA5621, and pMA562, respectively, in which the PGK promoter has been replaced by a hybrid PGK-GAL promoter that can be induced by galactose *(17)*.

2. Materials

1. Restriction enzymes: Klenow fragment, calf intestinal alkaline phosphatase, T4 DNA polymerase, T4 polynucleotide kinase, and T4 DNA ligase can be purchased from New England Biolabs (Beverly, MA).
2. Glusulase can be purchased from du Pont.
3. TE buffer: 10 m*M* Tris-HCl, 1 m*M* EDTA (pH 8.0).
4. 10× BamHI restriction enzyme buffer: 1.5*M* NaCl, 60 m*M* $MgCl_2$, 60 m*M* Tris-HCl (pH 7.8).
5. 10× Klenow buffer: 0.5*M* Tris-HCl (pH 7.2), 0.1*M* $MgSO_4$.
6. 10× T4 polymerase buffer: 0.33*M* Tris acetate (pH 7.9), 0.66*M* potassium acetate, 0.1 *M* magnesium acetate, 5 m*M* dithiothreitol (DTT), 1 mg/mL bovine serum albumin.
7. 10× Kinase buffer: 0.5*M* Tris-HCl (pH 7 .6), 0.1*M* $MgCl_2$, 50 m*M* DTT, 1 m*M* spermidine, 1 m*M* EDTA.
8. 10× Ligase buffer: 0.5*M* Tris-HCl (pH 7.5), 0.1*M* $MgCl_2$, 10 m*M* spermidine.
9. 10× AD buffer: 0.1 *M* DTT, 20 m*M* ATP.
10. STET buffer: 8% (w/v) sucrose, 5% (v/v) Triton X-100, 50 m*M* EDTA, 50 m*M* Tris-HCl (pH 8.0).
11. Sarkosyl solution: 0.1% (v/v) Triton X-100, 62.5 m*M* EDTA, 50 m*M* Tris-HCl (pH 8.0).
12. STC buffer: 1*M* sorbitol, 10 m*M* $CaCl_2$, 10 m*M* Tris-HCl (pH 7.5).
13. L broth: 1% (w/v) tryptone, 0.5% (w/v) yeast extract, 1% (w/v) NaCl. Add 2% (w/v) agar for plates.
14. YEPD medium: 2% (w/v) peptone, 1% (w/v) yeast extract, 2% (w/v) glucose.
15. Regeneration agar: 1*M* sorbitol, 0.67% (w/v) yeast nitrogen base without amino acids, 1% (w/v) glucose, 3% (w/v) agar. Autoclave, cool, and add appropriate amino acids depending on the auxotrophy of the yeast strain used. In order to maintain selection in culture, leucine should be omitted. Amino acid stocks can be made as a 50× solution and filter-sterilized. Dispense the regeneration agar into 50-mL Falcon tubes and keep warm (<48°C) in a water bath until required.
16. Synthetic complete-glucose (SC-glc) medium: 0.67% (w/v) yeast nitrogen base without amino acids, 1% (w/v) glucose. Add appropriate amino acids after autoclaving. For plates add 2% (w/v) agar.
17. Phenol/chloroform: Add 8-hydroxyquinoline (to 0.1% [w/v]) to molecular biology grade phenol (this can be obtained from GIBCO-BRL). Extract the phenol twice with 1*M* Tris-HCl (pH 8.0) to neutralize.

Mix the phenol 1:1 with chloroform and equilibrate with TE buffer. Store at 4°C in a dark bottle.

18. Water-saturated chloroform.
19. Water-saturated butanol.
20. Solution containing 2 m*M* of each of dCTP, dATP, dGTP, TTP.
21. 10 m*M* DTT.
22. 3*M* sodium acetate (pH 7.0).
23. Ethidium bromide, 10 mg/mL.
24. 0.5*M* EDTA (pH 8.5).
25. 100 m*M* $CaCl_2$.
26. Lysozyme, 10 mg/mL.
27. 25% (w/v) sucrose dissolved in 0.05*M* Tris-HCl (pH 8.0).
28. 1*M* sorbitol.
29. 44% (w/v) polyethylene glycol (PEG) 4000.
30. 40% glycerol.

3. Methods

In this protocol, DNA fragments or synthetic oligomers are cloned into a VLP expression vector and transformed into yeast.

3.1. Vector Preparation

1. Digest 5 µg of pMA5620 (or appropriate derivative) by mixing the DNA with 5 U of BamHI, 5 µL of 10× BamHI restriction enzyme buffer, and water to bring the final vol to 50 µL. Incubate at 37°C for 60 min.
2. If a blunt-ended vector is required, mix 1 µg of digested DNA, 1 µL of 10× Klenow buffer, 1 µL of 10 m*M* DTT, 0.2 U of Klenow fragment, 0.5 µL of a mixture of 2 m*M* dATP, dCTP, dGTP and TTP, and water to bring the final vol to 10 µL. Incubate at room temperature for 30 min. Add 40 µL of water, extract twice with 100 µL of phenol/chloroform and twice with 50 µL of chloroform, and precipitate the DNA on ice with 0.1 vol of 3*M* sodium acetate and 2.5 vol of ethanol for 10 min. Microfuge for 15 min, remove the supernatant, and wash the pellet with 150 µL of 75% ethanol. Resuspend the pellet in 10 µL of TE buffer.
3. Dephosphorylate the digested vector by diluting 1 µg of DNA to 50 µL with restriction enzyme buffer, add 1 µL of calf intestinal alkaline phosphatase (1 U/µL), and incubate at 37°C for 30 min. Extract twice with 100 µL of phenol/chloroform and twice with 50 µL of chloroform, and precipitate the DNA on ice with 0.1 vol of 3*M* sodium acetate and 2.5 vol of ethanol for 10 min. Microfuge for 15 min, remove the supernatant,

and wash the pellet with 150 µL of 75% ethanol. Resuspend the pellet in 10 µL of TE buffer.

3.2. Fragment Preparation

1. Cut the vector containing a DNA fragment of interest with appropriate restriction enzyme(s). Run fragments on a low-melting-point (LMP) agarose gel containing 5 µL of 10 mg/mL ethidium bromide in 50 mL of gel, and excise the band. Put the gel piece into a weighed microfuge tube and calculate the volume of the gel (100 mg = 100 µL). Add 10 µL of 10 mg/mL tRNA in water, 50 µL of 3*M* sodium acetate, and TE buffer to bring the final vol to 500 µL. Melt the gel at 65°C for 15 min and mix well. Extract twice with 500 µL of phenol/chloroform, twice with 250 µL of chloroform, and precipitate and microfuge the DNA as described above. Resuspend the pellet in 10 µL of TE buffer.
2. Fill in the 5' overhanging ends with Klenow fragment as described above, if necessary.
3. If the fragment has protruding 3' ends, these can be converted to blunt ends using T4 DNA polymerase. Mix 1 µg of DNA fragment, 2 µL of 10× T4 polymerase buffer, 1 µL of a 2 m*M* solution of all four dNTPs, 2.5 U of T4 polymerase and water to bring the final vol to 20 µL. Incubate at 37°C for 5 min. Add 1 µL of 0.5*M* EDTA and 29 µL of water. Extract twice with 100 µL of phenol/chloroform and twice with 50 µL of chloroform, and precipitate and microfuge the DNA as described above. Resuspend the pellet in 10 µL of TE buffer.

3.3. Phosphorylation of Synthetic Oligomers

If synthetic oligomers are to be inserted into the VLP vectors, they must be phosphorylated and annealed prior to ligation into dephosphorylated vector.

1. Lyophilize 10 pmol of each strand from aqueous solution. Phosphorylate each strand separately by resuspending 10 pmol of DNA in 2 µL of 10× kinase buffer, 1 µL 15m*M* ATP, 16 µL water, and 1 µL T4 polynucleotide kinase. Incubate at 37°C for 30 min.
2. Anneal the two complementary strands by mixing together 10 µL of each, floating in a boiling water bath, and then allowing to cool to <40°C over at least 1 h.

3.4. Ligation of Fragments

1. A three- to fivefold molar excess of insert over vector should be used. Mix 1 µL of dephosphorylated vector (100 ng/µL), 0.5–5 µL of insert

fragment, 2 µL of 10× ligase buffer, 2 µL of 10× AD buffer, 1 µL of T4 DNA ligase, and water to bring the final vol to 20 µL. Incubate at 16°C overnight. The ligation mixture can be used directly to transform *E. coli*. Alternatively, add 30 µL of water, extract once with 100 µL of phenol/chloroform and twice with 50 µL of chloroform, and precipitate and microfuge the DNA as described above. Resuspend the pellet in 20 µL of TE buffer.

3.5. *Transformation of* E. coli

1. Inoculate 100 mL of L broth with 1 mL of a fresh overnight culture of *E. coli*, such as HB101. Grow at 37°C to an optical density of OD_{600} = 0.4. Chill the culture on ice for 5 min and then pellet the cells for 5 min at 3500*g*. Resuspend the pellet in 50 mL of chilled 100 m*M* $CaCl_2$ and stand it on ice for 15 min. Pellet the cells as above and resuspend them in 6.7 mL of 100 m*M* $CaCl_2$.
2. Add 10 µL of ligation mixture to 100 µL of competent cells in a sterile microfuge tube. Stand the tube on ice for 20 min. Incubate at 42°C for 2 min and then add 1 mL of L broth. Incubate at 37°C for 30 min. Pellet the cells for 30 s in a microfuge, resuspend in 100 µL of L-broth, and plate onto L-broth agar plates containing 100 µg/mL carbenicillin. Incubate overnight at 37°C.

3.6. *Screening of Transformants*

1. Inoculate 1-mL aliquots of L broth with transformed cells from single colonies. Grow overnight at 37°C.
2. Pellet the cells and resuspend in 0.35 mL of STET buffer. Add 20 µL of 10 mg/mL lysozyme and place in a boiling water bath for 40 s. Spin in a microfuge for 10 min and remove the pellet with a wooden cocktail stick. Extract the supernatant once with 400 µL of phenol/chloroform, and precipitate DNA by adding an equal volume of isopropanol. Place the tubes at –20°C for 10 min, microfuge for 10 min, drain the tubes, and resuspend the DNA in 20 µL of TE buffer.
3. Digest 5 µL of "miniprep" DNA with appropriate restriction enzymes to determine the orientation of the inserted DNA fragment.

3.7. *Preparation of Plasmid DNA*

There are several methods available for purifying plasmid DNA. Since the VLP vectors plus inserts are >10 kb, the preferred procedure is by lysis in SDS.

1. Inoculate 500 mL of L broth with a single colony using a sterile inoculating loop. Incubate with vigorous shaking overnight at 37°C.
2. Pellet the cells at 3500*g* for 10 min, wash in 50 mL of cold TE buffer, and resuspend in 6 mL of cold 25% sucrose in 0.05*M* Tris-HCl, pH 8.0. Add 2 mL of fresh 10 mg/mL lysozyme solution and swirl on ice for 10 min. Add 2 mL of cold 0.5*M* EDTA and swirl on ice for 5 min. Add 10 mL of cold Sarkosyl solution, mix quickly, and stand the mixture on ice for 10 min. Spin the mixture in a Sorvall SS34 rotor (or equivalent) for 60 min at 15,000 rpm at 4°C.
3. Transfer the supernatant to a 50-mL Falcon tube. Add 6 mL of phenol, mix well, and transfer to 30-mL Corex tubes. Spin in a Sorvall SS34 rotor (or equivalent) for 10 min at 9000 rpm. Remove the aqueous phase and precipitate with 0.1 vol of sodium acetate and 2.5 vol of ethanol as described above.
4. Pellet the DNA in a Sorvall SS34 rotor (or equivalent) for 30 min at 9000 rpm. Dry the pellet and resuspend in 10 mL of TE buffer.
5. Add 10.5 g of CsCl and 0.3 mL of a 10 mg/mL ethidium bromide solution. Stand the mixture on ice for 10 min, and then pellet the precipitate in a Sorvall HB-4 rotor (or equivalent) for 6 min at 8000 rpm. Transfer the supernatant into two 4.5-mL Quickseal™ tubes and spin in a Beckman VTi65 (or equivalent) rotor for >6 h at 15°C at 54,000 rpm.
5. Remove the band of plasmid DNA by side-puncturing the tube with a 19-gage needle and 1-mL syringe. Extract with an equal volume of water-saturated butanol until the pink color disappears (the DNA is in the lower phase). Add 3 vol of TE buffer and precipitate with ethanol. Pellet the DNA in a Sorvall HB-4 rotor at 9000 rpm for 30 min. Resuspend the pellet in 300 µL of TE buffer, precipitate with ethanol again, and resuspend in 200 µL of TE buffer.

3.8. Transformation of Yeast

All manipulations must be carried out aseptically.

1. Inoculate 10 mL of YEPD medium with a single colony of untransformed yeast. (The strain of *Saccharomyces cerevisiae* used should be auxotrophic for leucine biosynthesis.) Grow overnight at 30°C. Inoculate 100 mL of YEPD medium from the overnight culture to give a cell density of 1.5×10^7 cells/mL after 16 h of growth. This will be dependent on the doubling time of the strain used and should be determined empirically.
2. When the culture is at 1.5×10^7 cells/mL, harvest the cells at 3500*g* in sterile 50-mL Falcon tubes. Wash the cells with 20 mL of 1*M* sorbitol. Pool

the suspension into one tube, pellet the cells again, and resuspend in 10 mL of 1*M* sorbitol.

3. Add 200 µL of glusulase and incubate at 30°C for 2 h, with occasional gentle shaking. Determine the extent of spheroplasting achieved by mixing 100 µL of cells with 1 mL of water and observing under a light microscope. Spheroplasted cells will be lysed.
4. Wash spheroplasts twice with 10 mL of 1*M* sorbitol and once with 10 mL of STC solution. Treat the cells gently at this point and from now on, since they are very fragile. Finally, resuspend the spheroplasts in 1 mL of STC.
5. Add DNA (2–3 µg) to 100 µL of competent cells. Mix by flicking gently. The input of DNA must be <15 µL, or the sorbitol concentration will become too dilute and the cells will burst. Incubate at room temperature for 15 min.
6. Add 1 mL of sterile 44% PEG 4000. Mix by inversion and gentle flicking. Leave at room temperature for 10 min.
7. Gently, spin down the cells by using the pulse button on a microfuge for 45 s. The cells will form a ribbon down the side of the tube. Decant the supernatant and gently resuspend the cells in 1 mL of 1*M* sorbitol. Add the cells to 20 mL of warm (not >48°C) regeneration agar prealiquoted into 50-mL Falcon tubes. Amino acids and growth factors, as appropriate, should be added to the regeneration agar after autoclaving. Mix the cells and agar gently and pour into a petri plate. Allow the agar to set, invert the plates, and then incubate at 30°C. Colonies should be visible after 3–4 d.
8. Pick single transformants with a sterile inoculating needle and streak for single colonies onto selective SC-glc plates containing the appropriate amino acids. Incubate plates at 30°C for 2–3 d.

3.9. Storage of Yeast Transformants

All manipulations must be carried out aseptically.

1. Inoculate 50 mL of SC-glc medium (plus appropriate amino acids) with a single colony. Incubate the culture at 30°C, with vigorous shaking, until a cell concentration of $2–4 \times 10^7$ cells/mL is achieved. This should take approx 48 h, depending on the yeast strain used.
2. Mix together equal volumes of culture and 40% sterile glycerol and transfer 1-mL aliquots to sterile cryotubes. Place the tubes at –20°C for 2 h and then transfer to –70°C for storage.

4. Notes

1. Calculation of volume containing 10 pmol of oligonucleotide: Determine the optical density of single-stranded oligonucleotide solution at 260 nm (OD260). The volume, *x*, containing 10 pmol is then calculated as follows (where *n* is the number of bases):

$$10 \text{ pmol} = (n \times 4.39) / (OD_{260} \times 50) = x\ \mu L$$

2. Determination of yeast cell density: Cell density can be determined using a cell counter, such as an improved Neubauer hemocytometer (available from BDH). Remove an aliquot of cells from the culture aseptically and dilute in such a way that single cells can be observed under a light microscope. Load the hemocytometer and count the cells contained within the chamber (a minimum of 100 cells should be counted). Calculate the number of cells/mL in the original culture—the equation will depend on the type of hemocytometer used. Cell densities can be determined spectroscopically after determining a standard curve for a particular yeast strain. However, direct counting of cells is preferable, since it provides an opportunity to check the culture for contamination and serves as a visual monitor of cell condition.
3. Resuspension of spheroplasts: Following removal of the cell wall, yeasts become extremely fragile. Transformations can fail because the spheroplasts have not been handled gently. The damage usually occurs during washing of the spheroplasts, particularly when they are resuspended following centrifugation. The spheroplasts can be resuspended gently by initially adding only 1 mL of the wash solution and using a wide-bore plastic pipet to genetly agitate the solution over the surface of the pellet. Following resuspension in a small volume, the additional wash solution should be added down the side of the tube and mixed by gentle swirling.
4. Temperature of regeneration agar: If the regeneration agar temperature is >48°C, the number of yeast transformants will be drastically reduced. It is therefore important to check that the temperature of the water bath or incubator used to keep the agar molten is maintained at 48°C.
5. Glycerol stocks: Glycerol stocks stored at –70°C should be examined for viability every 6–9 mo and replaced as necessary.

References

1. Adams, S. E., Mellor, J., Gull, K., Sim, R. B., Tuite, M. F., Kingsman, S. M., and Kingsman, A. J. (1987) The functions and relationships of Ty-VLP proteins in yeast reflect those of mammalian retroviral proteins. *Cell* **49,** 111–119 .

2. Kingsman, A. J. and Kingsman, S. M. (1988) Ty: A retroelement moving forward. *Cell* **53,** 333–335.
3. Adams, S. E., Dawson, K. M., Gull, K., Kingsman S. M., and Kingsman, A. J. (1987) The expression of hybrid Ty virus-like particles in yeast. *Nature* **329,** 68–70.
4. Kingsman, S. M. and Kingsman, A. J. (1988) Polyvalent recombinant antigens: A new vaccine strategy. *Vaccine* **6,** 304–307.
5. Mellor, J., Malim, M. H., Gull, K., Tuite, M. F., McCready, S. M., Dibbayawan, T., Kingsman, S. M., and Kingsman, A. J. (1985) Reverse transcriptase activity and Ty RNA are associated with virus-like particles in yeast. *Nature* **318,** 583–586.
6. Boeke, J. D., Garfinkel, D. J., Styles, C. A., and Fink, G. R. (1985) Ty elements transpose through an RNA intermediate. *Cell* **40,** 491–500.
7. Kingsman, A. J., Gimlich, R. L., Clarke, L., Chinault, A. C., and Carbon, J. A. (1981) Sequence variation in dispersed repetitive sequences in *Saccharomyces cerevisiae. J. Mol. Biol.* **145,** 619–632.
8. Mellor, J., Fulton, A. M., Dobson, M. J., Wilson, W., Kingsman, S. M., and Kingsman, A. J. (1985) A retrovirus-like strategy for expression of a fusion protein encoded by the yeast transposon, Ty1. *Nature* **313,** 243–246.
9. Mellor, J., Fulton, A. M., Dobson, M. J., Roberts, N. A., Wilson, W., Kingsman, A. J., and Kingsman, S. M. (1985) The Ty transposon of *Saccharomyces cerevisiae* determines the synthesis of at least three proteins. *Nucleic Acids Res.* **13,** 6249–6263.
10. Clare, J. and Farabaugh, P. (1985) Nucleotide sequence of a yeast Ty element: Evidence for an unusual mechanism of gene expression. *Proc. Natl. Acad. Sci. USA* **82,** 2829–2833.
11. Roeder, G. S. and Fink, G. R. (1983) Transposable elements in yeast, in *Mobile Genetic Elements* (Shapiro, J. A., ed.), Academic, New York, pp. 299–328.
12. Elder, R. T., Loh, E. Y., and Davis, R. W. (1983) RNA from the yeast transposable element Tyl has both ends in the direct repeats, a structure similar to retrovirus RNA. *Proc. Natl. Acad. Sci. USA* **80,** 2432–2436.
13. Wilson, W., Malim, M. H., Kingsman, A. J., and Kingsman, S. M. (1986) Expression strategies of the yeast retrotransposon Ty: A short sequence directs ribosomal frameshifting. *Nucleic Acids Res.* **14,** 7001–7015.
14. Malim, M. H., Adams, S. E., Gull, K., Kingsman, A. J., and Kingsman, S. M. (1987) The production of hybrid Ty:IFN virus-like particles in yeast. *Nucleic Acids Res.* **15,** 7571–7580.
15. Braddock, M., Chambers, A., Wilson, W., Esnouf, M. P., Adams, S. E., Kingsman, A. J., and Kingsman, S. M. (1989) HIV-1 TAT "activates" presynthesized RNA in the nucleus. *Cell* **58,** 269–279.
16. Dobson, M. J., Tuite, M. F., Roberts, N. A., King, R. M., Burke, D. C., Kingsman, A. J., and Kingsman S. M. (1982) Conservation of high efficiency promoter sequences in *Saccharomyces cerevisiae. Nucleic Acids Res.* **10,** 2625–2637.
17. Kingsman, S. M., Cousens, D., Stanway, C. A., Chambers, A., Wilson, W., and Kingsman, A. J. High efficiency expression vectors based on the promoter of the phosphoglycerate kinase gene. *Methods Enzymol.,* in press.

CHAPTER 25

Production and Purification of Hybrid Ty-VLPs

Nigel R. Burns, Jacqueline E. M. Gilmour, Susan M. Kingsman, Alan J. Kingsman, and Sally E. Adams

1. Introduction

The self-assembly properties of a protein encoded by the *TYA* gene of the yeast Ty element can be exploited to produce hybrid Ty-VLPs (virus-like particles) *(1,2)*. There has been developed a series of expression vectors that allow the construction of Ty fusion genes containing protein coding sequences of interest (*see* Chapter 24). Many different hybrid Ty-VLPs have now been produced that carry additional proteins that range in size from 3 to 42 kDa. These include regions from human immunodeficiency virus-1 (HIV-1) *env, pol, tat, rev, nef,* and *vif* genes; influenza virus hemagglutinin; human α-interferon, feline leukemia virus *env*; and bovine papillomavirus E1 and E2 (*1–5* and unpublished data).

An important feature of the Ty-VLP system is the ease with which pure VLPS can be prepared as a result of their particulate nature (Fig. 1). The sedimentation properties of different hybrid Ty-VLPs are similar, and this characteristic has been exploited to develop a simple purification process that can be used for any VLP, irrespective of the sequence of the additional protein. The system is therefore extremely versatile, allowing rapid production of a variety of recombinant proteins. Such ease of purification would, for

From: *Methods in Molecular Biology, Vol. 8: Practical Molecular Virology: Viral Vectors for Gene Expression*
Edited by: M. Collins © 1991 The Humana Press Inc., Clifton, NJ

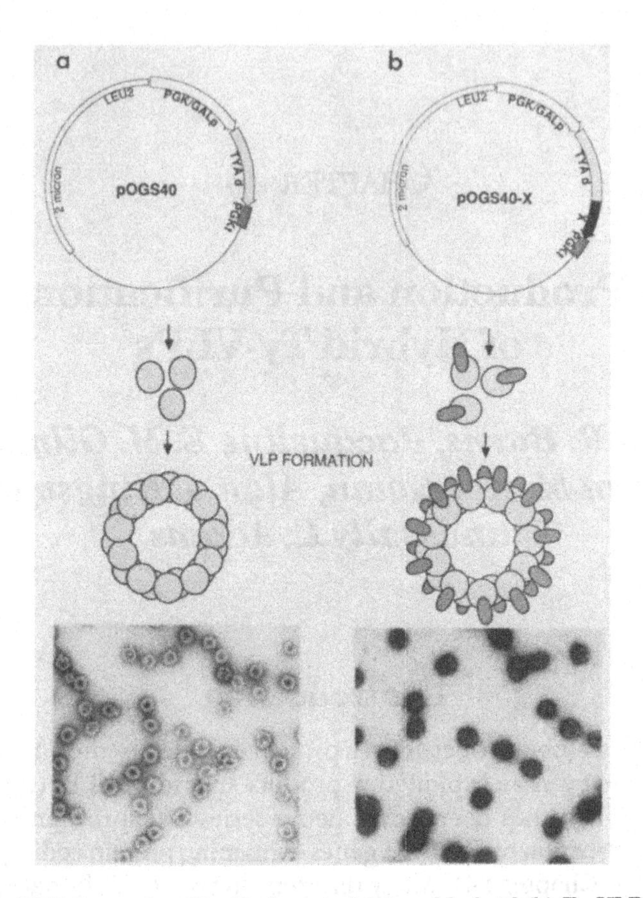

Fig. 1. Ty-VLP formation. Purified control (a) and hybrid (b) Ty-VLPs are shown below their relevant expression vectors and a schematic diagram of VLP formation. The vectors contain a hybrid PGK/GAL promoter (PGK/GALp; open arrow), the first 381 codons of the *TYA* gene (TYAd; lightly shaded arrow), a PGK terminator region (PGKt; darkly shaded box), selection and replication modules for yeast (open box) and *E. coli* (thin line). The purified VLPs in panel b contain regions of the HIV core p24 and p17 proteins fused to the Ty particle-forming protein.

example, make it feasible to survey a genome for regions that encode important antigenic determinants, such as those able to induce a protective immune response against a particular pathogen.

Hybrid VLPs can also be used in many other laboratory applications. The most obvious use is the production of defined polyclonal and monoclonal antisera as research reagents by selecting particular regions of a protein as the added antigen. These antibodies can then be used as research tools themselves, perhaps to purify native protein, as immunodetection reagents,

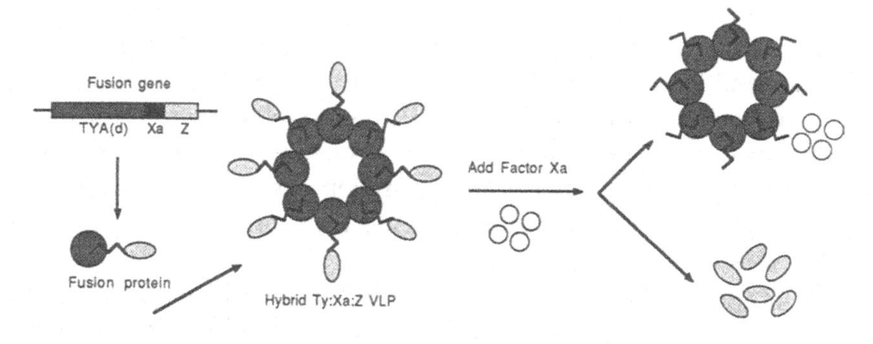

Fig. 2. Purification of proteins following factor Xa cleavage of hybrid Ty-VLPs. The recognition sequence for factor Xa is inserted between the coding sequence of *TYA*(d) and the protein of interest (Z). The resulting fusion protein assembles into VLPs. These are purified and cleaved with factor Xa (open cirles). The enzyme and residual particulate material are then purified from protein Z.

or to map functional domains of a protein. Conversely, hybrid VLPs can be used as a rapid primary screen to map monoclonal antibodies raised against non-VLP antigens. For example, a series of VLPs containing overlapping fragments of the HIV envelope protein gp120 have been used to map a monoclonal antibody to a 30 amino acid sequence *(1)*. The cost of fine mapping with peptides can therefore be decreased significantly by reducing the length of sequence to be covered by overlapping peptides.

The VLP system can also be manipulated to produce nonparticulate proteins or protein domains. By engineering a protease cleavage site at the C-terminus of the particle-forming protein p1, hybrid VLPs can be produced with the general structure of p1:cleavage site:added antigen. Hybrid VLPs have been constructed that contain various antigens downstream of the recognition sequence for the blood coagulation factor Xa. Purification of the particles followed by factor Xa cleavage results in a mixture of three proteins from which the protein of interest can be purified (Fig. 2). This technology has been used successfully to purify several HIV antigens, providing sufficient material for structural and functional analyses *(5,6)*. Such systems could result in a better understanding of disease pathology and the development of novel screening systems for antiviral compounds directed at virus-specific targets.

2. Materials

1. Synthetic complete-glucose (SC-glc) medium: 0.67% (w/v) Yeast nitrogen base without amino acids, 1% (w/v) glucose. Add appropriate amino

acids after autoclaving. The amino acids required will be dependent on the auxotrophy of the yeast strain used. The yeast strain used must be auxotrophic for leucine biosynthesis, and leucine is omitted from the media to ensure selection for the VLP plasmid following transformation. Amino acids can be prepared as a 50× stock solution and filter-sterilized.

2. SC-glc plates: 0.67% (w/v) Yeast nitrogen base without amino acids, 1% glucose (w/v), 2% Bacto-agar. Add appropriate amino acids after autoclaving.
3. SC-glc/gal medium: 0.67% (w/v) Yeast nitrogen base without amino acids, 0.3% (w/v) glucose, 1% (w/v) galactose. Add amino acids after autoclaving.
4. TEN buffer: 10 m*M* Tris-HCl (pH 7.4), 2 m*M* EDTA, 140 m*M* NaCl. This buffer is usually made up as a 10× stock solution.
5. Protease inhibitors: Separate solutions of 25 mg/mL chymostatin, antipain, leupeptin, and pepstatin A are made in dimethyl sulfoxide and stored in 25-µL aliquots. A 25-mg/mL solution of aprotinin is made in water and stored in the same manner. Then 25 µL of each of the above-mentioned solutions is added to 1 L of TEN buffer plus 1 mL of fresh 5 m*M* phenylmethylsulfonyl fluoride dissolved in ethanol. The protease inhibitors can be obtained from Sigma.
6. Acid-washed glass beads: Glass beads, 40-mesh, can be obtained from BDH Ltd. The beads are washed in concentrated sulfuric acid, rinsed 10 times in tap water, 10 times in distilled water, dried, and baked at 150°C for 2 h.
7. 60% (w/v) Sucrose dissolved in TEN buffer.
8. 5–20% Linear sucrose gradients: Prepare a 12.5% stock sucrose solution in TEN buffer and autoclave. Add 30 mL to Beckman SW28 tubes (or equivalent) and freeze at –20°C. The day before the gradients are needed, remove the required number and thaw slowly at 4°C for 16 h. Add 2 mL of 60% sucrose to the bottom of the tube using a Pasteur pipet before loading the gradients.
9. Sephacryl S1000 superfine can be obtained from LKB. The column is equilibrated in TEN buffer before use.
10. Ultrafiltration units with a 30,000-dalton cutoff can be obtained from Millipore (Immersible CX-30 units, catalog number PTTK11K25).
11. 5× DB: 20% (v/v) Glycerol, 10% (v/v), β-mercaptoethanol, 10% (w/v) SDS, 0.125*M* Tris-HCl (pH 6.8), 0.1% (w/v) bromophenol blue.
12. SDS-polyacrylamide gels:
    a. Separating gel: 10% stock acrylamide solution (acrylamide:*bis*-

acrylamide ratio of 30:0.8), 0.375*M* Tris-HCl (pH 8.8), 0.05% (wv) ammonium persulfate, 0.1% (w/v) SDS, 0.0005% (v/v) TEMED.

b. Stacking gel: 5% stock acrylamide solution, 0.125*M* Tris-HCl (pH 6.8), 0.1% (v/v) ammonium persulfate, 0.1% (w/v) SDS, 0.001% (v/v) TEMED.

13. Coomassie Blue stain: 0.25% (w/v) Coomassie Blue in 40% (v/v) methanol and 10% (v/v) acetic acid.
14. Destaining solution: 20% (v/v) Methanol, 10% (v/v) acetic acid.

3. Methods

In this protocol, 16-L cultures containing transformed yeast are grown, and Ty-VLPs are purified from them. Procedures are described for the production of Ty-VLPs using constitutive expression from the yeast phosphoglycerate kinase (PGK) promoter and using galactose-induced expression from a hybrid PGK-GAL promoter (*see* Chapter 24).

3.1. Growth of Cultures—Constitutive Expression

The procedures for transforming yeast with recombinant plasmids and preparing glycerol stocks of transformants are described in Chapter 24. Large cultures of transformed yeast should be grown from glycerol stocks.

1. Inoculate 100 mL of SC-glc medium (plus amino acids) with a 1-mL glycerol stock of the appropriate yeast transformant. Incubate, with vigorous shaking, at 30°C until a cell density of at least 2×10^7 cells/mL is achieved. This will take approx 2 d.
2. Put 1 L of SC-glc medium (plus amino acids) into each of two 2-L flasks, and inoculate each with 50 mL of preculture. Incubate overnight as before until the cell density is 5×10^7 cells/mL.
3. Split the 2 L of culture among 16 individual liters of SC-glc medium (plus amino acids). Grow overnight as before. Harvest the cells when the cell density is 4×10^7 cells/mL by centrifugation at 3500*g* for 20 min. Resuspend the pellets in 10 mL of water/L of culture. Transfer to 50-mL Falcon tubes (pool the cells from 2 L of culture into one tube) and centrifuge at 3500*g* for 5 min in a bench-top centrifuge. Repeat the washing procedure twice more. Finally, wash each pellet with 20 mL of TEN buffer plus protease inhibitors. Remove the supernatant and freeze the cell pellets at –20°C until required.

For a smaller-scale preparation, the cells can be harvested and washed at Step 2.

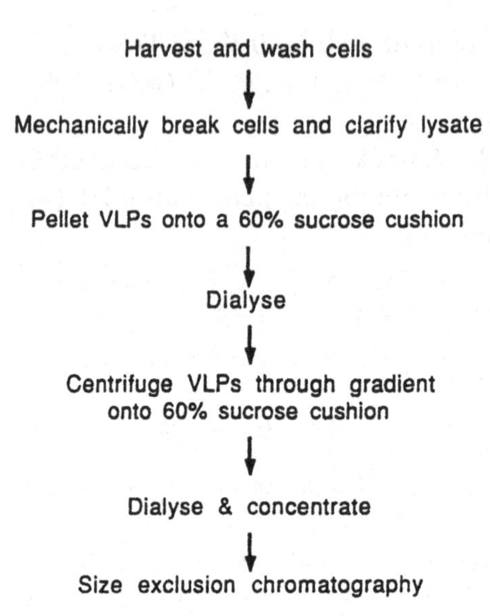

Fig. 3. Flowchart for the purification of hybrid Ty-VLPs.

3.2. Growth of Cultures—Inducible Expression

1. Inoculate 100 mL of SC-glc medium (plus amino acids) with a 1-mL glycerol stock of the appropriate yeast transformant. Incubate, with vigorous shaking, at 30°C until the cell density is 2–4 × 10^7 cells/mL. This will take approx 2 d.
2. Put 1 L of SC-glc medium (plus amino acids) into each of two 2-L flasks, and inoculate each with 50 mL of the preculture. Grow for approx 20 h, until the cell density is 4–6 × 10^7 cells/mL.
3. Split the 2 L of culture among 16 individual liters of SC-glc/gel medium (plus amino acids). Grow for 24 h.
4. Harvest the cells when a density of 4–8 × 10^7 cells/mL is reached. Wash the cells with water and TEN buffer as described above. Freeze the cell pellets at –20°C until required.

For a smaller-scale preparation, inoculate 50 mL of the preculture into 1 L of SC-glc medium (plus amino acids) at Step 2 and 150 mL from this culture into each of 2 × 1 L of SC-glc/gel medium (plus amino acids) at Step 3.

3.3. Purification of Ty-VLPs (see Note 1)

The following procedure is used for the purification of hybrid Ty-VLPs from 16 L of culture. The procedure is summarized in Fig. 3. Yields are generally in the range of 20–100 mg/16 L, and the VLPs constitute >90% of the final preparation. For smaller-scale preparations, the procedure can be scaled down proportionally, and a smaller column can be used for the final purification step. All procedures are performed at 4°C using prechilled buffers.

1. Thaw cells (in eight 50-mL Falcon tubes). Add 4 mL of TEN buffer containing protease inhibitors to each tube and resuspend the cells. Add 5 mL of acid-washed glass beads to each tube.
2. Vortex the tubes for 10, 30-s periods interspersed with 30-s periods of cooling in ice. Centrifuge the suspension at 2000*g* for 5 min. Remove and retain the supernatant on ice (*see* Note 2).
3. Add 4 mL of fresh TEN buffer (containing protease inhibitors) to the cell pellet and repeat the vortexing and centrifugation as above. Remove and retain the supernatant.
4. Add 3 mL of fresh TEN buffer (containing protease inhibitors) to the cell pellet and again repeat the vortexing and centrifugation. Pool the supernatants and centrifuge at 13000*g* (Sorvall HB4 rotor or equivalent) for 20 min.
5. Centrifuge the cleared supernatant at 100,000*g* (Beckman SW40 rotor, 30,000 rpm for 1 h; SW28 for 1.5 h at 28,000 rpm; or equivalent) onto a 2-mL 60% sucrose cushion (in TEN buffer). The use of a sucrose cushion prevents pelleting of the VLPs, which can result in a decrease in yield. Collect the cushions with a Pasteur pipet and dialyze overnight against 1 L of TEN buffer plus protease inhibitors (*see* Notes 3,4).
6. Centrifuge the dialysate at 13,000*g* for 20 min and retain the supernatant. Add 2 mL of 60% sucrose (in TEN buffer) to the bottom of each of six 5–20% sucrose gradients (Beckman SW28 tubes or equivalent) with a Pasteur pipet. Load each gradient with up to 2 mL of the supernatant and centrifuge at 53,000*g* (25,000 rpm Beckman SW28 rotor, or equivalent) for 6 h.
7. Following centrifugation, remove the sucrose cushions containing the VLPs using a Pasteur pipet. Remove the sucrose from this crude particle preparation by dialyzing against TEN buffer for 36 h.
8. Centrifuge the dialysate at 13,000*g* for 20 min, filter the supernatant through a 0.45-μm filter, and concentrate the sample to 16 mL using a Millipore ultrafiltration device. Load the sample onto a Sephacryl S1000 superfine 5 × 100 cm column, and develop at a rate of 16 mL/h with TEN buffer (*see* Notes 5,6).
9. Fractions containing hybrid Ty-VLPs are identified by running aliquots of the fractions on 10% SDS—polyacrylamide gels. Add 10 μL of 5 × DB to 40 μL of column fraction and boil the mixture for 5 min in a water bath. The quantity of sample loaded will depend on the protein gel ap-

paratus used. If a minigel (e.g., the Bio-Rad Mini-Protean) is used, then 10 μL is sufficient. Run the gel according to the manufacturer's instructions for the apparatus, stain with 0.25% Coomassie Blue solution, and destain.

10. Pool the fractions containing the VLPs, concentrate to approx 1 mg/mL, if necessary, and store in aliquots at –70°C.

4. Notes

1. Confirmation of the presence of fusion protein prior to starting the VLP purification: The presence of fusion protein can be detected by either Coomassie Blue staining, a total protein extract run on a 10% SDS-polyacrylamide gel or by Western blotting using either an anti-Ty antibody, or an antibody directed against the non-Ty component. As a control, a total protein extract of nontransformed yeast should be included. Total yeast protein extracts can be prepared as follows: Remove 5×10^8 cells at the end of the growth period and centrifuge at 2000*g* for 5 min in a 15-mL Falcon tube. Resuspend the cells in 1 mL of TEN buffer. Add 1 g of acid-washed and baked glass beads. Disrupt cells by vortexing the sample three times for 1 min each, interspersed with 1-min cooling periods on ice. Remove beads by centrifuging at 2000*g* for 5 min. Determine the protein concentration of the supernatant using a Bradford dye-binding assay (available from Bio-Rad). Load 30–50 μg of total protein for detection by Coomassie Blue staining or 3–5 μg for detection by Western blotting.
2. Breakage of yeast cells: In order to maximize yields of VLPs, it is essential to obtain efficient breakage of the yeast cells. When vortexing the cells with glass beads, it is important to hold the Falcon tube in such a manner that the liquid and beads are forced as far as possible up the sides of the tube. This can be achieved by keeping the tube vertical, holding the tube at the top, and pressing into the whirlimixer. Breakage can be evaluated by phase-contrast microscopic examination. Damaged cells will appear phase-dark, whereas unbroken cells will remain phase-bright. An alternative is to use a mechanical homogenizer, such as a Bead Beater (available from Biospec Products, Bartlesville, OK).
3. Harvesting VLPs from sucrose cushions: It is important to avoid mixing the supernatant (containing nonparticulate contaminants) with the sucrose cushion. This can be ensured by drawing off the supernatant with a Pasteur pipet before removing the sucrose cushion.

4. Dialysis: Extended dialysis of the crude VLP preparation often results in the formation of a precipitate. This does not contain the VLPS and is removed by centrifugation (13,000*g* for 20 min).
5. Filtration of VLPs: Hybrid VLPs have a diameter of 50–80 nm, depending on the size of the non-Ty component. They should therefore pass through 0.45- or 0.22-µm filters. However, prior to Sephacryl S1000 column chromatography, some contaminants are still present and filtration through a 0.45-µm filter may be difficult. It is therefore often advisable to prefilter using a 3-µm filter.
6. Alternative to size exclusion chromatography: If column chromatography facilities are not available, it is possible to replace this step with 15–45% sucrose gradients. However, it should be noted that resolution is improved by using the size exclusion column, rather than density gradient centrifugation. Layer 8-mL aliquots of 45, 35, 25, and 15% sucrose (in TEN buffer) into Beckman SW28 tubes (or equivalent) and leave to equilibrate overnight at 4°C. Load a maximum of 3 mL of the centrifuged, filtered, and concentrated dialysate onto each gradient and centrifuge at 53,000*g* (25,000 rpm in Beckman SW28, or equivalent) for 3 h at 4°C. Separate the gradients into 2-mL fractions. Identify the fractions that contain hybrid Ty-VLPs by running aliquots on 10% SDS-polyacrylamide gels as described above.

References

1. Adams, S. E., Dawson, K. M., Gull, K., Kingsman, S. M., and Kingsman, A. J. (1987) The expression of hybrid Ty virus-like particles in yeast. *Nature* **329,** 68–70.
2. Kingsman, S. M. and Kingsman, A. J. (1988) Polyvalent recombinant antigens: A new vaccine strategy. *Vaccine* **6,** 304–307.
3. Adams, S. E., Senior, J. M., Kingsman, S. M., and Kingsman, A. J. (1988) Induction of HIV antibodies by Ty:HIV hybrid virus-like particles, in *Technological Advances in Vaccine Development* (Lasky, L., ed.), Liss, New York, pp. 117–126.
4. Malim, M. H., Adams, S. E., Gull, K., Kingsman, A. J., and Kingsman, S. M. (1987) The production of hybrid Ty:IFN virus-like particles in yeast. *Nucleic Acids Res.* **15,** 7571–7580.
5. Braddock, M., Chambers, A., Wilson, W., Esnouf, M. P., Adams, S. E., Kingsman, A. J., and Kingsman, S. M. (1989) HIV-1 TAT "activates" presynthesized RNA in the nucleus. *Cell* **58,** 269–279.
6. Gilmour, J. E. M., Senior, J. M., Burns, N. R., Esnouf, M. P., Gull, K., Kingsman, S. M., Kingsman, A. J., and Adams, S. E. (1989) A novel method for the purification of HIV-1 p24 protein from hybrid Ty-VLPs. *AIDS* **3,** 717–723.

Chapter 26

Baculovirus Expression Vectors

Choice of Expression Vector

Vincent C. Emery

1. Introduction

Baculoviruses have been used for many years as effective pest-control agents; however, their interest to molecular virologists stems from their exploitation as helper-independent viral expression vectors for the production of proteins in a eukaryotic environment. The pioneering work that led to the development of the system came from the laboratory of Max Summers at Texas A & M, USA. Since its inception, the system has been increasingly utilized to express a broad range of eukaryotic proteins (reviewed in ref. *1*) and has been extensively modified *(2)*, in particular by David Bishop and Bob Possee at the Institute of Virology and Environmental Microbiology, Oxford, UK. This work has led to the construction of single and multiple baculovirus expression vectors. Extensive reviews on the subject of baculoviruses are available *(1,3–5)*, including a manual of methods specific for their use as expression systems *(6)*. Chapters 26–28 aim to provide a practical account of baculoviruses as expression vectors, along with the experimental methodologies required to produce a baculovirus transfer vector containing the gene to be expressed, the generation and selection of a recombinant expressing this gene product, and the methods available for processing downstream the baculovirus-produced protein.

From: *Methods in Molecular Biology, Vol. 8:*
Practical Molecular Virology: Viral Vectors for Gene Expression
Edited by: M. Collins © 1991 The Humana Press Inc., Clifton, NJ

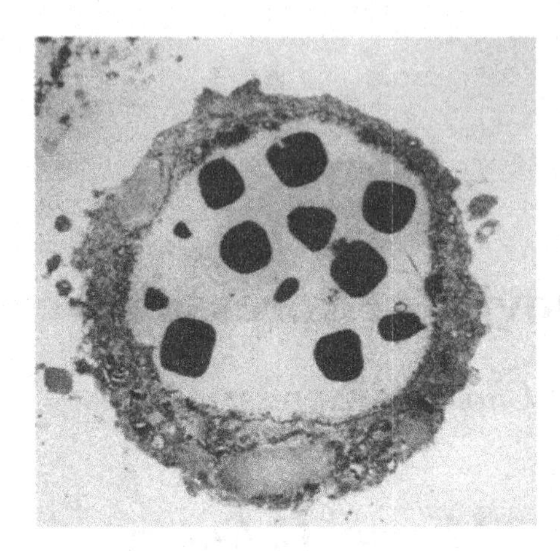

Fig. 1. Electron micrograph of a thin section of a *S. frugiperda* cell infected with *Autographa californica* nuclear polyhedrosis virus.

1.1. Baculovirus Life Cycle

Baculoviruses (family Baculoviridae) have been isolated from a number of arthropods that represent their natural hosts. The genome consists of a single molecule of circular, double-stranded DNA between 80 and 220 kb in length, depending on the viral trophism. The most highly studied subgroup of baculoviruses are the nuclear polyhedrosis viruses (NPV). The representative virus of this group is the *Autographa californica* nuclear polyhedrosis virus (AcMNPV or AcNPV; the latter abreviation will be used in these review chapters). AcNPV naturally infects *Autographa californica*, a lepidopteran commonly known as the alfalfa looper. Because of the wealth of information that has accumulated concerning the life cycle, replication, and molecular biology of AcNPV, it has been the baculovirus of choice for exploitation as an expression system. The less-characterized *Bombyx mori* NPV has also been used as an expression system *(5)*.

An important aspect of the life cycle of the NPV subgroup of baculoviruses that is particularly relevant to their exploitation as expression vectors is the presence of two forms of the virus: first, an occluded virus in which the virions are embedded within a crystalline matrix composed of the major late 29-kD polyhedrin protein; second, an extracellular budded virus that is not occluded. The former crystalline matrix of polyhedrin produces occlusion bodies approx 1–5 µm in diameter, with polyhedron symmetry, within the nucleus of infected cells (Fig. 1). In addition, since insects acquire infection *in natura* by consuming food contaminated with occluded virus, the polyhedrin is alkali-labile and dissolves in the midgut of the insect, releasing the virus particles in close proximity to the site of initial infection. Following infection of the midgut cells, systemic infection is mediated through extracellular nonoccluded virus that buds into the hemolymph of the insect.

AcNPV gene expression during infection can be broadly classified into three phases: early, late, and very late *(4)*. The most important phase for baculovirus expression vectors is the very late phase of gene expression. This phase begins at approx 24 h postinfection, continues up to 72 h postinfection, and is characterized by the production of polyhedrin, which accumulates within the nucleus to compose some 35% of total cell protein. The polyhedrin protein is therefore produced in large quantities very late in infection and is dispensible (*see above;* i.e., the ability for cell–cell transmission via nonoccluded virus). The majority of the expression vector systems have exploited the polyhedrin transcriptional machinery to promote and terminate the foreign gene that is placed in lieu of the polyhedrin gene. In addition, the absence of the polyhedrin phenotype in recombinant viruses yields a plaque morphology distinct from wild-type ACNPV and allows the visual selection of the recombinant virus.

1.2. Single and Multiple Vectors Available for Producing Recombinant Baculoviruses

A schematic representation of the production of a recombinant baculovirus is shown in Fig. 2. In essence, the baculovirus expressing the gene of interest is produced by homologous recombination between wild-type AcNPV DNA and transfer-vector DNA that contains the EcoR1 "I" fragment of the AcNPV viral genome, but with the polyhedrin gene replaced by the gene of interest.

One careful study *(2)* concludes that vectors containing the entire 5' leader sequence, up to and including the "A" of the normal polyhedrin ATG, produced the highest expression levels. A transfer vector (pAcYM1) based on these observations has been produced by these workers; subsequently, the data has been used to produce new vectors such as p36C *(7)*. It is interesting that the boundaries of the polyhedrin promoter appear to be small, encompassing a region from –1 to –60 from the translation start site of polyhedrin. This data has been exploited to produce a multiple expression vector system based on a duplication of the polyhedrin transcriptional machinery *(9)*. Naturally, polyhedrin fusion vectors are also invaluable for the production of peptides or proteins that are otherwise unstable.

1.2.1. Single Expression Vectors

The high-level expression vectors pAcYM1 or pAcRP23 are available from, respectively, D. H. L. Bishop and R. D. Possee (Institute of Virology and Environmental Microbiology, Mansfield Road, Oxford, OX1 3SR, UK). A physi-

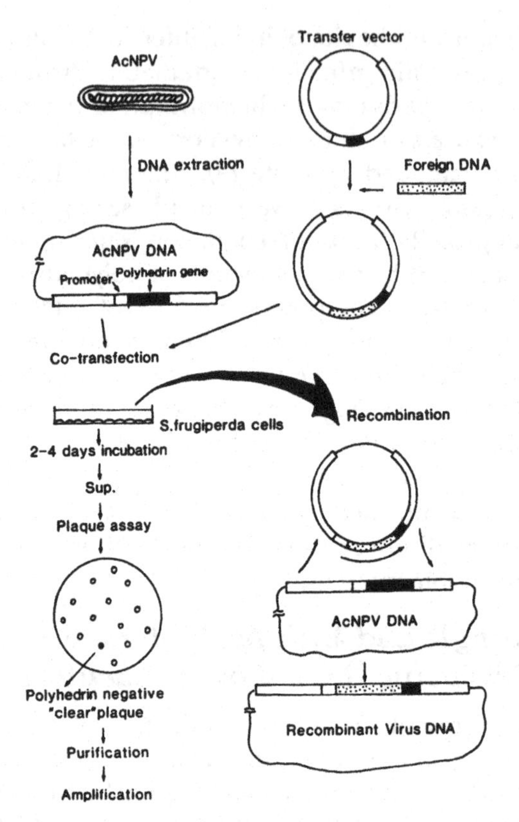

Fig. 2. A flowchart for the steps involved in generating a recombinant baculovirus.

cal map of pAcYM1 is shown in Fig. 3. It should be noted that all the "pAcRP" vectors (except pAcRP61 *(2)*) from the Institute of Virology possess a portion of the 3' coding sequence of the polyhedrin gene, whereas in pAcYM1 the entire polyhedrin gene coding sequences have been removed.

Two other "late" expression vectors are available from the Institute of Virology. Of these, one (pAcUW1; ref. *8*) utilizes the P10 promoter and the other (pAcMP1; ref. *10*) the basic protein promoter to transcribe the inserted gene. In addition, a vector based on pAcYM1 that allows singlestranded DNA to be produced for mutagenic experiments is also available from I. M. Jones *(11)* at the above address. The vector most commonly used to produce recombinant baculoviruses prior to the development of pAcYM1 was pAc373 *(12)*.

Miller and colleagues *(13)* have produced pEV55 that contains a multiple cloning site (BglII, XhoI, EcoR1, Xba1, Cla1, and Kpn1) inserted at the end of the intact polyhedrin leader sequence. The Kpn1 site will provide an "ATG" for genes without their own initiation codon.

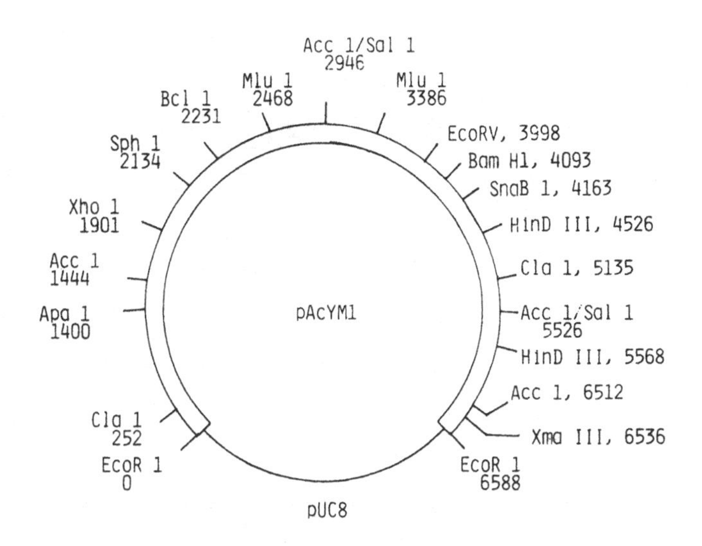

-1 +752
AGTTTTGTAATAAAAAAACCTATAAATA CGGATCCG GTTTATTAGTACAC

polyhedrin 5′ leader Bam H1 polyhedrin 3′ untranslated sequence

Fig. 3. Restriction map of pAcYMl. The sequence around the unique BamH1 cloning site is detailed. Nucleotide positions quoted are relative to the polyhedrin "ATG."

1.2.2. Fusion Vectors

For genes lacking their own "ATG" initiation codon, a range of vectors are available that use the polyhedrin gene initiation codon and varying amounts of the *N*-terminal amino acids. Summers and colleagues (Texas A & M) have developed pAc401, pAc436, pAc360, and pAc311 *(6)* to fulfill this role, whereas Possee (Institute of Virology) has produced fusion vectors, such as pAcRP14 *(14)* and derivatives thereof, that allow insertion in three reading frames.

1.2.3. Multiple Expression Vectors

For the controlled synthesis of two proteins under the control of the polyhedrin transcriptional machinery, two vectors are available from D. H. L. Bishop *(15,16)*. These vector types are shown in Fig. 4. The first, pAcVC1, consists of an intact polyhedrin gene and a duplicated polyhedrin promoter/terminator containing a unique Bgl11 site for cloning the gene for expression. Recombinants derived from this vector produce both polyhedrin and the gene of interest, and the occluded virus produced is very efficient at infecting *Trichoplusia ni* larvae *(9)*. A second vector, pAcVC3, replaces the

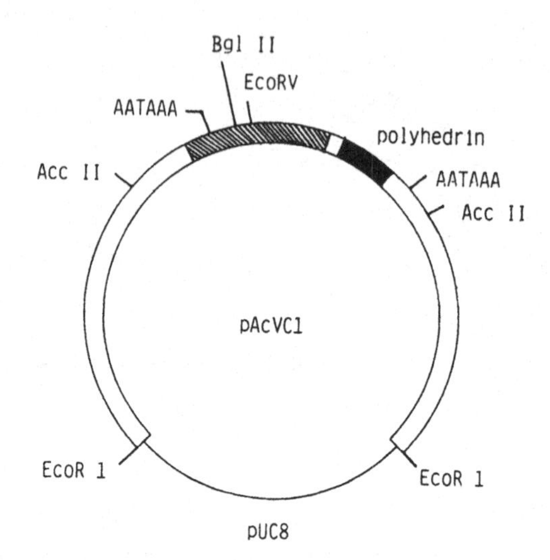

Fig. 4. Partial restriction map and orientation of the duplicated polyhedrin transcriptional control sequences within pAcVc1.

polyhedrin gene present in pAcVC1 with a unique BamH1 cloning site, thus allowing two different genes to be cloned into the same vector *(15,16)*. An alternative strategy to produce the dual expression system (especially if pAcYM1 has been utilized for the expression of the inividual proteins) is to follow the cloning procedures detailed in the original description of the multiple expression system *(9)*.

1.3. Generation and Characterization of the Transfer Vector

For expression of genes using such vectors as pAcYM1, the 5' leader sequence should be as short as possible (*see* Note 7). The amount of extraneous 3' sequence following the translation termination codon appears to be unimportant. Genes can be prepared for cloning using established procedures, such as deletion of unwanted sequences with Bal31; alternatively, they can be amplified directly with appropriate oligonucleotide primers and the polymerase chain reaction *(17)*. Once cloned into the transfer vector, DNA sequencing of the 5' and 3' junctions between vector and insert is highly recommended. The method of choice to perform these analyses is plasmid-sequencing using specific oligonucleotide primers complementary to the baculovirus DNA. The plasmid-sequencing methods described below are a modification *(18)* of those developed by Chen and Seeburg *(19)*.

2. Materials

1. T.0.1E: 10 m*M* Tris-HCl, pH 8.0, 0.1 m*M* EDTA.

2. 10× TM: 100 m*M* Tris-HCl, pH 8.0, 100 m*M* Magnesium chloride.
3. Relaxation solution (freshly made): 1 *M* sodium hydroxide, 1 m*M* EDTA.
4. Sepharose CL-6B (Sigma Chemical Co.).
5. Oligonucleotide primers (10 μg/mL): For any vector, the following 5' sequencing primer may be used: 5'-TAAGTATTTTACTGTTT. For vectors, such as pAcYM1, that do not contain the polyhedrin gene sequences, the following primer can be used to determine the sequence at the 3' junction: 5'-AAATCAACAACGCACAG. For vectors such as pAc373 and the "pAcRP" series, the following 3' junction sequencing primer should be used. This primer anneals within the residual polyhedrin gene sequence present in these vectors: 5'-TACGGATTTCCTTGAAG.
6. 5× BAL31 buffer: 3 *M* soduim chloride, 60 m*M* calcium chloride, 60 m*M* magnesium chloride, 100 m*M* Tris-HCl, pH 8.0, 1 m*M* EDTA, pH 8.0.
7. 200 m*M* EGTA, pH 8.0.

3. Methods

3.1. BAL 31 Deletion

1. Mix linearized DNA (1 μg/μL; 4 μL), water (48 μL) and 5× BAL 31 buffer, and dispense 9 μL of this mixture into six separate Eppendorf tubes.
2. Make a series of two- or fivefold dilutions of the BAL 31 enzyme in 1× BAL 31 buffer and add 1 μL of the enzyme to each tube containing the DNA substrate.
3. Incubate the tubes at 37°C for 30 min.
4. Add 200 m*M* EGTA (1 μL) to stop the reaction, and then heat the tubes at 65°C for 5 min.
5. Analyze the digested products by agarose gel electrophoresis to determine the optimum concentration of BAL 31 required to remove the unwanted sequences. The reaction can then be performed on a larger scale to produce the quantities of digested DNA required for subsequent cloning and sequencing.

3.2. Plasmid Sequencing

The following sequencing method works equally well with either mini-plasmid preparation DNA or cesium chloride purified DNA.

1. Dissolve approx 3–5 μg of plasmid DNA in 20 μL of water.
2. Add 5 μL of the relaxation solution and allow to stand at room temperature for 5 min.
3. Prepare the columns for spin dialysis by piercing a 0.6-mL Eppendorf tube with an 18-G syringe needle and plugging the hole with size 9 glass

beads (10–20 are usually sufficient). Then add a mixture of 2 parts Sepharose CL-6B to 1 part T.0.1E (550 µL) to the Eppendorf tube without disturbing the glass beads. Place the 0.6-mL Eppendorf tube in a 1.6-mL Eppendorf tube that has been completely punctured with an 18-G syringe needle and place the assembly in a neck of a 10-mL test tube (Sterilin). Centrifuge the assembly at 1200 g for 4 min to compact the column then remove the pierced 1.6-mL Eppendorf and replace it with an intact 1.6-mL Eppendorf tube.

4. Place the reaction mixture from step 2 onto the column prepared in step 3 and spin at 1200 g for 4 min.
5. Remove the column eluate containing the relaxed plasmid DNA and anneal the oligonucleotide primers as follows:
   a. column eluate 8.5 µL
   b. 10× TM 1 µL
   c. add oligonucleotide primer (10 µg/mL) 1 µL
   d. Incubate at 37°C for 15 min.
6. Use primed templates immediately in a conventional dideoxy sequencing reaction using ^{35}S-dATP as the radioactive deoxynucleotide triphosphate *(20)*.

4. Notes

1. Although there is broad variation between expression levels, it is generally expected that expression of intracellular proteins in baculoviruses exceeds the amounts produced in the majority of currently available mammalian expression systems, and certain proteins have been expressed to levels equating to that of polyhedrin itself, i.e., 35% of total cell protein *(2)*.
2. Proteins produced in baculoviruses usually possess full biological activity. Thus antibodies raised against the expressed protein are able to recognize the native antigens in appropriate assays. Conversely, the expressed protein can be used with some degree of certainty to detect the presence of antibodies for the specific protein in biological samples. These factors have contributed to the extensive use of baculovirus-expressed proteins as diagnostic reagents in which they generally produce products that possess a low cut-off value and a high signal-to-noise ratio *(21–23)*.
3. The insect cell can perform a variety of the posttranslational modifications exhibited by higher eukaryotes (reviewed in ref. *24*). Hence, mammalian signal peptides are cleaved at the appropriate site to yield the mature protein, nuclear transport signals are recognized, intracellu-

lar transport pathways result in the cell-surface display of expressed glycoproteins that normally reside on the cell surface, and proteins that are ultimately destined for secretion are processed and secreted accordingly. Similarly, phosphorylation, acylation, and sulphation have all been shown to occur to specific protein products expressed in the system. Insect cells will perform both *N*- and *O*-linked glycosylation, and it occurs at authentic sites. A major difference between insect and mammalian cells is the apparent absence of galactose and sialic acid transferases in the former. Thus, the extensive modification of the core oligosaccharide to produce the complex carbohydrates seen in a number of glycosylated mammalian and viral glycoproteins, e.g., HIV gp120 *(25)*, does not occur in insect cells, and carbohydrate sequences remain as a partially processed form of the initial high-mannose moiety, $GlcNAc_2Man_3$. It is interesting that, despite these differences, the biological activity of such proteins as the infuenza hemagglutinin *(26)* and interlukin-3 *(27)* were identical to their natural counterparts containing complex carbohydrate patterns.

4. Apart from considerations of posttranslational modifications, an important factor in the area of protein expression is the ability to purify the product. Because of the high level of synthesis observed for certain gene products in the baculovirus system, downstream processing is greatly facilitated. Also, removal of the transmembranous sequences of membrane glycoproteins, e.g., CD4, yields proteins that are processed and transported through the golgi apparatus and secreted into the medium, whence they can be purified. Scale-up facilities are available for insect cell culture (*see* Chapter 28), and exploitation of susceptible larva, such as *Trichoplusia ni*, is a plausible alternative for large-scale production.
5. As does any expression system, baculoviruses possess certain drawbacks. The absolute level of synthesis of a particular protein cannot be predicted prior to expression; however, as a general rule, intracellular DNA/RNA binding proteins and structural proteins are expressed in larger quantities than glycoproteins, which in turn are expressed in higher quantities than certain enzymes, e.g., proteases. The capacity of the system to secrete proteins appears to be limited, and expression levels of 1–5 mg/L are to be expected. These levels compare favorably with a number of mammalian expression systems (readers should compare the consecutive papers detailing CD4 expression *(28–31)*.
6. The majority of proteins expressed to date in the baculovirus system, together with information concerning posttranslational processing and biological activity, is shown in Table 1. Readers who wish to use the sys-

Table 1

Protein	Type	Reference
A: Viral proteins		
Bluetongue virus (BTV) nonstructural-1	Polymerizes to form microtubules	*32*
BTV VP1 protein	RNA-directed RNA polymerase	*33*
BTV VP2 protein	Used to raise neutralizing antibody to BTV-10 and also serotype 1,17 but not BTV13	*34*
BTV VP3 protein		*35*
BTV nonstructural-3	Two related products produced	*36*
Dengue virus nonstructural-1,E pre-M, and C proteins	Protected rabbits from fatal dengue encephalitis	*37*
Hantaan virus nucleocapsid protein	High expression, used as diagnostic antigen	*38*
Hepatitis B virus surface antigen	Acylated, glycosylated, secreted	*39,40*
Hepatitis B virus core antigen	High expression, diagnostic reagent	*23*
Herpes simplex virus (HSV) UL9	Sequence-specific DNA binding	*41*
HSV Vmw65		*42*
HSV glycoprotein D	Used to elicit high-titer neutralizing antibodies; protected mice	*43*
Human immunodeficiency virus (HIV) tat protein	Transactivator of HIV-LTR in cell fusion assay	*44*
HIV gag	p55 gag-pol processed,	*45*
	coinfection with p55 and protease yields authentic processing; p17 myristylate and p24 phosphorylated Retroviral-like core particles observed	*14,46*

HIV env	Elicits neutralizing antibodies; gp160 not cleaved to gp41 and gp120	*47*
HTLV-1 p40x	Phosphorylated, expressed in *Bombyx mori* 200 mg/2 5 × 10^8 cells	*48,49*
Influenza virus polymerase components	PB1, PB2, and PA expressed; PB1 and PB2 complex formed, but not tripartite complex including PA	*50*
Influenze virns hemagluttinin	Elicits protective immunity, cleaved to HA1 and HA2, glycosylated	*26,51*
Lymphocytic choriomeningitis virus		
Nucleocapsid protein	Highly expressed	*2*
Glycoprotein precursor	Glycosylated, not cleaved to G1 and G2, high expression	
	Nucleocapsid coexpressed with polyhedrin	*9*
Human papillomavirus (HPV) type 18 E6	Zinc finger protein, binds to DNA	*52*
HPV type 6 E2	DNA-binding properties	*53,54*
Bovine papillomavirus E2	Fusion with polyhedrin, used to infect *Bombyx mori*	*53,54*
Bovine papillomavirus E6	Located in golgi and plasma membrane carboxyl terminus extracellular	*55*
Human parainfluenza 3 fusion protein	Not cleaved to F1 and F2 subunits, elicits protective immunity	*56*
Human parainfluenza 3 hemaglutinin/neuraminidase	Glycosylated, biologically active	*57*
Poliovirus	Complete genome epressed VP0, VP1 and VP3 detected. No RNA encapsidated	*58*
Polyomavirus large T antigen	Polyomavirus origin-specific DNA binding	*59*

(continued)

Table 1 *(continued)*

Punta toro virus nucleocapsid	High-level expression	
Punta toro virus nonstructural (NS_s)		*60*
Punta toro virus glycoproteins	Intracellularly located	*61*
Rift valley fever virus M segment glycoproteins	Elicit protective immunity in mice; G2 sufficient for neutralization	*62*
Rotavirus VP4	Functional as a hemagglutinin	*63*
Rotavirus major capsid antigen (VP6)	Immunogenic in guinea pigs, not protective	*21*
Bovine rotavirus VP1	Expressed as 10% of cell protein	*64*
RSV fusion glycoprotein	Soluble product elicits protective immunity	*65*
Segment W of *Campletis sonorensis* virus	Glycosylated	*66*
Sindbis 26S cDNA	Glycosylated proteolytically cleaved	*67*
Simian immunodeficiency virus gag	Retroviral-like core particles produced	*68*
SV40 T antigen	Spliced inefficiently; specific DNA binding noted	*69,70*
SV40 T antigen and mouse p53	Complex formed after coinfection; individual proteins oligomerize	*71*
Snowshoe hare virus nucleocapsid protein	Overlapping reading frames produced the nucleocapsid and nonstructural proteins	
Snowshoe hare nonstructural protein		*72*
Vesicular stomatitis virus glycoprotein	Glycosylation not required for activity . Cell fusion at low pH	*73*
B: Miscellaneous proteins		
Human erythropoietin	Active; secreted 50-fold higher expression than COS cells	*74*
Human tissue plasminogen activator	Secreted, *N*-glycosylated to a endo-β-*N*-acetyl-D glucosamine H-resistant form *N*-terminal processed	*75,76*

Maize Ac element	Nuclear phosphoprotein	*77*
Bovine opsin	Biologically active, binds 11-*cis* retinal, expressed at 3 μg/10^6 cells	*78*
Human glucocerebrosidase	40% secreted, *N*-terminus processed, glycosylated	*79*
Human epidermal growth factor receptor	155-kD transmembrane protein, EGF stimulated tyrosine kinase activity	*80*
Epidermal growth factor receptor tyrosine kinase domain	Autophosphorylated; expressed at 0.5 mg/L	*81*
N. crassa qaIF activator protein	Mediates quinic acid induction of transcription	*82*
Insect toxin 1 of scorpion	High-level expression as polyhedrin fusion protein	*83*
5-Lipoxygenage	Enzymically active	*84*
Human transferrin receptor	Acylated, glycosylated, 70% of protein forms disulfide bridges. Expression 5.8×10^5 receptors/cell	*85*
T 11 (CD2)	Surface expression; did not mediate calcium influx	*86*
Human terminal transferase	Biologically active, phosphorylated; 200-fold higher expression than lymphoblastoid cells	*87*
Human alpha-interferon	Expressed in *Bombyx mori* (50 μg/larva)	*88,89*
IGF-II	Fusion with polyhedrin	*90*
Mouse interlukin-3	Glycosylated, biologically active, higher expression than COS cells	*27*
c-*myc* Oncogene	Major phosphoprotein component of infected cells. Present in nucleus	*91*
Drosophila Kruppel gene	Nuclear phosphoprotein	*92*
E coli β galactosidase	Fusion with polyhedrin	*93*
Human IL-2	Secreted, biologically active	*94*

(continued)

Table 1 *(continued)*

Human β-interferon	95% secreted. Active. Expressed at $(5 \times 10^6 U)/10^6$ cells	*12*
Patatin, major potato tuber protein	Glycosylated with lipid acylhydrolase and acyltransferage activity	*95*
RSV-LTR CAT	Polyhedrin promoter only active in expression in *Drosophila* and mosquito cells	*96*
Human acid β glucosidase	Biologically active	*97*
Antistasin	Anticoagulant and metastatic Factor Xa inhibitor	*98*
v-*sis* Growth factor	50- to 100-fold higher expression than other expression systems. Disulfide linked dimer formation. Biologically active	*99*
Human plasminogen	Secreted, biologically active	*100*
P2109 BCR-ABL	Tyrosine protein kinase; autophoshorylated. Expression level 4–5 mg/L	*101*
Prion protein	Failed to induce scrapie in hamsters. Not active	*102*
CD4	Both soluble and native CD4 produced. Soluble product 45% of secreted protein	*30,103*

tem are recommended to identify within the list proteins similar to those of their own individual interest and then to consult the cited references to ascertain whether the baculovirus expression system offers the most appropriate vechicle for production of the desired gene product.

7. The only factor that may consistently affect expression is the length of the untranslated leader of the gene to be expressed. Generally, the gene should contain the minimum amount of 5' untranslated sequence possible given the contraints of the cloning procedure, and the presence of "GC"-rich sequences should be minimized.

References

1. Luckow, V. A. and Summers, M. D. (1988) Trends in the development of baculovirus expression vectors. *Bio/Technology* **6,** 47–55.
2. Matsuura, Y., Possee, R. D., Overton, H. A., and Bishop, D. H. L. (1987) Baculovirus expression vectors: The requirements for high level expressicln of proteins, including glycoproteins. *J. Gen. Virol.* **68,** 1233–1250.
3. Kang, C. Y. (1988) Baculovirus vectors for expressing foreign genes. *Adv. Virus Res.* **35,** 177–192.
4. Kelly, D. C. (1982) Baculovirus replication. *J. Gen. Virol.* **63,** 1–13.
5. Maeda, S. (1989) Expression of foreign genes in insect cells using baculovirus vectors. *Annu. Rev. Entomol.* **34,** 351–372.
6. Summers, M. D. and Smith, G. E. (1987) *Manual of Methods for Baculovirus Vectors and Insect Cell Culture Procedures* (Tex. Agric. Exp. Stn. Bull., No. 1555).
7. Page, M. J. (1989) p36C: An improved baculovirus expression vector for producing high levels of mature recombinant proteins. *Nucleic Acid Res.* **17,** 451.
8. Weyer, U., Knight, S., and Possee, R. D. (1990) Analysis of very late expression by *Autographa californica* nuclear polyhedrosis virus and the further development of multiple expression vectors. *J. Gen. Virol.* **71,** 1525–1534.
9. Emery, V. C. and Bishop, D. H. L. (1987) The development of multiple expression vectors for the high level synthesis of eukaryotic proteins: Expression of the LCMV-*N* and AcNPV polyhedrin protein bv a recombinant baculovirus. *Protein Eng.* **1,** 359–366.
10. Hill-Perkins, M. S. and Possee, R. D. (1990) A baculovirus expression vector derived from the basic protein promoter of *Autographa californica* nuclear polyhedrosis virus. *J. Gen. Virol.* **71,** 971–976.
11. Livingstone, C. and Jones, I. M. (1989) Baculovirus expression vectors with single stranded capability. *Nucleic Acid Res.* **17,** 2366.
12. Smith, G. E., Summers, M. D., and Fraser, M. J. (1983) Production of human beta interferon in insect cells infected with a baculovirus expression vector. *Mol. Cell. Biol.* **3,** 183–192.
13. Miller, D. W., Safer, P., and Miller, L. K. (1986) An insect baculovirus host-vector for high-level expression of foreign genes, in *Genetic Engineering,* vol. 8, (Setloh, J. K. and Hollaender, A., eds.) Plenum, New York, pp. 277–298.
14. Overton, H. A., Fujii, Y., Price, I. R., and Jones, I. M. (1989) The protease and gag gene products of the Human immunodeficiency virus: Authentic cleavage and posttranslational modification in an insect cell expression system. *Virology* **170,** 107–116.

15. Bishop, D. H. L. and Emery, V. C. (1988) Expression vectors for the synthesis of proteins and plasmid replicons and sequence cassettes for use in constructing such vectors. Eur. Patent 88307452.8.
16. Bishop, D. H. L. and Emery, V. C. (1989) Expression vectors for the synthesis of proteins and plasmid replicons and sequence cassettes for use in constructing such vectors. US Patent PCT/GB88/00663.
17. Saiki, R. K., Gelfand, D. H., Stoffel, S., Scharf, S. J., Higuchi, R., Horn, G. T., Mullis, K. B., and Erlich, H. A. (1988) Primer directed enzymic amplification of DNA with a thermostable DNA polymerase. *Science* **239,** 487–491.
18. Murphy, G. and Kavanagh, T. (1988) Speeding up the sequencing of double-stranded DNA. *Nucleic Acid Res.* **16,** 5198.
19. Chen, E. Y. and Seeburg, P. H. (1985) Supercoil sequencing: A fast and simple method for sequencing plasmid DNA. *DNA* **4,** 165–170.
20. Sambrook, J., Fritsch, E. F., and Maniatis, T. (1989) *Molecular Cloning: A Laboratory Manual,* 2nd Ed. (Cold Spring Harbor Laboratory, Cold Spring Harbor, NY).
21. Estes, M. K., Crawford, S. E., Penaranda, M. E., Petrie, B., Burns, J. W., Chan, W., Ericson, B., Smith, G. E., and Summers, M. D. (1987) Synthesis and immunogenicity of the rotavirus major capsid antigen using a baculovirus expression system. *J. Virol.* **61,** 1488–1494.
22. Hu, S., Kowoski, S. G., and Schaaf, K. F. (1987) Expression of the envelope glycoproteins of human immunodeficiency virus by an insect virus vector. *J. Virol.* **61,** 3617–3620.
23. Takehara, K., Ireland, D., and Bishop, D. H. L. (1988) Coexpression of the hepatitis B surface and core antigens using baculovirus multiple expression vectors. *J. Gen. Virol.* **69,** 2763–2777.
24. Miller, L. K. (1988) Baculoviruses for foreign gene expression in insect cells. *Biotechnology* **10,** 457–465.
25. Mizouchi, T., Spellman, M. W., Larkin, M., Solomon, J., Basa, L. J., and Feizi, T. (1988) Carbohydrate studies of the human immunodeficiency virus (HIV) recombinant envelope glycoprotein gp120 produced in Chinese hamster ovary cells. *Biochem. J.* **254,** 599–603.
26. Kuroda, K., Hauser, C., Rott, R., Klenk, H., and Doerfler, W. (1986) Expression of the influenza virus haemagglutinin in insect cells by a baculovirus vector. *EMBO J.* **5,** 1359–1365.
27. Miyajima, A., Schreurs, J., Otsu, K., Kondo, A., Arai, K., and Maeda, S. (1989) Use of the silkworm *Bombyx mori,* and an insect baculovirus vector for high level expression and secretion of biologically active mouse interlukin-3. *Gene* **58,** 273–281.
28. Deen, K. C., McDougal, J. S., Inacker, R., Folena-Wasserman, G., Arthos, J., Rosenberg, J., Maddon, P. J., Axel, R., and Sweet, R. W. (1988) A soluble form of CD4 (T4) protein inhibits AIDS virus infection. *Nature* **331,** 82–84.
29. Fischer, R. A., Bertonis, J. M., Meier, W., Johnson, V. A., Costopoulus, D. S., Liu, T., Tizard, R., Walker, B. D., Hirsch, M. S., Schooley, R. T., and Flavell, R. A. (1988). HIV infection is blocked in vitro by recombinant soluble CD4. *Nature* **331,** 76–78.
30. Hussey, R. E., Richardson, N. E., Kowalski, M., Brown, N. R., Chang, H.-S., Siliciano, R. F., Dorfman, T., Walker, B., Sodroski, J., and Reinherz, E. L. (1988) A soluble CD4 protein selectively inhibits HIV replication and sycytium formation. *Nature* **331,** 7881.
31. Traunecker, A., Luke, W., and Karjalainen, K. (1988) Soluble CD4 molecules neutralize human immunodeficiency. *Nature* **331,** 84–86.
32. Urakawa, T. and Roy, P. (1988) Bluetongue virus tubules made in insect cells by

recombinant baculoviruses: Expression of the NS1 gene of bluetongue virus serotype 10. *J. Virol.* **62,** 3919–1927.

33. Urakawa, T., Ritter, D.G., and Roy, P. (1989) Expression of the largest RNA segment and synthesis of VP1 protein of bluetongue virus in insect cells by recombinant baculoviruses: Association with RNA polymerase activity. *Nucleic Acid Res.* **17,** 7395–7402.
34. Inamaru, S. and Roy, P. (1987) Production and characterisation of the neutralisation antigen VP2 of bluetongue virus serotype 10 using a baculovirus expression vector. *Virology* **157,** 472–479.
35. Inamaru, S., Ghiasi, H., and Roy, P. (1987) Expression of bluetongue virus group-specific antigen Vp3 in insect cells by a baculovirus vector: Its use for the detection of bluetongue virus antibodies. *J. Gen. Virol.* **68,** 1627–1635.
36. French, T.J., Inamaru, S., and Roy, P. (1989) Expression of two related nonstructural proteins of bluetongue virus (BTV) type 10 in insect cells by a recombinant baculovrirus: Production of polyclonal ascitic fluid and characterisation of the gene product in BTV-infected BHK cells. *J. Virol.* **63,** 3270–3278.
37. Zhang, Y. M., Haves, E. P., McCarty, T. L., Dubois, D. R., Summers, P. L., Eckels, K. H., Chanock, R. M., and Lai, C. J. (1988) Immunisation of mice with dengue structural proteins and nonstructural protein NS1 expressed by baculovirus recombinant induces resistance to dengue virus encephalitis. *J. Virol.* **62,** 3027–3031.
38. Schmaljohn, C. S., Sugiyama, K., Schmaljohn, A. L., and Bishop, D. H. L. (1988) Baculovirus expression of the small genome segment of Hantaan virus and potential use of the expressed nucleocapsid protein as a diagnostic antigen. *J. Gen. Virol.* **69,** 777–786.
39. Kang, C. Y., Bishop, D. H. L., Seo, J. S., Matsuura, Y., and Choe, M. H. (1987) Secretion of human hepatitis B virus surface antigen from *Spodoptera frugiperda* cells. *J. Gen. Virol.* **68,** 2607–2613.
40. Landford, R. E., Luckow, V. A., Kennedy, R. C., Dreesman, G. R., Notvall, L., and Summers, M. D. (1989) Expression and characterisation of hepatitis B virus surface antigen polypeptides in insect cells with a baculovirus expression system. *J. Virol.* **63,** 1549–1557.
41. Olivo, P.D., Nelson, N.J., and Challberg, M.D. (1988) Herpes simplex virus DNA replication: The UL9 gene encodes an origin-binding protein. *Proc. Natl. Acad. Sci. USA* **85,** 5414–5418.
42. Capone, J. (1989) Sceening recombinant baculovirus plaques *in situ* with antibody probes. *Genet. Anal. Tech.* **6,** 62–66.
43. Krishna, S., Blacklaws, B. A., Overton, H. A., Bishop, D. H. L., and Nash, A. A. (1989) Expression of glycoprotein D of herpes simplex virus type 1 in a recombinant baculovirus: Protective responses and T cell recognition of the recombinant cell extracts. *J. Gen. Virol.* **70,** 1805–1811.
44. Jeang, K. T., Shank, P. R., Rabson, A. B., and Kumar, A. (1988) Synthesis of functional human immunodeficiency virus tat protein in baculovirus as determined by a cell-cell fusion assay. *J. Virol.* **62,** 3874–3878.
45. Madisen, L., Travis, B., Hu, S. L., and Purchio, A. F. (1987) Expression of the human immunodeficiency gag gene in insect cells. *Virology* **158,** 218–250.
46. Gheysen, D., Jacobs, F., de Feresta, F., Thiriart, C., Francotte, M., Thines, D., and De Wilde, M. (1989) Assembly and release of HIV-1 precursor Pr55gag virus-like particles from recombinant baculovirus infected insect cells. *Cell* **59,** 103–112.

47. Rusche, J. R., Lynn, D. L., Robert-Guroff, M., Langlois, A. J., Lyerly, H. K., Carson, H., Krohn, K., Ranki, A., Gallo, R. C., and Bolognesi, D. P. (1987) Humoral immune response to the entire human immunodeficiency virus envelope glycoprotein made in insect cells. *Proc. Natl. Acad. Sci. USA* **84**, 6924–6928.
48. Nyunoya, H., Akagi, T., Ogura, T., Maeda, S., and Shimotohno, K. (1988) Evidence for phosphorylation of transactivator p40x of human T-cell leukamia virus type 1 produced in insect cells with a baculovirus expression vector. *Virology* **167**, 538–544.
49. Jeang, K.T., Holmgren-Konig, M., and Khoury, G. (1987) Abundant synthesis of fuctional human T—cell leukemia virus type 1 p40x protein in eucaryotic cells by using a baculovirus expression vector. *J. Virol.* **61**, 1761–1764.
50. St. Angelo, C., Smith, G. E., Summers, M. D., and Krug, R. M. (1987) Two of the three infuenza viral polymerase proteins expressed by using baculovirus vectors form a complex in insect cells. *J. Virol.* **61**, 361–365.
51. Possee, R. D. (1986) Cell surface expression of influenza virus haemagglutinin in insect cells using a baculovirus vector. *Virus Res.* **5**, 43–59.
52. Grossman, S. R., Mora, R., and Laimins, L. A. (1989) Intracellular localisation and DNA-binding properties of human papillomavirus type 18 E6 protein expressed with a baculovirus vector. *J. Virol.* **63**, 366–374.
53. Tada, A., Fuse, A., Sekine, H., Simizu, B., Kondo, A., and Maeda, S. (1988) Expression of the E2 open reading frame of papillomaviruses BPV1 and HPV6b in silkworm by a baculovirus vector. *Virus Res.* **9**, 357–367.
54. Sekine, H., Fuse, A., Tada, A., Maeda, S., and Simuzu, B. (1988) Expression of human papillomavirus type 6b E2 gene product with DNA-binding activity in insect (*Bombyx mori*) cells using a baculovirus expression vector. *Gene* **65**, 187–193.
55. Burkardt, A., Willingham, M., Gay, C., Jeang, K.T., and Schlegel, R. (1989) The E5 oncoprotein of bovine papillomavirus is orientated asymmetrically in Golgi and plasma membranes. *Virology* **170**, 334–339.
56. Ray, R., Galinski, M. S., and Compans, R. W. (1989) Expression of the fusion glycoprotein of human parainfluenza type 3 virus in insect cells by a recombinant baculovirus and analysis of its immunogenicity. *Virus Res.* **12**, 169–180.
57. Van Wyke-Coelingh, E. L., Murphy, B. R., Collins, P. L., Lebacq-Verheyden, A. M., and Battey, J. F. (1987) Expression of biologically active and antigenically authentic parainfluenza type 3 virus hemagglutinin-neuraminidase glycoprotein by a recombinant baculovirus. *Virology* **160**, 465–472.
58. Urakawa, T., Ferguson, M., Minor, P. D., Cooper, J., Sullivan, M., Almond, J. W., and Bishop, D. H. L. (1989) Synthesis of immunogenic, but non-infectious, poliovirus particles in insect cells by a baculovirus expression vector. *J. Gen. Virol.* **70**, 1453–1463.
59. Rice, W. C., Lorimer, H. E., Prives, C., and Miller, L. K. (1987) Expression of polyomavirus large T antigen by using a baculovirus vector. *J. Virol.* **61**, 1712–1716.
60. Overton, H. A., Ihara, T., and Bishop, D. H. L. (1987) Identification of the *N* and SS proteins coded by the ambisense S RNA of the Punta Toro phlebovirus using monospecific antisera raised to baculovirus expressed *N* and SS protein. *Virology* **157**, 338–350.
61. Matsuoka, Y., Ihara, T., Bishop, D. H. L., and Compans, R.W. (1988) Intracellular accumulation of Punta Toro virus glycoproteins expressed from cloned cDNA. *Virology* **167**, 251–260.
62. Schmaljohn, C. S., Parker, D., Ennis, W. H., Dalrymple, J., Collett, M. S., Suzich, J. A., and Schmaljohn, A. L. (1989) Baculovirus expression of the M genome segment of

Rift Valley fever virus and examination of the antigenic and immunogenic properties of the expressed proteins. *Virology* **170,** 184–192.

63. Mackow, E. R., Barnett, J., Chan, H. and Greenberg, H. B. (1989) The rhesus rotavirus outer capsid protein VP4 functions as a hemagglutinin and is antigenically conserved when expressed by a baculovirus recombinant. *J. Virol.* **63,** 1661–1668.
64. Cohen, J., Charpillenne, A., Chilmanazyk, S., and Estes, M. (1989) Nucleotide sequence of bovine rotavirus-gene 1 and expression of the gene product in baculovirus. *Virology* **171,** 131–140.
65. Wathen, M. W., Brideau, R. J., and Thomsen, D. R. (1989) Immunisation of cotton rats with the human respiratory syncytial virus F glycoprotein produced using a baculovirus vector. *J. Infect. Dis.* **159,** 255–264.
66. Blissard, G. W., Theilmann, D. A., and Summers, M. D. (1989) Segment W of Campoletis sonorensis virus: Expression, gene products and organisation. *Virology* **169,** 78–89.
67. Oker-Blom, C. and Summers, M. D. (1989) Expression of Sindbis virus 26S cDA in *Spodoptera frugiperda* (Sf 9) cells, using a baculovirus expression vector. *J. Virol.* **63,** 1256–1264.
68. Delchambre, M., Gheysen, D., Thines, D., Thiriart, C., Jacobs, E., Verdin, E., Horth, M., Burny, A., and Bex, F. (1989) The GAG precursor of simian immunodeficiency virus assembles into virus-like particles. *EMBO J.* **8,** 2653–2660.
69. Murphy, C. I., Weiner, B., Bikel, I., Piwnica-Worms, H., Bradley, M. K., and Livingstone, D. M. (1988) Purification and functional properties of simian virus 40 large and small T antigens overproduced in insect cells. *J. Virol.* **62,** 2951–2959.
70. Jeang, K. T., Giam, C. Z., Nerenberg, M., and Khoury, G. (1987) Abundant synthesis of functional human T-cell leukemia virus type 1 p40x protein in eukaryotic cells by using a baculovirus expression vector. *J. Virol.* **61,** 701–713.
71. O'Reilly, D. R. and Miller, L. K. (1988) Expression and complex formation of simian virus 40 large T antigen and mouse p53 in insect cells. *J. Virol.* **62,** 3109–3119.
72. Urakawa, T., Small, D. A., and Bishop, D. H. L. (1988) Expression of snowshoe hare bunyavirus S RTA coding proteins by recombinant baculoviruses. *Virus Res.* **11,** 303–317.
73. Bailey, M. J., McCleod, D. A., Kang, C. Y., and Bishop, D. H. L. (1989) Glycosylation is not required for the fusion activity of the G protein of vesicular stomatitis virus in insect cells. *Virology* **169,** 323–331.
74. Wojchowski, D. M., Orkin, S. H., and Sytkowski, A. J. (1987) Active human erythropoietin epressed in insect cells using a baculovirus vector: A role for *N*-linked oligosaccharide. *Biochem. Biophys. Acta* **910,** 224–232.
75. Furlong, A. M., Thomsen, D. R., Marotti, K. R., Post, L. E., and Sharma, S. K. (1988) Active human tissue plasminogen activator secreted from insect cells using a baculovirus vector. *Biotechnol. Appl. Biochem.* **10,** 454–464.
76. Jarvis, D. L. and Summers, M. D. (1989) Glycosylation and secretion of human tissue plasminogen activator in recombinant baculovirus-infected insect cells. *Mol. Cell. Biol.* **9,** 214–223.
77. Hauser, C., Fusswinkel, H., Li, J., Oellig, C., Kunze, R., Muller-Neumann, M., Heinlein, M., Starlinger, P., and Doerfler, W. (1988) Overproduction of the protein encoded by the maize transposable element Ac in insect cells by a baculovirus vector. *Mol. Gen. Genet.* **214,** 373–378.
78. Janssen, J. J., Van-de-Ven, W. J., van-Groningen-Luyben, W. A., Roosien, J., Vlak, J. M.,

and de-Grip, W. J. (1988) Synthesis of functional bovine opsin in insect cells under control of the baculovirus polyhedrin promoter. *Mol. Biol. Rep.* **13,** 65–71.

79. Martin, B. M., Tsuji, S., LaMarca, M. E., Maysak, K., Eliasom, W., and Ginns, E. I. (1988) Glycosylation and processing of high levels of active human glucocerebrosidase in invertebrate cells using a baculovirus epression vector. *DNA* 7, 99–106 .
80. Greenfield, C., Patel, G., Clark, S., Jones, T., and Waterfield, M. D. (1988) Expression of the human EGF receptor with ligand-stimulatable kinase activity in insect cells using a baculovirus vector. *EMBO J.* **7,** 139–146.
81. Wedegaertner, P. B. and Gill, G. N. (1989) Activation of the purified protein kinase domain of the epidermal growth factor receptor. *J. Biol. Chem.* **264,** 11346–11353.
82. Baum, J. A., Geever, R., and Giles, N. H. (1987) Expression of the qa-1F activator protein: Identification of upstream binding sites in the qa gene cluster and localisation of the DNA binding domains. *Mol. Cell. Biol.* **7,** 1256–1266.
83. Carbonell, L. F., Hodge, M. R., Tomaiski, M. D., and Miller, L. K., (1988) Synthesis of a gene coding for an insect specific scorpion neurotoxin and attempts to express it using baculovirus vectors. *Gene* **73,** 409–418.
84. Funk, C. D., Gunne, H., Steiner, H., Izumi, T., and Samuelsson, B. (1989) Native and mutant lipoxygenase expression in a baculovirus/insect cell system. *Proc. Natl. Acad. Sci. USA* **86,** 2592–2596.
85. Domingo, D. L. and Trowbridge, I. S. (1988) Characterisation of the human transferrin receptor produced in a baculovirus expression system. *J. Biol. Chem.* **263,** 13386–13392.
86. Alcover, A., Chang, H. C., Sayre, P. H., Hussey, R. E., and Reinherz, E. L. (1988) The T11 (CD2) cDNA encodes a transmembrane protein which expresses T 11(1), T 11(2) alld T 11(3) epitopes but which does not independently mediate calcium influx: Analysis by gene transfer in a baculovirus system. *Eur. J. Immunol.* **18,** 363–367.
87. Chang, L. M. S., Rafter, E., Rusquet-Valerius, R., Peterson, R. C., White, S. T., and Bollum, F. J. (1988) Expression and processing of recombinant human terminal transferase in the baculovirus system. *J. Biol. Chem.* **263,** 12509–12513.
88. Horiuchi, T., Marumoto, Y., Saeki, Y., Sato, Y., Furusawa, M., Kondo, A., and Maeda, S. (1987) High level expression of the human alpha interferon gene through the use of an improved baculovirus vector in the silkworm. *Bombyx mori. Agric. Biol. Chem.* **51,** 1573–1580.
89. Maeda, S., Kawai, T., Ohinata, M., Fujiwara, H., Horiuchi, T., Saeki, Y., Sato, Y., and Furusawa, M. (1985) Production of human alpha interferon in silkworm using a baculovirus vector. *Nature* **315,** 592–594.
90. Marumoto, Y., Sato, Y., Fujiwara, H., Sakano, K, Saeki, Y., Agata, M., Furusawa, M., and Maeda, S. (1987) Hyperproduction of polyhedrin-IGF II fusion protein in silkworm larvae infected with recombinant *Bombyx mori* nuclear polyhedrosis virus. *J. Gen. Virol.* **68,** 2599–2606.
91. Miyamoto, C., Smith, G. E., Farrell-Towt, J., Chizzonite, R., and Summers, M. D. (1985) Production of human c-*myc* protein in insect cells infected with a baculovirus epression vector. *Mol. Cell. Biol.* **5,** 2860–2865.
92. Ollo, R. and Maniatis, T. (1987) Drosophila Kruppel gene product produced in a baculovirus expression system is a nuclear phosphoprotein that binds to DNA. *Proc. Natl. Acad. Sci. USA* **84,** 5700–5704.

93. Pennock, G. D., Shoemaker, C., and Miller, L. K. (1984) Strong and regulated expression of *Escherichia coli* beta galactosidase in insect cells with a baculovirus vector. *Mol. Cell. Biol*. **4**, 399–406.
94. Smith, G. E., Ju, G., Ericson, B. L., Moschera, J., Lahm, H. W., Chizzonite, R., and Summers, M. D. (1984) Modification and secretion of human interlukin 2 produced in insect cells by a baculovirus expression vector. *Proc. Natl. Acad. Sci. USA* **82**, 8404–8408.
95. Andrews, D. L., Beames, B., Summers, M. D., and Park, W. D. (1988) Characterisation of the lipid acly hydrolase activity of the major potato *(Solanum tuberosum)* tuber protein, patatin, by cloning and abundant expression in a baculovirus vector. *Biochem. J.* **252**, 199–206.
96. Carbonell, E. F., Klowden, M. J., and Miller, L. K. (1985). Baculovirus-mediated expression of bacterial genes in dipteran and mammalian cells. *J. Virol.* **56**, 153–160.
97. Grabowski, G. A., White, W. R., and Grace, M. E. (1989) Expression of functional human acid beta-glucosidase in COS-1 and *Spodoptera frugiperda* cells. *Enzyme* **41**, 133–142.
98. Han, J. H., Law, S. W., Keller, P. I., Kniskern, P. J., Siberklang, M., Gasic, T. B., Gasic, G. J., Friedman, P. A., and Ellis, R. W. (1989) Cloning and expression of cDNA encoding antistasin, a leech derived protein having anti-coagulant and anti-metastatic properties. *Gene* **75**, 47–57.
99. Giese, N., May-Siroff, M., LaRochelle, W. J., van Wyke Coelingh, K., and Aaronson, S. A. (1989) Expression and purification of biologically active v-sis/platelet-derived growth factor B protein by using a baculovirus vector system. *J. Virol.* **63**, 3080–3086.
100. Whitefleet-Smith, J., Rosen, E., McLinden, J., Ploplis, V. A., Fraser, M. J., Tomlinson, J. E., McClean, J. W., and Castellino, E. J. (1989) Expression of human plasminogen cDNA in a baculovirus vector-infected insect cell system. *Arch. Biochem. Biophys.* **271**, 390–399.
101. Pendergast, A. M., Clark, R., Kawsaki, E. S., McCormick, F. P., and Witle, O. N. (1989) Baculovirus expression of functional P210 BCR—ABL oncogene product. *Oncogene* **4**, 759–766.
102. Scott, M. R., Butler, D. A., Bredesen, D. E., Walchli, M., Hsiao, K. K., and Prusiner, S. B. (1988) Prion protein gene expression in cultured cells. *Protein Eng.* **2**, 69–76.
103. Webb, N. R., Madoulet, C., Tosi, P-F., Broussard, D. R., Sneed, L., Nicolau, C., and Summers, M. D. (1989) Cell surface expression and purification of human CD4 produced in baculovirus infected insect cells. *Proc. Natl. Acad. Sci. USA* **86**, 7731–7735.

CHAPTER 27

Baculovirus Expression Vectors

Generating a Recombinant Baculovirus

Vincent C. Emery

1. Introduction

This chapter aims to provide the reader with the experimental protocols required to produce a baculovirus containing the gene of interest, including transfection of *Spodoptera frugiperda* cells and the screening of progeny virus by plaque assay. In addition, an outline of the biochemical methods routinely used to characterize the expressed protein product within the baculovirus system will be given. Throughout the experimental procedures, areas that researchers generally find technically difficult will be highlighted in the hope that this may ameliorate some of the problems encountered in using the system *ab initio.*

2. Materials

1. *Spodoptera frugiperda* (Sf9) cells can be obtained from the American Type Culture Collection, 12301 Parklawn Drive, Rockville, MD 20852-1776; Accession No. CRL1171) or from D. H. L. Bishop, NERC Institute of Virology and Environmental Microbiology, Mansfield Road, Oxford, OX1 3SR UK. Cells have a doubling time of about 18 h in TC100 medium at 28°C.

From: *Methods in Molecular Biology, Vol. 8: Practical Molecular Virology: Viral Vectors for Gene Expression*
Edited by: M. Collins © 1991 The Humana Press Inc., Clifton, NJ

2. 2*M* hydrochloric acid.
3. 10*M* potassium hydroxide.
4. TC100 medium (powdered), supplied by GIBCO-BRL, Paisley, Scotland, UK, Cat. No. 074-03000 A. Ready-made medium is also available from GIBCO-BRL, but is commensurately more expensive.
5. Penicillin/streptomycin containing 10,000 U/mL of penicillin and 10,000 µg/mL of streptomycin.
6. Kanamycin, 5000 µg/mL.
7. Lysis buffer A: 10% w/v sodium *N*-lauryl sarcosinate, 10 m*M* EDTA.
8. 2× HEPES-buffered saline: 40 m*M* HEPES, pH 7.05, 2 m*M* Na_2HPO_4, 10 m*M* KCl, 280 m*M* NaCl.
9. 100 m*M* glucose (filter-sterilized).
10. 2.5*M* calcium chloride (filter-sterilized).
11. 0.2*M* NaOH.
12. 2*M* NaCl.
13. Lysis buffer B: 30 m*M* Tris-HCl, pH 7.5, 10 m*M* magnesium acetate, 1% Nonidet P-40.
14. RIPA immunoprecipitation buffer: 1% Triton X-100, 1% sodium deoxycholate, 0.15*M* NaCl, 50 m*M* Tris-HCl, pH 7.4, 10 m*M* EDTA, 1% SDS.
15. 2× Protein dissociation buffer: 2.3% SDS, 10% glycerol, 5% 2-mercaptoethanol, 62.5 m*M* Tris-HCl, pH 6.8, 0.01% bromophenol blue.

3. Methods

3.1. Preparation of TC100 Medium for Propagating *S. frugiperda* Cells

1. Add 19.95 g of GIBCO TC100 powdered medium to 800 mL of distilled water and stir for 30 min.
2. Adjust the pH of the solution to 5.02 with 2*M* HCl.
3. Stir the solution until it is clear.
4. Add sodium bicarbonate to bring the final concentration to 0.35 g/L.
5. Adjust the pH of the solution to 6.18 with 10*M* potassium hydroxide.
6. Adjust the volume to 900 mL and filter-sterilize.
7. Prior to use, add 10% (final volume) fetal calf serum, 20 mL of penicillin/streptomycin, and 10 mL of kanamycin.

3.2. Insect Cell Culture

3.2.1. Monolayer Culture

1. Resuspend the cells from a nearly confluent monolayer culture by gently tapping the side of the flask or by gently pipeting across

the monolayer. Excessive foaming should be avoided to maximize cell viability.

2. Transfer the cells into new flasks containing a suitable quantity of TC100 medium. Cells can be routinely split into the ratio 1:5 and are usually confluent within 3 d.
3. Incubate the cells at 28°C. Cells do not require CO_2.
4. Cell viability can be assayed by adding 0.1 mL of trypan blue to 1 mL of cells followed by microscopic examination. Cell viability should be at least 97% for lóg-phase cultures.

3.2.2. Suspension Culture

1. Seed the cells into spinner flasks at a density of about 1×10^6 cells/mL.
2. Incubate the spinner, with stirring, at 50–60 rpm at 28°C.
3. When the cell density reaches about 2.5×10^6 cells/mL, subculture by removing 80% of the suspension and replacing with fresh medium.
4. In order to prevent the accumulation of contaminants and metabolic byproducts, suspension cultures should be routinely concentrated by gentle centrifugation and resuspended in fresh medium in a new vessel.
5. For large spinner cultures, diffusion-aeration may be required.
6. Sf9 cells are not anchorage-dependent and may be transferred between monolayer and suspension cultures without undue stress.

3.3. Freezing and Storage of Insect Cells

3.3.1. Freezing Cells

1. 97% viable log-phase cell cultures should be used for freezing.
2. Concentrate the cells by gentle centrifugation and resuspend them in fresh medium at a density of $4–5 \times 10^6$ cells/mL.
3. Dilute the cell suspension with an equal volume of TC100 containing 20% DMSO (filter-sterilized) to yield a final DMSO concentration of 10%. Maintain the cells on ice.
4. Using suitable freezing vials, freeze the cells slowly either in liquid nitrogen vapor or insulated at –20°C, and then at –80°C. Transfer the vials to liquid-nitrogen storage.

3.3.2. Thawing Cells

1. Rapidly thaw the frozen cells by gentle agitation at 37°C.
2. Place the cells into a 25-cm^2 flask and add 5 mL of fresh TC100 medium.
3. Allow the cells to attach for 1 h at room temperature.

4. Incubate at 28°C. After 1 d, discard the old medium and replace with fresh TC100.

3.4. Infection of Sf9 Cells

3.4.1. Monolayer Culture

1. Count the cells using a hemocytometer and seed into flasks or dishes at the appropriate density (*see* Section 3.2.). Allow the cells to attach for at least 30 min. These procedures can be ignored if large tissue-culture flasks, e.g., >75 cm^2, of confluent cell monolayers are to be infected.
2. Following attachment, remove the medium and add the appropriate amount of virus, being particularly careful to cover the cells (i.e., a 75-cm^2 flask requires about 1 mL of inoculum).
3. Incubate the cells at 28°C or at room temperature for 1 h.
4. Remove the inoculum and replace with fresh TC100 medium.
5. Incubate at 28°C for the required time — usually 2–4 d. Examine the culture daily for cytopathic effects.
6. To collect the extracellular virus, centrifuge the infected cell medium to pellet residual cells and store the supernatant at 4°C. Virus stocks should be stored at 4°C, not frozen.

3.4.2. Suspension Cultures

1. Determine the cell number and viability as previously described (Section 3.2.).
2. Calculate the quantity of virus required to achieve the relevant multiplicity of infection (moi).
3. Pellet the suspension culture and resuspend it in the virus inoculum, and then add TC100 to yield an initial density of 1×10^7 cells/mL.
4. Incubate at 28°C or room temperature for 1 h.
5. Dilute the cells to a density of $1–5 \times 10^6$ cells/mL by adding fresh TC100.
6. Incubate at 28°C for 2–4 d and process as for the monolayer cultures.

3.5. Purification of Autographa californica Nuclear Polyhedrosis Virus (AcNPV) DNA

1. Infect confluent monolayers of Sf9 cells with AcNPV at an moi of about 0.5–1. Usually the virus obtained from five 175-cm^2 flasks is sufficient for extracting viral DNA.
2. At 48–72 h postinfection, remove the infected cells by centrifugation and subject the extracellular virus to ultracentrifugation at 100,000*g* (25,000 rpm in an 8×50-mL rotor) for 1 h at 4°C.
3. Discard the supernatant and resuspend the viral pellet in a small volume

of TE. Layer the virus onto a gradient consisting of equal volumes of 10 and 50 % sucrose in TE (w/w). Centrifuge in a swing-out rotor for 90 min at 100,000*g*. Remove the band of virus at the 10–50% sucrose interface.

4. Dilute the viral band with TE to 50 mL and pellet at 100,000*g* in a fixed-angle rotor for 1 h at 4°C. Resuspend the pellet in 1 mL of TE.
5. Add 0.6 mL of lysis buffer to the gradient-purified virus and incubate at 60°C for 20 min.
6. Layer the solution from step 5 onto a 54% (w/w) cesium chloride/TE gradient containing ethidium bromide. Centrifuge at 200,000*g* for 18 h at 20°C.
7. View the gradients under UV light and harvest the viral DNA bands corresponding to supercoiled and open circular DNA.
8. Remove ethidium bromide from the samples by gently extracting with water-saturated butan-1-ol.
9. Remove butan-1-ol from the DNA sample by dialysis overnight against *sterile* TE.
10. Store viral DNA at 4°C, not frozen.

This procedure yields intact AcNPV DNA that functions efficiently in transfections.

3.6. Transfection of Sf9 Cells

The following transfection procedure is a modification of that described by Summers and colleagues *(1)*.

1. Seed Sf9 cells onto 35-mm tissue-culture dishes at a density of 1.5–2.0 × 10^6 cells/dish and allow them to adhere for 30 min.
2. Mix the following constituents in an Eppendorf tube:

2× HEPES-buffered saline	0.475 mL
AcNPV DNA	1 μg
transfer vector DNA	25 or 50 μg
100 m*M* glucose	0.1 mL
water	to 0.95 mL

3. Add 50 μL of 2.5*M* calcium chloride solution, dropwise and with vortexing, to the above transfection mixture. Leave at room temperature for 30 min to coprecipitate the DNA. The precipitate should appear as a slight opaqueness in the transfection mixture, not as a heavy white precipitate.
4. Remove the TC100 medium from the Sf9 cells prepared in Step 1 and replace it with the transfection mixture from Step 3. Leave at room temperature for 1 h.

5. Remove the transfection mixture and replace it with fresh TC100.
6. Leave the transfected cells at 28°C in a humidified container for 2 d. Check the cells for the presence of polyhedra by light microscopy to confirm that transfection has been successful. Under normal circumstances, polyhedra should be visible within 2 d. The appearance of transfected Sf9 cells varies, but they invariably appear stressed.
7. After 2 d, harvest the extracellular virus by removing the supernatant from the transfected cells and briefly microfuging to remove cell debris. Store the transfection supernatants at 4°C prior to plaque assay.

3.7. Plaque Assay to Identify Recombinant Viruses

Identification of the polyhedrin-negative phenotype that typifies a recombinant baculovirus using plaque assay *(2)* is the most difficult facet of the baculovirus expression system to master.

1. Seed 35- × 10-mm tissue culture dishes with Sf9 cells to a density of 1.5×10^6 cells/dish. The cells should not be confluent in the dishes, since they will grow over the course of the plaque assay.
2. Prepare tenfold dilutions of the transfection mixture; usually 10^{-2} and 10^{-3} dilutions yield a convenient number of plaques from transfections performed according to the methods outlined in Section 3.6.
3. Remove the media from the attached cells and add 0.1 mL of the diluted virus to each plate. Twenty dishes at each dilution should be sufficient for the identification of recombinant viruses with a polyhedrin-negative phenotype.
4. Incubate the dishes for 1 h at room temperature.
5. While the dishes are incubating, prepare a 1.5% low-melting-point (Seaplaque) agarose overlay by mixing equal volumes of 3% low-gelling agarose (melted in a microwave oven) and complete TC100. The TC100 medium should not be added until the agarose has cooled to approx 45°C, nor should the final mixture be remicrowaved. Maintain the 1.5% agarose–TC100 mixture at 45°C in a water bath until required.
6. After 1 h of virus adsorption, remove the inoculum using a Pasteur pipet and add 2 mL of agarose overlay to the edge of each dish, ensuring an even spread.
7. Leave the dishes for at least 30 min to allow the overlay to solidify.
8. Add 1 mL of TC100 to each dish and incubate for 4–6 d in a humidified container at 28°C.
9. Three rounds of plaque purification are usually sufficient to isolate recombinant virus without contaminating wild-type polyhedrin-positive viruses.

3.8. Detection of Recombinant Viruses Following Plaque Assay

3.8.1. Visual Screening

1. In order to detect polyhedrin-negative viruses, discard the 1 mL of TC100 medium from the plaque-assay dish and invert the dish over a light box, to which a piece of black velvet or blackened X-ray film has been attached. Using this contrasting mechanism, wild-type virus plaques appear refractile, with a white or yellowish color.
2. Choose plaques that appear less refractile in nature and circle them with a marker pen as potential recombinant viruses.
3. View the circled plaques under an inverted phase microscope at 400× magnification. Appearance of polyhedra in any cells within the plaque means that the plaque is not a recombinant. All circled plaques should be rigorously viewed in this way to ensure that they are truly polyhedrin-negative.
4. Plaques that are definitely polyhedrin-negative are then picked using sterile glass Pasteur pipets. The agar plug above the plaque is removed from the overlay by gentle suction and ejected into fresh TC100 medium (0.4 mL), vortexed, and stored at 4°C. Virus isolated from the plaque is then subjected to three more rounds of plaque assay at dilutions of 10^{-1} and 10^{-2} to ensure that it is not contaminated with wild-type AcNPV.

3.9. Detection of Recombinants by Limiting Dilution and DNA Dot Blot Hybridization

Although it should by possible to identify recombinant baculoviruses by the visual screening procedures outlined above, alternative methods are available to identify recombinants. The simplest and most efficient of these is limiting dilution coupled with DNA dot blot hybridization. The method described below is that reported in ref. *(3)*.

1. Seed 1.5×10^4 Sf9 cells in 80 µL of TC100/well of a 96-well plate. Allow the cells to attach for 30 min.
2. Prepare 10-fold dilutions of the transfection mixture. Remove the TC100 medium from wells in the 96-well plate and replace with 50 µL of suitably diluted transfection mixture.
3. Incubate the plate at 28°C for 5–8 d in a humidified container.
4. Transfer the supernate of each well to a new plate (to serve as a master plate) and store the new plate at 4°C.

5. Add 0.2 mL of 0.2*M* sodium hydroxide to each well of the original plate to lyse the cells. Detach the cells in each well by mixing with a pipet and leave for 15 min at room temperature.
6. Prepare nylon or nitrocellulose blotting membranes according to the manufacturer's instructions (e.g., Hybond N from Amersham International, UK) and place into a dot blot apparatus.
7. Add 0.4 mL of 2*M* sodium chloride/well of the dot blot apparatus followed by 0.2 mL of the lysed cells. Aspirate the solution through the membrane with a vacuum pump.
8. Prehybridize and hybridize membranes with a radioactive probe representing the gene to be expressed using standard protocols *(4)*, and expose the filter to X-ray film.
9. Virus stocks from the wells that are hybridization-positive can then be subjected to further limiting dilution analysis to effectively enrich the recombinant virus. After three rounds of screening, the recombinant viral stock is usually homogeneous.

3.10. Preparation of Recombinant Viral DNA for Restriction Endonuclease Digestion

The most facile method for demonstrating that homologous recombination between transfer vector and AcNPV DNA has occurred authentically is via Southern blot analysis of restriction-digested viral DNA. Viral DNA suitable for this purpose may be derived from either extracellular virus or infected cells.

3.10.1. Extraction of Recombinant Viral DNA from Extracellular Virus

1. Follow steps 1–5 of Section 3.5.
2. Do not proceed to purify the viral DNA with cesium chloride; instead, phenol-extract and ethanol-precipitate the nucleic acid.
3. Viral DNA prepared by precipitation can then by digested with restriction endonuclease and subjected to Southern blot analysis *(4)*.

3.10.2. Extraction of Viral DNA from Infected Cells

Alternatively, viral DNA can be purified from the infected Sf9 cell nucleus. This is facilitated by the fact that, late in infection, approx 25% of the total nucleic acid in an Sf9-infected cell nucleus is AcNPV DNA. This method is less time-consuming than method 3.10.1., but viral DNA accounts for only 25% of the DNA isolated.

1. Infect confluent monolayers of Sf9 cells with a virus at an moi of 1 (two 75-cm^2 flasks should be sufficient for viral DNA isolation). At 72 h postinfection, decant the culture medium and discard.
2. Add 5 mL of lysis buffer B to each flask and leave for 5 min in a horizontal position, during which time the cells will detach.
3. Transfer the cell suspension to a corex tube and keep on ice for 15 min. Vortex 3–4 times during this period.
4. Centrifuge the suspension at 2000 rpm for 3 min and discard the supernatant containing cytoplasmic nucleic acid.
5. Wash the pelleted nuclei with cold PBS and repellet.
6. Add lysis buffer A (1 mL) and incubate at 60°C for 30 min.
7. Phenol-extract (0.5 vol) and ethanol-precipitate the nucleic acid.

3.11. Radiolabeling the Recombinant Proteins

When radiolabeling proteins synthesized in recombinant AcNPV-infected Sf9 cells, it is important to ascertain the appropriate radiolabeled amino acid to use. The default should always be ^{35}S-methionine, especially since the major late p10 protein of AcNPV does not contain methionine and thus does not interfere with the analysis of low-mol-wt recombinant protein products. Alternatively, [^3H]-leucine or ^{35}S-cysteine can be used. It is obviously imperative to perform similar parallel analysis of cells infected with wild-type AcNPV and of mock infected cells to provide the relevant control samples. Usually protein synthesis is assayed at 24 and 48 h postinfection.

1. Seed 35- × 10-mm dishes with 2 × 10^6 Sf9 cells and allow them to adhere.
2. Remove the TC100 medium from dishes and add virus (0.1 mL inoculum) to achieve an moi of 5–10.
3. Incubate at room temperature for 1 h.
4. Remove the viral inoculum and replace it with 1 mL of fresh TC100.
5. After 24 or 48h, remove the supernatant and gently wash the cell monolayer with TC100 depleted of the amino acid that is to be used as the radioactive label. The depleted TC100 does not require addition of fetal calf serum or antibiotics.
6. Add 1 mL of the depleted TC100 and incubate at 28°C for 1 h.
7. Remove the depleted TC100 medium, replace it with depleted medium (0.2 mL) containing the appropriate radiolabeled amino acid (10μCi/dish), and incubate the dishes at 28°C for at least 1 h. If the recombinant protein is secreted, labeling should proceed for about 6 h.
8. After the labeling period, remove cells by gently pipeting the labeling medium and microfuge to pellet the cells. The supernatant can be directly analyzed by SDS-PAGE after the addition of 2× protein dissocia-

tion buffer. The cell pellet should be washed twice with PBS and resuspend in RIPA buffer. Radiolabeled proteins can the be either immune-precipitated or subjected to SDS-PAGE analysis after addition of 2 protein dissociation buffer followed by boiling for 5 min. Protein extracts can be frozen for 2 wk without significant proteolytic degradation.

4. Notes

1. The condition of the insect cells is a consistent factor that affects the success (and failure) rate of using the baculovirus system. If the appearance of the cells indicates senescnece, then it is worth passaging the cells rather than using them for a critical experiment. Often, cells can be revitalized following passage, even if they appear severely distressed.
2. Insect cells obtained from the Institute of Virology and Environmental Microbiology usually give optimum results when used between passages 155–180. After passage 180, cells should not be used for plaque assay. In contrast, insect cells derived from the ATCC can be used immediately, i.e., at low passage numbers, although data concerning their efficiency for plaque assay after passage 180 is not available.

References

1. Smith, G. E., Summers, M. D., and Fraser, M. J. (1983) Production of human beta interferon in insect cells infected a baculovirus expression vector. *Mol. Cell. Biol.* **3,** 2156–2165.
2. Brown, M. and Faulkner, P. (1977) A plaque assay for nuclear polyhedrosis viruses using a solid overlay. *J. Gen. Virol.* **36,** 361–364.
3. Fung, M.-C., Chiu, K. Y. M., Weber, T., Chang, T.-W., and Chang, N. T. (1988) Detection and purification of a recombinant human B lymphotropic virus (HHV-6) in the baculovirus expression system by limiting dilution and DNA dot-blot hybridization. *J. Virol. Methods* **19,** 33–42.
4. Sambrook, J., Fritsch, E. F., and Maniatis, T. (1989) *Molecular Cloning: A Laboratory Manual,* 2nd Ed. Cold Spring Harbor, Cold Spring Harbor Laboratory, New York.

CHAPTER 28

Baculovirus Expression Vectors

Scale-Up and Downstream Processing

Vincent C. Emery

1. Introduction

The baculovirus system has the potential for the large-scale production of protein products by two methods. First, as a result of recent advances *(1,2)*, large-scale cell culture is now possible and, second, the recombinant baculovirus can be used to infect susceptible insect larvae *(3)*. Whether using small-scale cell culture or the aforementioned methods for scale-up, the requirement for downstream processing of the protein product is manifest. Purification of the product is greatly facilitated when expression is high (ca. 30% of total cell protein); however, when expression is relatively low or membrane proteins are required, the researcher faces the same problems encountered in ascertaining the optimum conditions for any *de novo* protein purification. Obviously, the degree of purity required for a given product is indicative of its final use; consequently, a number of antigens for use in diagnostic procedures have been relatively crude preparations *(4,5)*. This chapter highlights the methods that have been used to purify baculovirus-expressed protein products, and is aimed at providing general guidelines to the purification of certain types of protein product rather than a definitive guide to protein purification.

From: ***Methods in Molecular Biology, Vol. 8:***
Practical Molecular Virology: Viral Vectors for Gene Expression
Edited by: M. Collins

2. Materials

1. Fourth instar *Trichoplusia ni* caterpillars: Researchers interested in using these animals for scale-up should consult D. H. L. Bishop (Institute of Virology and Environmental Microbiology, Mansfield Road, Oxford, OX1 3SR, UK) for expert advice.
2. Phosphate buffered saline.
3. 0.2% solution of sodium diethylthiocarbamate.
4. Lipid mixture: Cod liver oil, 10 mg; Tween-80, 25 mg, cholesterol, 4.5 mg, α-tocopherol acetate, 2 mg, ethanol, 1 mL, filter-sterilized using a 0.22-μ*M* filter.
5. Pleuronic mixture: 10% Pleuronic polyol F-68 (BASF, Parsippany, NJ) in water (10 mL), filter-sterilized using a 0.22-μ*M* filter.
6. Polyethylene glycol 6000.
7. TNE: 10 m*M* Tris-HCl, pH 7.4, 50 m*M* sodium chloride, 100 m*M* EDTA.
8. Extraction buffer 1: 100 m*M* Tris-HCl, pH 7.6, 100 m*M* sodium chloride, containing 2% *n*-octyl-β-D-glucopyranoside.
9. 10 m*M* Tris-HCl, pH 7.6; 10 m*M* sodium chloride.
10. Extraction buffer 2: 10 m*M* HEPES, pH 7.4, containing 10 m*M* sodium chloride, 2 m*M* EGTA, 6 m*M* 2-mercaptoethanol, 4 m*M* benzamadine, 2 μg/mL aprotinin, 2 μg/mL hepeptin, 1 m*M* phenylmethylsulfonyl fluoride, 1% glycerol, and 0.05% Triton X-100.
11. 100 m*M* ATP.
12. 100 m*M* Sodium vanadate.
13. 100 m*M* Magnesium chloride.
14. Washing buffer 1: 10 m*M* HEPES, pH 7.4, 30 m*M* sodium chloride, 1 m*M* EDTA, 6 m*M* 2-mercaptoethanol, 0.05% Triton X-100, 1% glycerol.
15. Washing buffer 2: Same as washing buffer 1, but including 0.05*M* sodium chloride.
16. 50 m*M* Phenyl phosphate in washing buffer 1 without Triton X-100.
17. Extraction buffer 3: 0.15*M* sodium chloride, 10 m*M* Tris-HCl, pH 8, 1% Triton X-114, 0.2 m*M* phenylmethylsulfonyl fluoride.
18. Washing buffer 3: 20 m*M* Tris-HCl, pH 7, 1*M* sodium chloride.
19. 1*M* Sodium citrate, pH 6,5,4, and 3.

3. Methods

3.1. Infection of T.ni *Larvae*

1. Fourth instar *T.ni* caterpillars grown on semisynthetic growth media *(6)* are required for propagation of the baculovirus.
2. Infect the larvae by mixing either nonoccluded recombinants (4×10^5

PFU/larvae) or polyhedral inclusion bodies (4×10^4/larvae) with a small quantity of the synthetic growth media. Polyhedral inclusion bodies may be prepared from insect cells infected with dual expression vectors, such as VC2 *(3)*, either by freezing and thawing in liquid nitrogen or by a brief sonication (30 s). Cellular debris should be removed by low-speed centrifugation (1000 rpm for 5 min) (*see* Note 1).

3. After 4 d, harvest the caterpillars and place them at –70°C prior to extraction of the hemolymph containing the desired protein product. Best results are obtained when the caterpillars are harvested prior to death.
4. Extract the protein products by first resuspending the larvae in PBS containing 0.02% diethlydithiocarbamate (0.2 mL/larvae) and then subjecting the larvae to homogenization. This procedure usually extracts the hemolymph while maintaining the larval skin intact. The latter can then be removed by low-speed centrifugation and the supernatant used for subsequent processing.

3.2. Large-Scale Insect Cell Culture

Insect cells are unusually shear-sensitive, and this has hampered the development of suitable bioreactors. The cells are also damaged by gas sparging and have a high oxygen demand. The following method for large-scale culture is based on that described in refs. *(1)* and *(2)*, and relies on the use of a specially formulated lipid emulsion to overcome some of the aforementioned problems (*see* Note 2).

1. To produce the lipid emulsion, slowly add the aqueous pleuronic polyol solution to the lipid mixture at 37°C while vortexing. The mixture initially turns cloudy, but clears on further addition of the pleuronyl polyol. This microsuspension is then added to 1 L of basal medium.
2. A suitable airlift fermentor is Vessel type A, supplied by Chemap, South Plainfield, NJ. Agitation and dissolved oxygen (20% air saturation) are controlled by sparging at 0.02 vol of gas/vol of culture/min (VVM; i.e., 420 mL/min in a 21-L vessel) with a controlled mixture of nitrogen/air/oxygen using a fermentor controller (Cetus). Bubble diameter should be kept at 0.5–1 cm to minimize cell damage.
3. Initial cell densities for fermentation are subject to trial and error, but such densities as 0.1×10^6 cells/mL have been used. After 7 d of growth, densities of 5×10^6 cells/mL have been achieved. The latter figure is comparable with the densities achieved using a nonoxygen-limited 100-mL spinner flask.
4. Infection of cells with recombinant baculoviruses should occur when high cell densities have been reached, but the cells are still exponen-

tially growing. A multiplicity of infection (moi) of 1 is sufficient for efficient infection.

5. Recombinant protein expression usually occurs during the period 48 h postinfection, but may peak at periods up to 4–5 d postinfection. Obviously, such information must be determined empirically.

3.3. Purification of the Hepatitis B Virus Surface Antigen (HBsAg) (7)

1. The supernatant from cells infected with a recombinant baculovirus expressing the Hepatitis B surface antigen is recovered 4–5 d postinfection.
2. The supernatant is clarified by the addition of PEG 6000 (7%), followed by centrifugation at 20,000*g*, and the HBsAg recovered by precipitation with 9% PEG 6000.
3. The pelleted material is resuspended in TNE, loaded onto a 20–60% (w/w) sucrose gradient in TNE buffer, and centrifuged at 150,000*g* for 15 h, using a swing-out rotor (e.g., SW41, Beckman).
4. The gradient is split into fractions, and fractions containing HBsAg are identified by radioimmunoassay and concentrated by centrifugation at 100,000*g* for 15 h at 4°C.

3.4. Purification of Hepatitis B Virus Core Antigen (HBcAg) (7)

1. Cells are infected with recombinant viruses expressing HBcAg at an moi of 5.
2. After 4 d postinfection, the cells are harvested by low-speed centrifugation and the pellet resuspended in approx 1/100th of the original vol of PBS.
3. The cells are lysed either by three freeze–thaw cycles or by sonication. Cellular debris is removed by low-speed centrifugation and the supernatant fluids centrifuged through a 30% sucrose cushion (in TNE) at 106,000*g* for 16 h.
4. The pellet is resuspended in TNE (2 mL) and HBcAg purified by CsCl isopycnic centrifugation at 106,000*g* for 36 h. Peak fractions containing HBcAg are pelleted by centrifugation at 106,000*g* for 10 h.

3.5. Purification of Membrane Proteins

3.5.1. The Fusion Protein of Parainfluenza Virus

The method described below has been used to partially purify the fusion glycoprotein of human parainfluenza virus type 3 *(8)*.

1. Infect insect cells with the recombinant virus as described in Chapter 27, Section 3.5., at a moi of 5.
2. After 48 h, harvest the infected cells and wash them twice in PBS.
3. Resuspend the cell pellet in extraction buffer 1 and incubate at room temperature for 60 min.
4. Centrifuge the cell suspension at 300,000*g* for 30 min.
5. Harvest the clear supernatant and dialyze it against 0.01 *M* Tris-HCl, 0.01 *M* sodium chloride (pH 7.6) for 72 h at 4°C. The dialyzed supernatant can then be concentrated, using an appropriate Amicon filter if necessary.

In the study reported in *(8)*, the detergent-solubilized protein was mixed with Freund's complete adjuvant and used to immunize hamsters by intramuscular hindleg injection. However, further purification of the protein product could be achieved using established procedures, e.g., affinity chromatography (*see below*).

3.5.2. Immunoaffinity Purification of Phosphotyrosine Containing Proteins

The following method has been used to purify to homogeneity the tyrosine kinase domain of the epidermal growth factor receptor *(9)*. It is generally applicable to any protein containing phosphotyrosine residues, but also illustrates the use of immunoaffinity columns to purify recombinant baculovirus-expressed proteins.

1. Infect suspension cultures of insect cells with the recombinant baculovirus expressing the tyrosine kinase domain at a moi of 10.
2. At 72 h postinfection, harvest the cells by low-speed centrifugation and homogenize the pellet in 10 mL of extraction buffer 2.
3. Centrifuge the cell lysate for 5 min at 400*g* to remove cellular debris.
4. Subject the supernatant from Step 3 to ultracentrifugation at 115,000*g* for 1 h.
5. Add magnesium chloride, ATP, and sodium vanadate to the supernatant to yield a final concentration of 2 m*M*, 25 μ*M*, and 100 μ*M*, respectively, and allow the autophosphorylation to proceed for 10 min at 4°C.
6. Adsorb the protein solution onto an antiphosphotyrosine antibody column prepared by the covalent linkage of a monoclonal antiphosphotyrosine antibody to BIO-Gel A-15M beads (Bio-Rad) at a ratio of 4 mg of antibody/mL of beads, using the cyanogen bromide coupling reaction *(10)*. A 20-mL sample is loaded onto a 4-mL antibody column at a flow rate of 23 mL/h. The sample run-through is recycled through the column three times to ensure adsorption of the target protein.

7. The column is washed with 30 column vol of washing buffer 1 followed by 30 vol of washing buffer 2, and then a further 30 vol of washing buffer 1.
8. The kinase domain is eluted with phenyl phosphate. Two 5-mL elutions are sufficient to elute >90% of the adsorbed protein.
9. The eluates are dialyzed against washing buffer 1 without Triton X-100.

The tyrosine kinase domain was purified to virtual homogenity by the above procedure in a 23% yield (an approx 1400-fold purification) with a specific activity of 5000 U/mg. Approximately 180 µg of purified kinase domain can be obtained from 400 mL of infected insect cells, which is equivalent to 0.5 mg/L of *S. frugiperda* cells.

3.5.3. Purification of Baculovirus-Expressed CD4 (11)

1. Two days postinfection, harvest the insect cells expressing the CD4 molecule by low-speed centrifugation.
2. After washing the cells in TC100 medium, resuspend them in extraction buffer 3 and leave for 20 min on ice.
3. Clarify the cell suspension by low-speed centrifugation (2000 rpm, 10 min) at 4°C.
4. Incubate the supernatant at 37°C for 10 min, and then centrifuge (2000 rpm, 10 min) at room temperature to produce three phases.
5. Harvest the detergent phase and dilute with an equal vol of PBS.
6. Ultracentrifuge the diluted detergent phase at 100,000*g* for 1 h.
7. Following centrifugation, the supernatant is applied to an immunoaffinity column composed of a monoclonal anti-CD4 antibody linked to Affi-Gel 10 beads (4 mg/mL gel).
8. Wash the column with washing buffer 3 (20 column vol) and subsequently with 20 vol each of the sodium citrate solutions, pH 6, 5, and 4.
9. Elute the CD4 with sodium citrate, pH 3 (2 column vol), and adjust the solution to pH 7.

4. Notes

1. Infection of *T. ni* larvae is much more efficient using occluded virus rather than occlusion-negative virus *(3)*. Nonoccluded viruses, which at present form the majority of the recombinants generated, yeild varying efficiencies of infection of larvae. A plausible method for occluding such occlusion-negative recombinants is to coinfect insect cells with both the recombinant baculovirus and wild-type AcNPV, on the assumption that the viral DNA of the recombinant will be occluded *(12)*. The efficiency

of this procedure has not been fully investigated and, for reproducibility, such baculoviruses as VC2 *(3,13,14)* or their equivalents are still the viruses of choice for efficient infection of larvae.

2. The lipid emulsion containing pleuronyl polyol F-68 not only reduces bubble-induced cell damage and foaming, but also is a substitute for fetal calf serum in the propagation of insect cells. Large-scale cell culture is therefore less expensive using this serum substitute, but it may be used equally well during small-scale growth. Three passages are usually required for the cells to adapt to the serum-free medium. Obviously, the absence of serum proteins from the medium facilitates the purification of secreted proteins expressed using recombinant baculoviruses.

References

1. Maiorella, B., Inlow, D., Shauger, A., and Harano, D. (1988) Large scale insect cell culture for recombinant protein production. *Bio/Technology* **6,** 1406–1410.
2. Murhammer, D. W. and Goochee, C. F. (1988) Scaleup of insect cell cultures: Protective effects of Pleronic F-68. *Bio/Technology* **6,** 1411–1418.
3. Emery, V. C. and Bishop, D. H. L. (1987) The development of multiple expression vectors for the high level synthesis of eukaryotic proteins: Expression of theLCMV-N and AcNPV polyhedrin protein by a recombinant baculovirus. *Protein Eng.* **1,** 359–366.
4. Madisen, L., Travis, B., Hu, S. L., and Purchio, A. F. (1987) Expression of the human immunodeficiency virus gag gene in insect cells. *Virology* **158,** 248–250.
5. Schmaljohn, C. S., Sugiyama, K., Schmaljohn, A. L., and Bishop, D. H. L. (1988) Baculovirus expression of the small genome segment of Hantaan virus and potential use of the expressed nucleocapsid protein as a diagnostic antigen. *J. Gen. Virol.* **69,** 777–786.
6. Hoffman, J. D., Lawson, F. R., and Yamamoto, R. (1966) Synthetic growth media for propagating insect larvae, in *Insect Colonisation and Mass Production* (Smith, C. N., ed.), Academic, NY, pp. 155–190.
7. Takehara, K., Ireland, D., and Bishop, D. H. L., (1988) Coexpression of the Hepatitis B surface and core antigens using baculovirus multiple expression vectors. *J. Gen. Virol.* **69,** 2763–2777.
8. Ray, R., Galinski, M. S., and Compans, R. W. (1989) Expression of the fusion glycoprotein of human parainfluenza type 3 virus in insect cells by a recombinant baculovirus and analysis of its immunogenicity. *Virus. Res.* **12,** 169–180.
9. Wedegaertner, P. B. and Gill, G. N. (1989) Activation of the purified protein kinase domain of the epidermal growth factor receptor. *J. Biol. Chem.* **264,** 11346–11353.
10. Cuatrecasas, P. (1970) Protein purification by affinity chromatography. *J. Biol. Chem.* **245,** 3059–3065.
11. Webb, N. R., Madoulet, C., Tosi, P.F., Broussard, D. R., Sneed, L., Nicolau, C., and Summers, M. D. (1989) Cell surface expression and purification of human CD4 produced in baculovirus infected cells. *Proc. Natl. Acad. Sci. USA* **86,** 7731–7735.
12. Price, P. M., Reichelderfer, C. F., Johansson, B. E., Kilbourne, E. D., and Acs, G. (1989) Complementation of recombinant baculoviruses by coinfection with wild type

virus facilitates production in insect larvae of antigenic proteins of hepatitis B virus and influenza virus. *Proc. Natl. Acad. Sci. USA* **86**, 1453–1457.

13. Bishop, D. H. L. and Emery, V. C. (1988) Expression vectors for the synthesis of proteins and plasmid replicons and sequence cassettes for use in constructing such vectors. Eur. Patent 88307452.8.
14. Bishop, D. H. L. and Emery, V. C. (1988) Expression vectors for the synthesis of proteins and plasmid replicons and sequence cassettes for use in constructing such vectors. US Patent PCT/GB88/00663.

Index